Graduate Texts in Contemporary Physics

Series Editors:

Joseph L. Birman
Helmut Faissner
Jeffrey W. Lynn

Graduate Texts in Contemporary Physics

Heinz Oberhummer (Ed.)

Nuclei in the Cosmos

With Contributions by
J. H. Applegate   J. J. Cowan   F. Käppeler
H. V. Klapdor-Kleingrothaus   K. Langanke
R. A. Malaney   H. Oberhummer   B. E. J. Pagel
C. Rolfs   G. Schatz   G. Staudt   F.-K. Thielemann
J. Vervier   M. Wiescher

With 97 Figures

Springer-Verlag

Berlin Heidelberg New York
London Paris Tokyo
Hong Kong Barcelona
Budapest

Professor Dr. Heinz Oberhummer
Institut für Kernphysik, Technische Universität Wien
Wiedner Hauptstraße 8–10/142, A-1040 Wien, Austria

Series Editors

Joseph L. Birman
Department of Physics
The City College of the
City University of New York
New York, NY 10031, USA

Helmut Faissner
Physikalisches Institut
RWTH Aachen
W-5100 Aachen
Fed. Rep. of Germany

Jeffrey W. Lynn
Department of Physics and Astronomy
University of Maryland
College Park, MD 20742, USA

ISBN-13: 978-3-642-48842-9   e-ISBN-13: 978-3-642-48840-5
DOI: 10.1007/978-3-642-48840-5

Library of Congress Cataloging-in-Publication Data. Nuclei in the cosmos / edited by Heinz Oberhummer: with contributions by J. H. Applegate ... (et al.). p.   cm. – (Graduate texts in contemporary physics) Includes bibliographical references and index. 1. Nuclear astrophysics. I. Oberhummer, Heinz. II. Applegate, J. H. (James H.) III. Series. QB464.N78   1991   523.01'97–dc20 91-27070

© Springer-Verlag Berlin Heidelberg 1991
Softcover reprint of the hardcover 1st edition 1991

Typesetting: Camera ready by Author
56/3140-543210 – Printed on acid-free paper

Preface

Nuclear astrophysics as it stands today is a fascinating science. Even though, compared to other scientific fields, it is a young discipline which has developed only in this century, it has answered many questions concerning the understanding of our cosmos. One of these great achievements was the concept of nucleosynthesis, the creation of the elements in the early universe in interstellar matter and in stars. Nuclear astrophysics has continued to solve many riddles of the evolution of the myriads of stars in our cosmos.

This review volume attempts to provide an overview of the current status of nuclear astrophysics. Special emphasis is given to the interdisciplinary nature of the field: astronomy, nuclear physics, astrophysics and particle physics are equally involved. One basic effort of nuclear astrophysics is the collection of observational facts with astronomical methods. Laboratory studies of the nuclear processes involved in various astrophysical scenarios have provided fundamental information serving both as input for and test of astrophysical models. The theoretical understanding of nuclear reaction mechanisms is necessary, for example, to extrapolate the experimentally determined reaction rates to the thermonuclear energy range, which is relevant for the nuclear processes in our cosmos. Astrophysical models and calculations allow us to simulate how nuclear processes contribute to driving the evolution of stars, interstellar matter and the whole universe. Finally, elementary particle physics also plays an important role in the field of nuclear astrophysics, for instance through weak interaction processes involving neutrinos.

The purpose of this book is to present in ten contributions written by experts in their specific fields our understanding of different astrophysical environments and their connection to nuclear processes. The volume starts with new techniques in experimental nuclear astrophysics and their results (C. Rolfs; J. Vervier). Four papers follow, concerning theoretical models for the explanation of nuclear processes and their astrophysical implications (H. Oberhummer and G. Staudt; K. Langanke) and observational facts relevant for nuclear astrophysics (B. Pagel; G. Schatz). Finally different astrophysical scenarios ranging from primordial nucleosynthesis (R. Malaney; F.-K. Thielemann, J.H. Applegate, J.J. Cowan and M. Wiescher) to the s-process (F. Käppeler) and nuclei far from stability (H.V. Klapdor-Kleingrothaus) are discussed.

I would like to express my thanks to all colleagues who have contributed their expertise to this book. The excellent collaboration with G. Wolschin, U. Beiglböck and F. Holzwarth of Springer-Verlag is gratefully acknowledged. I also want to thank my coworkers K. Grün, H. Krauss and T. Rauscher for their help in editing the book. Last but not least I am especially indebted to G. Raimann who answered all the questions and solved all the problems that arose in the application of TEX.

Vienna, August 1991   *H. Oberhummer*

Contents

List of Contributors

- James H. Applegate
  Department of Astronomy, Columbia University, New York, NY 10027, USA

- John J. Cowan
  Department of Physics and Astronomy, University of Oklahoma, Norman, OK 73019, USA

- Franz Käppeler
  Kernforschungszentrum Karlsruhe, Institut für Kernphysik, Postfach 3640, D-7500 Karlsruhe, Germany

- Hans V. Klapdor-Kleingrothaus
  Max-Planck-Institut für Kernphysik,
  Postfach 10 39 80, D-6900 Heidelberg, Germany

- Karlheinz Langanke
  Institut für Theoretische Physik I, Westfälische Wilhelms Universität Münster, Wilhelm-Klemm-Str. 9, D-4400 Münster, Germany

- Robert A. Malaney
  Institute of Geophysics and Planetary Physics, Box 808, University of California, Lawrence Livermore National Laboratory, CA 94550, USA

- Heinz Oberhummer
  Institut für Kernphysik, TU Wien, Wiedner Hauptstr. 8-10, A-1040 Wien, Austria

- Bernard E.J. Pagel
  NORDITA, Blegdamsvej 17, DK-2100 Copenhagen X, Denmark

- Claus Rolfs
  Institut für Experimentalphysik III, Ruhr-Universität Bochum, Universitätsstr. 150, D-4630 Bochum, Germany

- Gerd Schatz
  Kernforschungszentrum Karlsruhe, Institut für Kernphysik, Postfach 3640, D-7500 Karlsruhe, Germany

– Günter Staudt
Physikalisches Institut, Univ. Tübingen, Auf der Morgenstelle 14, D-7400 Tübingen, Germany
– Friedrich-Karl Thielemann
Harvard-Smithsonian Center for Astrophysics, 60 Garden Street, Cambridge, MA 02138, USA
– Jean Vervier
Université Catholique de Louvain, Institut de Physique Nucléaire, Chemin du Cyclotron, 2, B-1348 Louvain-la-Neuve, Belgium
– Michael Wiescher
Department of Physics, University of Notre Dame, Notre Dame, IN 46556, USA

Experimental Determination
of Stellar Reaction Rates

C. Rolfs

Institut für Experimentalphysik III, Ruhr-Universität Bochum,
Universitätsstr. 150, D-4630 Bochum, Germany

1. Introduction

An understanding of most of the critical stellar features, such as time scales, energy production, and nucleosynthesis of the elements, hinges directly on the magnitude of the reaction rate per particle pair [Rol88, Bar89],

$$\langle \sigma v \rangle = \left(\frac{8}{\pi \mu} \right)^{1/2} \frac{1}{(kT)^{3/2}} \int_0^\infty \sigma(E) \, E \, \exp \left(-\frac{E}{kT} \right) \, dE, \qquad (1)$$

where μ is the reduced mass, T the stellar temperature, E the center-of-mass energy and $\sigma(E)$ the cross section of the reaction. As a star evolves, its temperature changes, and hence the reaction rate $\langle \sigma v \rangle$ must be evaluated for each temperature of interest. This evaluation requires $\sigma(E)$ to be known in the range from nearly zero energy to high values (see below).

In charged-particle-induced reactions, the cross section drops rapidly (on an exponential scale) at low energies, and it becomes increasingly difficult to measure the relevant cross sections:

$$\sigma(E) = S(E) \, \frac{1}{E} \, \exp(-2\pi\eta) \qquad , \qquad (2)$$

where the exponential represents the tunneling through the Coulomb barrier (assuming s-wave particles) with $2\pi\eta$ as the Sommerfeld parameter,

$$2\pi\eta = \frac{2\pi Z_1 Z_2 e^2}{\hbar v} = 31.29 \, Z_1 \, Z_2 \, (\mu/E)^{1/2}, \qquad (3)$$

(E in units of keV, Z_i = nuclear charges). The quantity $S(E)$ defined by Eq. 2 is called the astrophysical $S(E)$ factor.

For nonresonant reactions, with $S(E)$ varying slowly with energy, the integrand of Eq. 1 forms a peak, the Gamow peak, at energy E_0

$$E_0 = 1.22 \, (Z_1^2 \, Z_2^2 \, \mu \, T_6^2)^{1/3} \text{ keV} \qquad (4)$$

(with T_6 in units of 10^6 K) and of width Δ

$$\Delta = 0.749 \, (Z_1^2 \, Z_2^2 \, \mu \, T_6^5)^{1/6} \text{ keV} \,, \tag{5}$$

and this nonresonant rate $\langle \sigma v \rangle$ reduces to the expression

$$\langle \sigma v \rangle_{\text{nr}} = \left(\frac{2}{\mu}\right)^{1/2} \frac{\Delta}{(kT)^{3/2}} \, S(E_0) \, \exp\left(-\frac{3E_0}{kT}\right). \tag{6}$$

Nuclear reactions take place predominantly over the energy window $E = E_0 \pm \Delta/2$. This energy window increases with temperature and with the Coulomb barrier E_C of the nuclei involved. It is over this energy range where information regarding $\sigma(E)$ must be obtained. The typical stellar temperature range of $T_6 \approx$ 10 to 5000 corresponds to $E_0/E_C \approx 0.01$ to 1; thus, these energies are at and far below the Coulomb barrier and are called subCoulomb energies.

For narrow resonances (total width $\Gamma_R \ll$ energy E_R) the reaction rate (Eq. 1) reduces to

$$\langle \sigma v \rangle_{\text{r}} = \left(\frac{2\pi}{\mu kT}\right)^{3/2} \hbar^2 \sum_i (\omega\gamma)_i \, \exp\left(-\frac{E_i}{kT}\right) \tag{7}$$

where $(\omega\gamma)_i$ and E_i are the strength and energy of a given resonance, respectively, and the sum extends over all resonances. The strengths $(\omega\gamma)_i$ also depend critically on the effects of the Coulomb barrier:

$$\omega\gamma = \omega \, \frac{\Gamma_a \Gamma_b}{\Gamma_R} \approx \omega \, \Gamma_a \qquad \text{for } E \ll E_C \tag{8}$$

(Γ_a and Γ_b = particle width of entrance and exit channels, respectively, and $\Gamma_R = \Gamma_a + \Gamma_b + \dots$). These strengths decrease very rapidly with decreasing resonance energy E_R due to the barrier penetration factor $P_\ell(E, R_n)$ contained in Γ_a:

$$\Gamma_a(E, \ell) = \frac{2\hbar}{R_n} \left(\frac{2E}{\mu}\right)^{1/2} P_\ell(E, R_n) \, \theta_\ell^2 \tag{9}$$

(R_n = interaction radius, ℓ = orbital angular momentum, θ_ℓ^2 = reduced width). The exponential term in Eq. 7 shows that for a given stellar temperature T resonances with energies E_R near kT will dominate $\langle \sigma v \rangle$. Therefore, for low stellar temperatures, it is extremely important to know the locations and strengths of low-energy resonances.

With improved experimental techniques (see below) direct measurements of $\sigma(E)$ can be extended toward lower energies, but in practise one hardly reaches the relevant energy regions $E_0 \pm \Delta/2$ for quiescent stellar burning. The observed energy dependence of $\sigma(E)$, or equivalently of $S(E)$, must therefore be extrapolated into the stellar energy region (essentially to zero energy). Of course, the basis for extrapolation will be improved if extremely low-energy data are available. However, these extrapolated data represent only lower limits of $\langle \sigma v \rangle$. If there are resonances near the particle threshold at $-E_R$ or $+E_R$, they can completely dominate $\langle \sigma v \rangle$ for low stellar temperatures. In contrast, for neutron-induced reactions, there is no Coulomb barrier, and the cross section is very large increasing with decreasing energy (resonances are superposed on a smooth nonresonant yield following often the $1/v$ law). Thus, experimental measurements of $\sigma(E)$ can be carried out directly at the relevant stellar energies, which is here near $kT \simeq 30\,\text{keV}$ (for details, see [Rol88] and references therein). In the following we will restrict our discussion to the case of charged-particle-induced reactions.

2. Direct Measurements

Because the cross section $\sigma(E)$ drops by many orders of magnitude for energies between the Coulomb barrier and the point at which the ratio E/E_C is about 0.01, experimental investigation of $\sigma(E)$ requires high ion-beam currents, stable and pure targets, and efficient detectors.

The need for high currents establishes the requirements for ion sources, accelerators, beam transport systems, and beam integration (for details, see [Rol88]). For example, a $100\,\text{kV}$ accelerator has been built at the Ruhr-Universität Bochum with beam currents (H^+ and He^+ ions) of the order of $1\,\text{mA}$ on target [Kra87]. This accelerator allowed direct measurements of several favorable reactions to energies as low as $E/E_C \approx 0.01$ (see article on electron screening by S. Engstler, [Eng90]).

The requirement of stable and pure targets can be fulfilled by windowless gas targets or specially prepared solid targets [Rol88]. In recent years the implantation technique has proven as an excellent means for producing isotopically enriched targets capable of withstanding high beam bombardments [Seu87]. Through proper preparation and choice of the implantation host material (the backing), such targets can be made sufficiently free of contaminants. The implantation technique also appears to be a useful tool for the production of radioactive targets, e.g. ^{22}Na [Seu90].

A perfect detector would have 100% detection effciency, 4π geometry, unique identification of the reaction products, insensitivity to background radiation, high energy and time resolution, and reasonable costs. Obviously, such perfection is unattainable; for each specific experiment, these requirements conflict, and compromises must be made in the choice and design of the detectors.

In the case of neutron detection with high efficiency, the target is surrounded in 4π geometry with a material (e.g. polyethylene, graphite or heavy water), which moderates the emitted high-energy neutrons. These neutrons are then detected with ^3He-filled proportional counters surrounding the material, leading to an efficiency of about 30% [Seu88]. With this detection technique several astrophysically important (α,n) reactions could be observed recently to low energies [Wan91]. Of course, the price is a loss of energy information of the emitted neutrons and thus extreme care must be taken to minimize the contribution of background reactions to the observed neutron yield. Even in the absence of such background reactions, the sensitivity of these detectors is limited by the yield contributions from cosmic rays: typically about 0.1 counts per sec. A reduction of the cosmic rays by nearly 3 orders of magnitude can be achieved by going deep underground.

When the reaction products are charged particles (p, α, etc.), they are usually observed with an array of surface barrier detectors around the target leading to an efficiency of nearly 100%. The limit of yield measurements is determined also by the contributions of cosmic rays. Surrounding the detectors with an anticoincidence shielding can reduce this background by about one order of magnitude; again, the best solution is to go underground.

Finally, the detection of γ-rays in capture reactions can be carried out with "crystal balls" made of NaI(Tl), BGO or BaF$_2$ crystals leading to nearly 100% efficiency for MeV γ-rays [Wis89]. A cheaper (but similar) solution is to surround the target by a thick scintillator (e.g. a NaI(Tl) crystal 30 cm thick), where the energies of the emitted γ-rays (e.g. cascades) are summed leading to a peak in the spectrum at an energy $E_\gamma = Q + E_p$ (Q = reaction Q-value, E_p = projectile energy). If background reactions have Q values differing from that of the reaction of interest, their contributions to the spectrum appear at different sum energies. If this summing spectrometer is surrounded by an anticoincidence shield to reject cosmic-ray events, the later background could be reduced by one order of magnitude at $E_\gamma \approx 5 - 10$ MeV [Sch90]. Further significant reduction requires to go underground.

Aside from further development of ion beams, targets, and detectors, measurements at energies far below the Coulomb barrier could be significantly improved by a "quantum jump" by going underground; the installation of a low-energy accelerator at the underground Gran Sasso laboratory in Italy is underway.

3. Indirect Approaches

As discussed in the previous section, resonances near the particle threshold are very important and their contribution to $\langle \sigma v \rangle$ is proportional to their strength $\omega\gamma = \omega\Gamma_a$, with the partial width given by Eq. 9. Note that this width Γ_a scales with the reduced width θ_b^2, where the latter quantity could be measured e.g. via stripping reactions. The existence of possible states near the particle threshold of the compound nucleus has to be studied experimentally using several other reactions, which form this compound nucleus at the relevant excitation energies. Measurement of the appropriate quantities (i.e. E_R, ℓ, J, and θ_ℓ^2) needed to calculate $\omega\gamma$ and thus $\langle \sigma v \rangle$ frequently demands nuclear studies that have only subtle connections to the reaction of interest to astrophysics. A celebrated example is the $3\alpha-$process [Rol88].

Sometimes it is easier to determine the cross section $\sigma_{ab}(E)$ of a reaction X(a,b)Y via measurement of $\sigma_{ba}(E)$ of the inverse reaction Y(b,a)X, where both cross sections are related by the principle of detailed balance [Rol88]. A recent example of this approach is the reaction $^8\mathrm{Li}(\alpha,\mathrm{n})^{11}\mathrm{B}$, which is difficult to measure directly due to the instable $^8\mathrm{Li}$ ($T_{1/2} \approx 1\,\mathrm{sec}$). The inverse reaction $^{11}\mathrm{B}(\mathrm{n},\alpha)^8\mathrm{Li}$ was thus studied recently [Par90]. However, due to the high Q-value several excited states in $^{11}\mathrm{B}$ can be fed via the direct reaction, while the inverse reaction always starts with $^{11}\mathrm{B}$ in its ground state. Thus, such studies provide only partial information on $\sigma_{\mathrm{ab}}(E)$, i.e. a lower limit on $\sigma_{\mathrm{ab}}(E)$.

Another example are radiative capture reactions, $X(\mathrm{a},\gamma)Y$, where the inverse reaction $Y(\gamma,\mathrm{a})X$ is also called photodisintegration. Due to phase space factors, $\sigma_{\gamma\mathrm{a}}$ is much larger than $\sigma_{\mathrm{a}\gamma}$. This might be taken as indicating that, instead of directly studying the capture reaction, it would be easier to investigate the photodisintegration reaction, at a given E. However, photodisintegration reactions are not easy to carry out from a technical point of view. Another approach was suggested recently [Reb87]. When a charged nuclear projectile Y moves with moderately high kinetic energy through the Coulomb field of a high-Z nucleus such as lead, strong electromagnetic fields are present for a short time. The effect of the time-varying electromagnetic field is equivalent to a high-flux spectrum of virtual photons. These photons can lead to Coulomb dissociation of the projectiles, $Y + Z \to X + \mathrm{a} + Z$. The copious source of virtual photons offers a promising way to study the photodisintegration process. However, it is not yet clear how far this method can be pushed in practice. Of course, the method provides again only partial information on the capture cross section, namely on σ_{cap} to the ground state of Y only (see above). Fortunately, for light nuclides this capture process is often the dominant process.

4. Reactions Involving Radioactive Nuclides

In the hot and explosive burning phases of stars, where E_0 approaches E_C, nuclear burning times can be measured in seconds. If the lifetime of a radioactive nucleus is longer than or of the same order as the burning time, that nucleus will be involved in the nuclear burning processes. A quantitative understanding of the observed nuclear ashes from these stars requires a knowledge of $\langle \sigma v \rangle$, and this in turn requires measuring (p,γ), (α, γ), and other reactions involving these radioactive nuclides. If the half-life of the nuclides is longer than a day or so, they may be made into a radioactive target (Sect. 2). However, in a great majority of interesting cases, the half-life is shorter (e.g. $T_{1/2} = 10\,\mathrm{min}$ for $^{13}\mathrm{N}$). In this case, the only direct method is to produce the radioactive nuclides in an accelerator, separate them, accelerate them in a second accelerator, and finally allow the radioactive ion beam (RIB) to interact with a H_2 or He target. All of this must be achieved in a time shorter than the decay lifetime of the radioactive nuclides.

To develop RIB several proposals (e.g. TRIUMF and CERN) have been made. A promising project is already well under way at Louvain-la-Neuve to study the reaction ^{13}N(p,γ)^{14}O via ^{1}H(^{13}N,γ)^{14}O (for technical details of this project as well as of earlier proposals, see [Ver91]). It is clear that additional research and technical development are needed on all aspects of the problem (ion sources, accelerators, targets, detectors, and radiation safety), before reactions involving short-lived nuclides can be routinely studied in the laboratory. In what follows, I will concentrate on the problem of detectors in radiative capture reactions.

The capture process A(x,γ)B can be studied by observation of either the capture γ-rays or the product nuclei B, or both. Detection of the γ-rays with standard detectors (NaI(Tl), BGO or Ge) is hampered by the intense radiation from the RIB (note that a 1 nA particle beam of ^{13}N corresponds to an activity of about 1 Ci). An ideal detector for such studies should have (in addition to the characteristics discussed earlier) a threshold property, i.e. zero efficiency for the main γ-ray flux of the RIB (predominantly 0.51 MeV) and a high and nearly uniform efficiency for (say) $E_\gamma = 3 - 9$ MeV capture γ-rays. A detector developed recently is based on the photodisintegration of deuterium (2.225 MeV threshold) with subsequent counting of the moderated neutrons [Seu88]. The system consists of a cylindrical tank filled with 242 liters of D_2O and surrounded by 30 ^{3}He-filled proportional counters. The detector measures angle-integrated γ-ray fluxes with a rather low efficiency, $\epsilon_\gamma \approx 4 \cdot 10^{-4}$ for $E_\gamma = 3 - 9$ MeV. A subthreshold source strength of about 1 Ci is the limit of the detector because of the γ-ray pileups in the ^{3}He counters. The (current) sensitivity limit arises from a background rate of 0.7 c/sec produced by cosmic rays. Alternatively, radiative capture γ-rays might be observed with a detector based on e^+e^- pair production, provided that the quantum energies of all γ-rays from the radioactivity were below the 1.022 MeV threshold.

Since all the recoiling nuclei B produced in the reaction A(x,γ)B travel in essentially the same direction as the beam, their observation with detectors having 100% efficiency would greatly improve the experimental sensitivity. In this approach, thin targets (with no backings) or windowless gas targets are needed. However, there are some obvious problems if a detector is placed in the beam direction: one observes not only the capture products B, but also the incident beam (including any beam contaminants), small angle elastic scattering products, and background events (e.g. from multiple scattering processes). These problems can be solved by various techniques, such as combining a Wien filter and magnetic and electrostatic analyzers with particle identification in detector telescopes. One such system has been built at Caltech and used in the study of ^{4}He(^{12}C,γ)^{16}O [Kre88]. Recent work on ^{1}H(^{12}C,γ)^{13}N using this system has also shown its usefulness to purify the beam (i.e. to reduce the ratio of beam particles to recoiling nuclei B) even though the mass difference between the beam and residual nuclei is here only one mass unit. With a beam purification of the order 10^{10} one could also detect the residual nuclei B via their characteristic radioactivity signatures [Smi91].

References

Bar89   C.A. Barnes: In *Nuclear Astrophysics*, ed. M. Lozano et al. (Springer, 1989) p.197
Eng90   S. Engstler et al.: In *Proceedings of the International Symposium on Nuclear Astrophysics, Baden/Vienna 1990*, ed. by H. Oberhummer and W. Hillebrandt (Max-Planck Institut for Physics and Astrophysics, Garching 1990) p. 185
Kra87   A. Krauss et al.: Nucl. Phys. **A467**, 273 (1987)
Kre88   R.M. Kremer et al.: Phys. Rev. Lett. **60**, 1475 (1988)
Par90   T. Paradellis et al.: Zeitsch. Phys. **A337**, 211 (1990)
Reb87   H. Rebel: Lect. Notes Phys. **287**, 38 (1987)
Rol88   C. Rolfs, W.S. Rodney: *Cauldrons in the Cosmos* (University of Chicago Press, 1988)
Sch90   Th. Schange: Diplomarbeit, Universität Münster (1990)
Seu87   S. Seuthe et al.: Nucl. Instr. Meth. **A260**, 33 (1987)
Seu88   S. Seuthe et al.: Nucl. Phys. **A272**, 814 (1988)
Seu90   S. Seuthe et al.: Nucl. Phys. **A514**, 471 (1990)
Smi91   M. Smith et al.: Nucl. Instr. Meth. (in press) (1991)
Ver91   J. Vervier: In *Nuclei in the Cosmos*, ed. by H. Oberhummer (Springer, Berlin 1991) this volume
Wan91   T.R. Wang, R.B. Vogelaar, R.W. Kavanagh: Phys. Rev. (1991) in press
Wis89   K. Wisshak et al.: Tech. Rep. KfK **4652** (1989) (Kernforschungszentrum Karlsruhe)

Radioactive Ion Beams in Nuclear Astrophysics

Jean Vervier

Université Catholique de Louvain, Institut de Physique Nucléaire, Chemin du Cyclotron, 2, B-1348 Louvain-la-Neuve, Belgium

1. Introduction

The development of Radioactive Ion Beams (RIB), and their uses in Physics in general - and in particular in Nuclear Astrophysics -, represent a field of Science which is still in its infancy, but which is currently attracting an increasing interest [Buc85a, Kub89, Mye90, McC90, Obe90]. Very few RIB with the ranges of energies and intensities needed for studying nuclear reactions of astrophysical interest have been produced so far [Dar90a, Dar90b], but many RIB projects have recently been elaborated (see Subsec. 3.3 below), most of which include Nuclear Astrophysics as one of their possible applications.

In Sect. 2 of the present article, we shall outline the interest of RIB for the measurement of quantities useful in Nuclear Astrophysics. Sect. 3 will be devoted to the methods used to produce RIB, and will underline what has been achieved so far and what is likely to be in operation within a few years. The experimental techniques which are available or which will have to be developed in order to measure cross sections for nuclear reactions of astrophysical interest induced by RIB will be described in Sect. 4. We shall present some concluding remarks in Sect. 5.

2. Interest of Radioactive Ion Beams in Nuclear Astrophysics

2.1 Nuclear Reactions of Astrophysical Interest Involving Radioactive Nuclei

Nuclear reactions which produce energy and synthetise new elements at various astrophysical sites in the cosmos often include radioactive nuclei in their entrance channels. This general statement is illustrated in several other articles of the present book [Kla91, Mal91, Thi91], and a well-known example is the production of neutron-rich heavy nuclei by the rapid neutron capture r-process [Kla91]. Nuclear reactions involving proton-rich light radioactive nuclei may also be of significant importance, for instance in primordial nucleosynthesis for inhomogeneous Big Bang models [Mal91, Thi91], in massive star nucleosynthesis and explosive events such as novae and supernovae [Woo90]. In general, for large stellar temperatures and densities, the rates for nuclear reactions may become larger than the rates for nuclear (mainly beta) decays ; in these conditions, radioactive nuclei produced in nuclear reactions do not have enough time to decay before being involved in other nuclear reactions, thereby contributing to the production of energy and to the synthesis of new elements. This is illustrated by the following three examples.

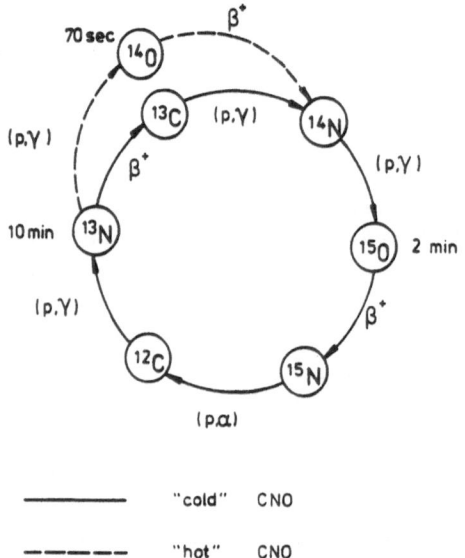

Fig. 1. "Cold" and "hot" CNO cycles [Rol88]. The half-lifes of the radioactive nuclei ^{13}N, ^{14}O and ^{15}O are also indicated

The production of energy in hydrogen-burning stars of the main sequence occurs, either by the proton-proton (p-p) chain or by the CNO cycle [Rol88]. The latter effectively fuses four protons into a ^{4}He nucleus using ^{12}C as a catalyst, as represented in figure 1, and it is dominant in second- or later-generation stars with temperatures $T_6 > 20$. At very large stellar temperatures ($T_6 > 100$) and densities ($\rho > 10^{-4}$ g/cm^3), this "ordinary" or "cold" CNO cycle could be replaced by the "hot" CNO cycle, also represented in Fig. 1, in which the rate of the ^{13}N(p,γ)^{14}O reaction is larger than the rate for the β^+ decay of ^{13}N with $T_{1/2} = 10$ min. The stellar conditions under which the transition from the "cold" to the "hot" CNO cycles could occur depend on the cross section for the ^{13}N(p,γ)^{14}O reaction. For the relevant range of stellar temperatures, the proton capture reaction on ^{13}N is expected to be dominated

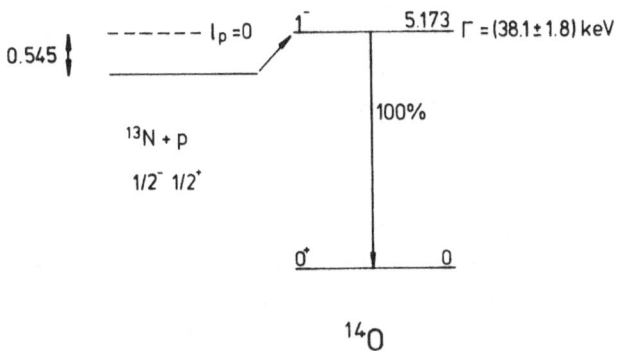

Fig. 2. Partial decay scheme of ^{14}O and of the ^{13}N(p,γ)^{14}O resonance capture at low energies [Ajz86]. Γ is the total width of the 5.173 MeV level [Chu85]. Energies are in MeV

(Fig. 2) by a resonance at an energy of 0.545 MeV in the p-^{13}N center of mass, which corresponds to the first excited level of ^{14}O, located at an energy of 5.173 MeV, with spin and parity $J^\pi = 1^-$ [Ajz86], and with a total width Γ equal to 38.1 (1.8) keV [Chu85]. The partial γ-width Γ_γ of this state has been calculated by various authors using different models [Mat84, Lan85, Bar85, Fun87, Des89] ; their results, summarized in Table 1, span the 1-10 eV range. A value of 2.7 (1.3) eV has been proposed for Γ_γ by Fernandez et al.[Fer89], on the basis of an indirect measurement wherein this 1^- level has been populated via the ^{12}C(^{3}He,n)^{14}O* reaction and its γ-ray-to-total decay branching ratio has been determined. Using these data, the phase diagramme delimiting, in the density versus temperature plane, the regions of the "cold" or "normal" and of the "hot" CNO cycles can be calculated [Fer89]. Astrophysical sites where the "hot" CNO cycle could be important include (still hypothetical) supermassive (10^5 to 10^8 solar masses) stars, novae and supernovae outbursts and accreting neutron stars [Rol88, Arn90]. One of the reasons why the relative importances of the "cold" and "hot" CNO cycles are of great interest is the following. The "hot" CNO cycle effectively "bypasses" the nucleus ^{13}C (Fig. 1), which would thereby be underproduced ; the latter is a possible source of neutrons for the slow neutron capture s-process [Käp91], through the ^{13}C(α,n)^{16}O reaction, and its relative depletion could affect the importance of this process in the relevant stars [Jor89].

Table 1. Calculated values of the partial γ-decay width Γ_γ of the 1^- level at 5.173 MeV in ^{14}O

Γ_γ (eV)	Ref.
2.44	[Mat84]
1.5	[Lan85]
1.2	[Bar85]
7.8	[Fun87]
4.1	[Des89]

The burning of hydrogen in main sequence stars with $T_6 < 20$ mainly occurs through the p-p chain. The latter, represented in Fig. 3 [Rol88], includes various branches, denoted I to III, whose relative importances, for the sun where T_6 = 15, are also given in Fig. 3. It should be noticed that chain III involves the $^7Be(p,\gamma)^8B$ nuclear reaction with the radioactive nucleus ^7Be ($T_{1/2}$ = 53 days), and that the high-energy neutrinos from the β^+ decay of ^8B are at the heart of the so-called missing solar neutrino problem [Rol88]. At very high stellar temperatures ($T_6 > 100$), another p-p chain, also shown in Fig. 3, the "hot" p-p chain, could be of importance [Wie89]. It includes nuclear reactions with the radioactive nuclei ^7Be and ^{11}C, and in particular $^{11}C(p,\gamma)^{12}N$, whose rate could be larger than the rate for the β^+ decay of ^{11}C with $T_{1/2}$ = 20 min. The relative importance of the hot p-p chain depends, within other quantities, on the cross section for the latter reaction. In the temperature range of interest, the proton capture reaction on ^{11}C should be mostly determined by two resonances, at center-of-mass energies of 0.36 and 0.59 MeV, corresponding to the 2^+ and 2^- first and second excited levels of ^{12}N at 0.96 and 1.19 MeV, respectively [Ajz85]. The theoretical properties of these states and resonances have recently been recalculated using the microscopic cluster model [Des90a]. The 0.59 MeV resonance should be the dominant one, and the computed partial γ and proton widths of the corresponding 2^- level in ^{12}N are : $\Gamma_\gamma = 0.14$ eV and $\Gamma_p = 130$ keV, respectively. The $^{11}C(p,\gamma)^{12}N$ reaction could be of importance in massive (M > 20 solar masses) first-generation zero-metal stars, since the hot p-p chain could then lead to the synthesis of ^{12}C, through the β^+ decay of ^{12}N [Des90a] ; this production mechanism for ^{12}C would then compete with the standard triple-α process [Rol88].

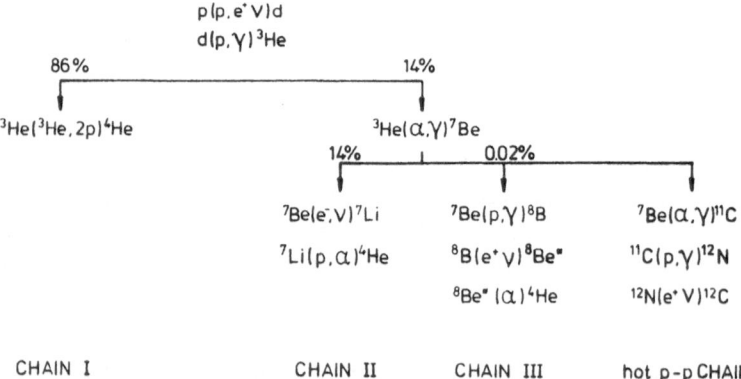

Fig. 3. Classical p-p chains I, II and III in main sequence stars, with their branchings in the sun [Rol88], and "hot" p-p chain [Wie89]

It has been suggested [Wal81] that "breakout" or "leakage" from the CNO cycle could occur through the $^{15}O(\alpha,\gamma)^{19}Ne(p,\gamma)^{20}Na$ reactions. One of the conditions for this process to be effective is that the rate of the $^{19}Ne(p,\gamma)^{20}Na$ reaction should be larger than the rate for the β^+ decay of ^{19}Ne with $T_{1/2}$ = 17 sec. According to recent results [Kub88, Kub90], the cross section for the latter reaction should be dominated by a resonance at a center-of-mass energy of 0.44 MeV, corresponding to a 1^+ level at 2.637 MeV in ^{20}Na [Ajz87]. A theoretical study of the ^{19}Ne$(p,\gamma)^{20}$Na resonances and of the ^{20}Na levels has recently been carried out [Des90b]. This breakout from the CNO cycle would lead to the production of ^{20}Ne, through the β^+ decay of ^{20}Na, and thereby to the so-called rapid proton capture rp process [Wal81], which could synthetize nuclei with A $\geq$ 20.

The ^{13}N$(p,\gamma)^{14}$O, ^{11}C$(p,\gamma)^{12}$N and ^{19}Ne$(p,\gamma)^{20}$Na reactions are only three out of many nuclear reactions which include radioactive nuclei in their entrance channels, and which could be of importance in Nuclear Astrophysics. Other examples are ^{8}Li$(\alpha,n)^{11}$B, which could play a significant role in the recently proposed inhomogeneous Big-Bang nucleosyntheses [Mal91, Thi91], ^{15}O$(\alpha,\gamma)^{19}$Ne, which comes about in the breakout from the CNO cycle mentioned above, as well as many others [Arn90]. In almost all cases, the only informations available on the cross sections for these reactions arise from theoretical calculations, as illustrated for the three examples considered earlier. These are affected by large uncertainties, as shown in Table 1 for the ^{13}N$(p,\gamma)^{14}$O reaction. It is thus of considerable importance to strengthen these indications by experimental data, preferentially by direct measurements of these cross sections in the relevant energy ranges.

2.2 Indirect Methods

Useful experimental informations on the reactions considered in Subsect. 2.1 can be obtained by indirect methods. The first of them [Par85] is the determination of the properties of the levels which are important for these reactions, by populating them through reactions involving only stable nuclei in their entrance channels. Examples of this method mentioned in Subsect. 2.1 are : the study of the branching ratio in the decay of the 1^- level at 5.173 MeV in ^{14}O through the ^{12}C$(^3$He,n$)^{14}$O* reaction [Fer89] ; the determination of the excitation energy, spin and parity of the 1^+ level at 2.637 MeV in ^{20}Na through the $(^3$He,t$)$ and (p,n) reactions on ^{20}Ne [Kub88, Kub90]. This method often requires measurements of very small quantities (for example, the γ-ray-to-total decay branching ratio for the 1^- level in ^{14}O, which is the order of 10^{-4}), which are very sensitive to the presence of tiny amounts of impurities in the target [Agu89]. A second indirect method is based on the study of the inverse reactions and on the detailed balance theorem. The ^{11}B$(n,\alpha)^8$Li reaction, for example, has recently been reexamined at low neutron energies [Kos90], and the results have yielded useful information on the ^{8}Li$(\alpha,n)^{11}$B reaction, which has implications for primordial nucleosynthesis as noted above. These data were however limited to the ground-state to ground-state transition, and hence give only a lower limit to the cross section for the production of particle-stable levels in ^{11}B through the ^{8}Li$(\alpha,n)^{11}$B reaction. A third indirect method aims at the study of radiative capture reactions such as (p,γ) and (α,γ) on radioactive targets, through the desintegration of the final nuclei in the Coulomb field of a heavy nucleus, using the virtual photon spectrum present in this interaction and the detailed balance theorem [Bau86]. An example could be the study of the desintegration of ^{14}O into ^{13}N + p, which is related to the ^{13}N(p,γ)^{14}O reaction, and which could be investigated with high energy (60-80 MeV/nucleon) ^{14}O radioactive ion beams on high-Z targets [Agu90]. These experiments require the solution of several problems, both in their experimental techniques (to ensure that a pure Coulomb interaction is taking place between the projectile and the target, and to identify unambiguously the reaction products) and in their theoretical interpretation (to deduce the radiative capture cross sections of interest from the measured Coulomb breakup cross section). One can thus conclude that the three indirect methods just described could yield useful informations on the nuclear reactions considered in Subsect. 2.1. However, in view of their difficulties and limitations mentioned above, they should be strengthened and completed by direct measurements of the cross sections for the relevant reactions. This requires the development of RIB, as will now be shown.

2.3 Radioactive Ion Beams Versus Radioactive Targets

The measurement of the cross sections for nuclear reactions which include a radioactive nucleus together with a (generally stable) partner in their entrance channel can be carried out in two ways (Fig. 4) : either by sending a stable beam, generally protons or alpha particles, on a radioactive target, prepared off-line or on-line (Figs. 4a and b) ; or by producing a RIB and directing it on a stable target, generally hydrogen or helium (Fig. 4c). The comparison between these two methods has been carried out in detail [Rol88]. It has been concluded (Fig. 4d) that the most advantageous way of studying these reactions depends on the halflife $T_{1/2}$ of the radioactive nuclei : if $T_{1/2} \gg 1$ hour, the radioactive target approach is the best ; if $T_{1/2} \ll 1$ hour, RIB should be used. This conclusion is independent of the cross section of the nuclear reaction of interest and of the intensity of the primary radioactive beam which is used, either to prepare the radioactive target (figure 4a and b) or to irradiate the stable target (Fig. 4c).

Nuclear reactions of astrophysical interest involving long halflife radioactive targets have already been investigated, using proton beams on ^{7}Be ($T_{1/2} = 53$ days) [Fil83] and ^{22}Na ($T_{1/2} = 2.6$ years) [Seu90], or neutron beams on ^{7}Be and ^{22}Na [Koe88a, Koe88b] or on ^{36}Cl ($T_{1/2} = 3 \times 10^5$ years), ^{40}K ($T_{1/2} = 1.3 \times 10^9$ years) and ^{41}Ca ($T_{1/2} = 1.0 \times 10^5$ years) [Wag90]. In the present article, we concentrate on the radioactive ion beam approach, i.e. on short halflife radioactive nuclei such as ^{8}Li($T_{1/2} = 0.84$ sec), ^{11}C ($T_{1/2} = 20$ min), ^{13}N ($T_{1/2} = 10$ min), ^{15}O ($T_{1/2} = 2$ min), ^{19}Ne ($T_{1/2} = 17$ sec), which clearly fulfil the above-mentioned criterium for the RIB approach.

(a) RADIOACTIVE TARGET (OFF-LINE)

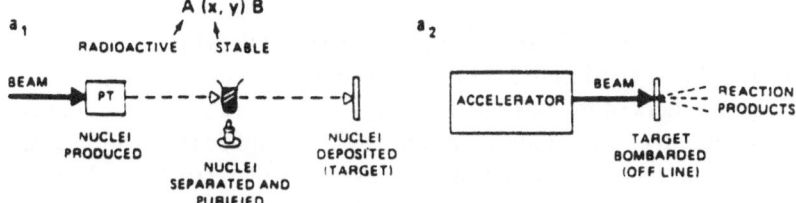

(b) RADIOACTIVE TARGET (ON-LINE)

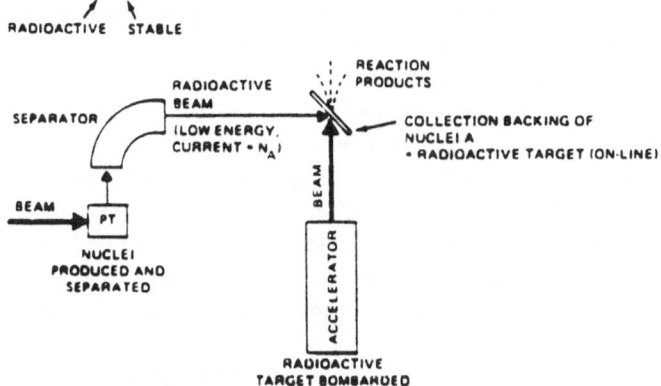

(c) RADIOACTIVE ION BEAM (RIB)

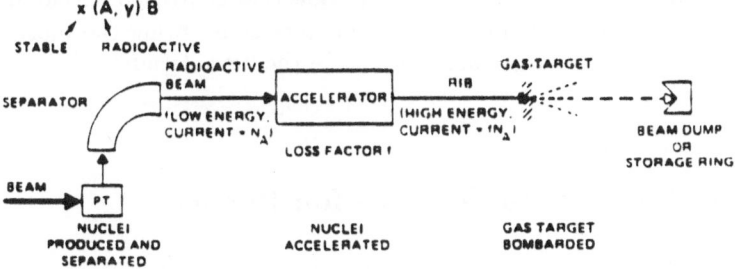

(d) RADIOACTIVE BEAM VERSUS RADIOACTIVE TARGET
(CHARGED PARTICLE REACTIONS)

RADIOACTIVE BEAM

RADIOACTIVE TARGET
(ON-LINE)

RADIOACTIVE TARGET
(OFF-LINE)

1 s   1 mm   1 h   1 d   1 y

HALF-LIFE OF RADIOACTIVE NUCLEI

Fig. 4. Comparison between radioactive targets, either off-line (a) or on-line (b), and radioactive ion beam (c), for the study of charged particle reactions involving radioactive nuclei [Rol88]

2.4 Remarks

As will be shown in Sect. 3, the intensities of the RIB which have been or will be produced and accelerated (in the pico- to nanoampere range) are much lower, by many orders of magnitude, than those of the stable beams currently available (in the micro-to milliampere range). This implies that the measurements of the cross sections for the nuclear reactions of astrophysical interest will require very long counting times and will impose the development of refined techniques to deal with very low signal-to-background ratios, as described in Sect. 4. These drawbacks are at least partly compensated by the following features. Nuclear reactions of astrophysical interest for "normal" (i.e. "low" temperatures) stars or "quiescent" (i.e. very slow) events have to be studied at very low energies [Rol88], much lower than the Coulomb barrier between the partners in the entrance channel (for charged particle reactions), where the relevant cross sections are extremely low (in the pico- to nanobarn range) : this represents a formidable task, even with very high intensity (in the milliampere range) stable beams. When dealing with "high" temperature stars and "fast" events, where nuclear reactions involving radioactive nuclei start to be important, the energy ranges of interest for nuclear astrophysics are higher, in the vicinity of the Coulomb barrier, and the relevant cross sections are larger (in the micro- to millibarn range), making their measurements (somewhat) less difficult.

The solution of the above-mentioned problems with RIB measurements represents a very difficult but very important challenge, as noted by W. Fowler in his 1983 Nobel lecture [Fow84] : "It is my view that continued development and application of radioactive ion beam techniques could bring the most exciting results in laboratory nuclear astrophysics in the next decade".

3. Production of Radioactive Ion Beams

3.1 The Two Methods

Two methods have been adopted so far to produce RIB [Nit90]. The first uses a single accelerator. Stable projectiles are sent on a stable target, wherein they induce a nuclear reaction whose secondary products include the radioactive nuclei of interest. The latter are separated from the primary beam by exploiting the kinematics of the production reaction, and the resulting secondary RIB is directed towards the experimental area. The second method involves two accelerators, separated by an ion source. The first accelerator produces, by a suitable nuclear reaction, the radioactive elements. They are extracted from the (primary) target, and transformed into ions by the ion source. These ions are brought to the required energies by the second accelerator, and sent on the user's (secondary) target.

The first method has been used in two different energy regimes. Low-energy heavy ions, for example 14-MeV ^{7}Li ions, are sent on a target containing a light element, for example a CD_2 foil; the production reaction, for example $^2H(^7Li,^8Li)^1H$, being exploited in the reverse kinematics, yields low-energy forward-peaked secondary RIB, for example 8-MeV ^{8}Li ions at 3°, which can be separated from the primary beam and focussed on the user's target [Boy89]. By this way, low-energy RIB ($\leq$ 3 MeV/nucleon) have been obtained [Boy89, Bec89, Kol90, Bec90], including ^{6}He, ^{7}Be, ^{8}Li and $^{18}F^m$ beams, with however rather low intensities ($10^3 - 10^6$ particles per second). In another approach, high-energy heavy ions (several tens to several hundreds of MeV/nucleon) are fragmented by suitable targets; the resulting secondary (radioactive) fragments are emitted very close to the direction of the primary beam, from which they can be separated by involved and often achromatic analysis techniques. This method has allowed to produce RIB at high energies (tens to hundreds of MeV/nucleon) and with high intensities (up to 10^8 particles per second) [Bim85, Duf86, Ann87, Yam89, Mue90, Alo90, Mun90].

The RIB produced by the single-accelerator method in the two energy regimes thus have, either energies in the range of interest for nuclear astrophysics ($\leq$ 3 MeV/nucleon) but too low intensities to allow useful measurements during reasonable counting times, or rather high intensities but much too high energies to be of interest for direct measurements of the cross sections for astrophysically important nuclear reactions. In the latter case, these beams could however be useful for indirect measurements, i.e. Coulomb desintegration in the field of a heavy target, as outlined in Subsect. 2.2 [Agu90]. The two-accelerator method, on the contrary, allows to produce RIB in the energy region of interest for nuclear astrophysics and with intensities suitable for useful experiments, as will now be shown.

3.2 Radioactive Ion Beams Produced by the Two-Accelerator Method

Radioactive ion beams produced by the two-accelerator method, with energies and intensities suitable for the study of astrophysically interesting nuclear reactions, are only available so far at Louvain-la-Neuve, Belgium, using the following scheme represented in Fig. 5 [Dar90a, Dar90b]. The first accelerator is the CYCLONE 30 cyclotron [Jon87], which is able to produce 15-30 MeV proton beams with very high intensities, up to 500 μA. These beams are bent by a 90° magnet, and focussed by a quadrupole doublet on the production target, which is located in the 3-m concrete wall of the cyclotron vault. This target contains ^{13}C, and very large quantities of ^{13}N nuclei are produced by the $^{13}C(p,n)^{13}N$ reaction [Sin89]. These ^{13}N nuclei are extracted from the target [Dar90c], purified by suitable traps and sent in the ion source, which is of the Electron Cyclotron Resonance (ECR) type [Dec90]. The $^{13}N^{1+}$ ions thereby produced are extracted out of the ECR source by an extraction voltage $\leq$ 10 kV, mass-analyzed by a 90° magnet, transported to the top of the second accelerator, the cyclotron CYCLONE, using two lenses, and axially injected into the machine by a 90° bending magnet followed by a focussing lens. The central region of CYCLONE has been modified to allow its use for acceleration in the sixth harmonic mode [Jon89], necessary to obtain the low energies required for nuclear astrophysics experiments, as for example 0.587 MeV/nucleon ^{13}N nuclei to study the 0.545 MeV resonance in the ^{13}N(p,γ)^{14}O reaction as described in Subsect. 2.1. The ^{13}N^{1+} beam extracted from CYCLONE is finally sent on the user's target by a switching magnet.

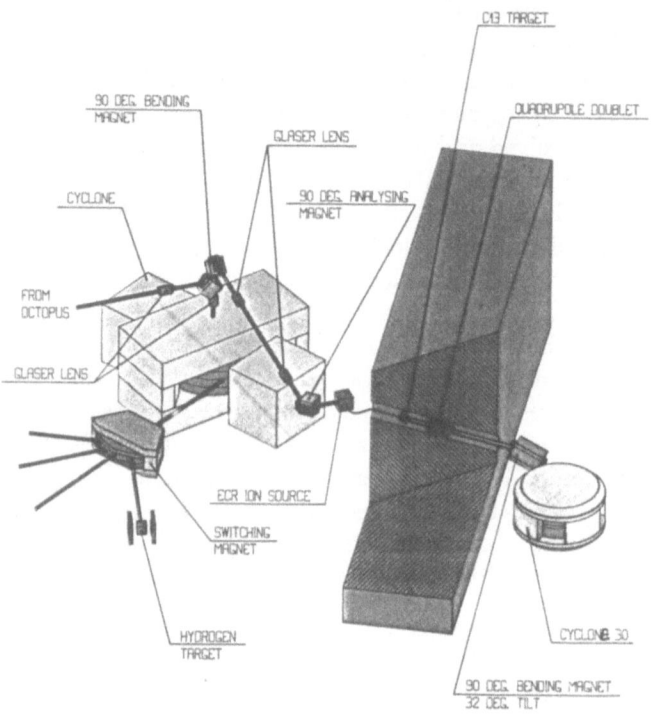

Fig. 5. The Louvain-la-Neuve scheme for the production of an intense beam of ^{13}N^{1+} radioactive ions [Dar90b]

The most recent performances of this equipment, for off-line separate tests of its various elements and on-line global tests of the whole set up, are summarized in Table 2. The thick target yield of the ^{13}C(p,n)^{13}N reaction has been measured for 5-30 MeV protons, its maximum value, at $E_p = 30$ MeV, is 1.6 x 10^{-3} ^{13}N nuclei per incident proton, and it is the highest one for producing ^{13}N with proton beams whose energies are limited to 30 MeV [Sin89]. Extensive tests of natural graphite targets (i.e. containing 1.1 % of ^{13}C), bombarded by 30-MeV proton beams with intensities up to 250 μA, have been carried out [Dar90c]. The ^{13}N activity has been extracted from these targets by a small ($\sim$ 0.1 standard cm^3 per hour) nitrogen support gas flow, which carries ^{13}N as ^{13}N^{14}N molecules. The extraction efficiency $\epsilon_{\text{extraction}}$, defined as the ratio between the number of ^{13}N nuclei extracted from and the number of ^{13}N nuclei produced into the target, is a very sensitive function of the beam power, i.e. of the target temperature, and a value $\epsilon_{extraction} \geq 80$ % has been reached in the most favourable conditions. Enriched ^{13}C targets have been obtained by graphitizing 99 % enriched ^{13}C powder under high temperatures and pressures; they were embedded into natural graphite rods in good thermal contact with a water-cooled copper cylinder. The proton beam current on such targets was 120-200 μA, and their extraction efficiencies were in the 10-30 % range.

Table 2. Radioactive Ion Beam production at Louvain-la-Neuve. I_p is the intensity of the 30-MeV proton beam. ^{13}N/p is the number of ^{13}N nuclei produced per incident 30-MeV proton [Sin89]. $\epsilon_{extraction}$ is the extraction efficiency of the ^{13}N activity out of the ^{13}C target [Dar90c]. $\epsilon_{ionization}$ is the ionization efficiency of the ECR source, defined as the ratio between the number of ^{13}N^{1+} ions produced per incident ^{13}N atom [Dec90]. $\epsilon_{acceleration}$ is the acceleration efficiency of the CYCLONE cyclotron for ^{13}N^{1+} ions at 8.4 MeV. The second column gives the results of the latest on-line run of the full system and the third column, those of separate off-line tests of its various elements. The values of N(^{13}N) and I(^{13}N) given in the third column represent what could be achieved if all the off-line test performances could be obtained simultaneously

Quantity	On-line run	Off-line tests
I_p	120 μA	500 μA
^{13}N/p	1.6×10^{-3}	1.6×10^{-3}
$\epsilon_{extraction}$	32 %	80 %
$\epsilon_{ionization}$	5.5 %	15 %
$\epsilon_{acceleration}$	2.5 %	10 %
N(^{13}N)	2.5×10^8 pps	(6×10^{10} pps)
I(^{13}N)	40 ppA	(10 pnA)

An ECR ion source, especially designed to optimize the ionization efficiency $\epsilon_{ionization}$, defined as the number of atomic ions per injected atom, has been constructed [Dec90]. Extensive tests of this source have been carried out, with different materials (copper or quartz) for the inner wall of the chamber, different types of ions and support gases ... The ionization efficiency is a very sensitive function of the source pressure, and a value $\epsilon_{ionization} = 15$ % has been achieved for ^{13}N^{1+} ions for a pressure of 1.5×10^{-5} mbar. The operation of CYCLONE in the sixth harmonic mode has been tested with ^{13}C^{1+} beams, and its optimum acceleration efficiency $\epsilon_{acceleration}$, defined as the fraction of the ions injected into CYCLONE which are extracted from it after acceleration, is currently about 5 %. This acceleration efficiency is a sensitive function of the residual pressure in the cyclotron vacuum box, which should be in the 10^{-7} mbar range, due to charge exchange collisions during the acceleration procedure [Jon89].

If all the above-mentioned optimal performances of the various elements of this equipment can be simultaneously achieved, an intensity in the range of several 10^{10} particles per second will be obtained for the ^{13}N^{1+} beam, as shown in the third column of Table 2. The latest on-line tests of the whole set up have allowed to reach a $^{13}N^{1+}$ intensity of 3×10^8 particles per second, i.e. 50 particlepicoamperes, with the conditions given in the second column of Table 2. These conditions were significantly worse than the optimum ones referred to above, obtained during separate tests and given in the third column. One of the reason is that the latter are to some extent contradictory. For example, a higher proton beam current produces more ^{13}N nuclei, a higher nitrogen carrier gas flow increases the extraction efficiency of this activity, but both lead to worse vacuum conditions in the ECR source, thereby reducing its ionization efficiency.

A major problem which has been and will be encountered during the production of RIB is the contamination of the RIB by isobaric stable atomic and molecular ions, in the present case $^{13}C^{1+}$ or $^{12}CH^{1+}$ contaminations in the $^{13}N^{1+}$ beam. Since the relative mass differences between ^{13}C and ^{13}N, and between ^{13}N and ^{12}CH, are $1.5 - 1.8 \times 10^{-4}$, separators with very high resolving power have to be used to eliminate these contaminations. The CYCLONE cyclotron is a reasonably good mass analyser, and it does completely discriminate between $^{13}N^{1+}$, $^{13}C^{1+}$ and $^{12}CH^{1+}$ ions.

3.3 Proposals for RIB Production by the Two-Accelerator Method

In the last few years, several proposals for the production of RIB by the two-accelerator method have been elaborated [Buc85b, Nom89, Nom90, Rol89, Haa90, Ilj90, Wal90, Huy90]. Their main characteristics are summarized in [Nit90]. The production accelerator is very often a high energy (600 MeV - 1 GeV) proton machine capable of delivering beams in the several μA to several 100 μA range. Radioactive nuclei are thereby produced by the spallation reaction on heavy targets, which yields very wide ranges of products both with respect to their charge Z and mass A. These nuclei are ionized by suitable sources, arc discharges, FEBIAD sources, surface ionizers or ECR sources, each of which being more favourable for specific chemical elements. An Isotope Separator On Line (ISOL) with a high resolving power is generally used to select the ions thereby produced, at least in mass number A, and, sometimes, to separate isobaric nuclei. The RIB accelerator is very often a combination of linear devices, including : a Radio Frequency Quadrupole (RFQ) accelerator, to bring the RIB at a modest energy (a few keV to a few tens of keV per nucleon) while focussing them in space and bunching them in time ; one or several heavy ion linacs, conventional or superconducting, to accelerate them to the desired energy range (from 1 to several tenths of MeV per nucleon). Strippers are sometimes inserted between the linacs, to increase the average RIB charge and thereby the acceleration efficiency of the subsequent linacs. Most of these proposals include Nuclear Astrophysics in their scientific programs.

4. Special Experimental Techniques in Nuclear Astrophysics Using Radioactive Ion Beams

4.1 Introduction

As stated in Sect. 2, RIB are interesting in Nuclear Astrophysics mainly for measuring the cross sections for nuclear reactions involving radioactive nuclei and other particles, most often protons or alpha particles. This is accomplished by sending the RIB on a target containing hydrogen or helium, and by detecting the products of the reaction, either on-line, i.e. the γ-rays, neutrons, light charged particles and recoil nuclei, or off-line, i.e. through the radioactivity of one of these products. Standard techniques developed to study astrophysically interesting nuclear reactions involving stable nuclei [Rol88] can be used, with however the following extra complications: (1) the present intensities of the RIB are, and they will probably remain, much lower than the ones of the stable beams, by many orders of magnitude (pico-to-nanoampere versus micro-to milli-ampere in some cases); (2) the radioactivity of the beam itself is a strong source of background for the detectors, mainly for the γ-ray detectors, emanating not so much from the beam dump (which can be located at a large distance from the counting area, and heavily shielded) but from the (even tiny) fractions of the beam lost on the beam pipes or collimators during the acceleration, or scattered in the target zone by the target itself; (3) the RIB may have nonnegligible contaminations from isobaric stable beams, atomic or molecular ($C_x H_y$ ions), which may produce "background" reactions in the target. The following experimental techniques have been, or will have to be, developed to overcome the difficulties associated with these complications.

4.2 Targets

Targets containing hydrogen or helium may be solid foils (for hydrogen), closed (i.e. with foil windows) or windowless (i.e. differentially pumped) gas cells (for both) [Rol88]. Titanium or zirconium foils with adsorbed hydrogen gas face two disadvantages: (a) the stopping cross section ϵ of the target, to which the yield Y for a given nuclear reaction is inversely proportional [Rol88], is large, due to the presence of the Ti or Zr "backing"; this reduces the value of Y with respect to pure hydrogen targets or plastic foils; (b) the mean opening angle of the RIB due to multiple scattering in the target is large, due to the high Z of Ti or Zr; this increases the background induced by the beam in the target zone. Plastic foils made of $(CH_2)_n$ molecules, polyethylene or polypropylene, used as hydrogen-containing targets, suffer less disadvantages due to these two points: (a) the loss in yield Y with respect to pure hydrogen targets is only a factor of 2; (b) the beam opening angle is much smaller than for Ti or Zr adsorbed targets, by at least a factor of 10. Their disadvantage lies in their progressive loss of hydrogen, even at beam intensities as low as a few nA, presumably due to beam heating of these thermally isolating materials; this can be overcome by aluminization and/or by the use of rotating targets, and monitored by using suitable proton recoil detectors.

The entrance and exit foil windows of closed gas targets induce the following problems: (i) the energy loss and straggling of the RIB, at the energies of interest in Nuclear Astrophysics and in windows thick enough to stand the target gas pressure, are often prohibitively large; (ii) the large beam opening angle in the two windows (generally made of aluminium or even heavier metals) increases the beam background problem. Windowless differentially pumped gas targets [Bec82] do not face these drawbacks provided the diameters of the entrance and exit holes can be made large enough to intercept a minimal fraction of the RIB, again to minimize the beam background problem. A special target of this type, including several entrance holes with decreasing diameters along the beam direction and differential pumping between these holes, has recently been developed, and optimized to match the (poor) emittance properties of cyclotron beams [Rol90a]. Such windowless gas targets will have to be used in the study of nuclear reactions involving RIB and alpha particles.

In practice, solid hydrogen-containing targets can most easily be made with rather large thicknesses (≥ 100 $\mu gr/cm^2$); the energy loss of the RIB in them is therefore larger than the typical widths of the resonances to be studied in Nuclear Astrophysics (below a few eV in the center-of-mass). As a consequence, they will mostly be used for measuring the cross sections of interest integrated over the investigated resonances. Windowless gas targets, on the other hand, can be made to work with much lower effective thicknesses. They will accordingly be useful to study the shape of the resonances and the nonresonant contributions to the cross section, provided the intensities of the RIB can be made large enough to yield reasonable signal-to-background ratios with thin targets.

One possible problem related to the use of natural hydrogen targets, solid or gaseous, lies in the presence of a fraction 1.5×10^{-4} of deuterium in them: when investigating (p,γ) reactions of low relative energies, with expected very low cross sections, the contribution of (d,n) reactions on the same initial nucleus leading to the same final nucleus, with possibly much higher cross sections, could represent a sizable "background" problem; the latter can be investigated by repeating the experiment with deuterium-containing targets, deuterated polyethylene foils or deuterium gas cells.

4.3 On-line Detectors

The astrophysically-interesting nuclear reactions with RIB include many capture reactions, (p,γ) or (α,γ). Therefore, the on-line detection of the capture γ-rays, which lie in the 4-10 MeV range, is a natural way to study these reactions. Due to the complications related to the use of RIB noted in Subsect. 4.1, these γ-ray detectors should have the following properties, which are unfortunately sometimes contradictory: (1) a high efficiency, to compensate the low beam intensity; (2) a low background rate, for the same reason; (3) some energy selection capability, to discriminate the capture γ-rays from those arising from the beam radioactivity, most often the 0.511 MeV annihilation radiation; (4) a high counting-rate capacity, to tolerate the high contribution from the direct or scattered radioactive beam in the target zone; (5) a good energy resolution to detect the capture γ-rays due to the RIB in the presence of those from nuclear reactions induced by the stable contaminant beams, in cases when the latter contribution is large. Good detector characteristics with respect to the latter property also increases the signal-to-background ratio, by concentrating the signal pulses in narrower peaks in the spectrum; (6) reasonable timing properties, if the RIB accelerator is pulsed. This also yields a better signal-to-noise ratio, if one registers the capture γ-rays in time coincidence with the beam pulses.

Standard γ-ray detectors fulfill some, but not all, of the above requirements. Arrays of several large NaI(Tl) scintillators are suitable for items (1) to (4) and (6) above, but not for (5) if the contaminant γ-ray spectrum has some complexity. Batteries of BaF_2 scintillators, which may be used as pair spectrometers [Top90], are even better than NaI(Tl) detectors for items (1), (4) and (6), due to their higher effective Z and fast light component, but worse for item (5). Large volume Ge diodes, with efficiencies close to 100 % of the ones of 3" × 3" NaI(Tl) scintillators, are definitely superior to the latter detectors for item (5), and somewhat worse for item (4) if the full energy resolution has to be preserved. Arrays of several such diodes have to be used in order to yield a sufficient detection efficiency; this is especially important if their distance to the target has to be increased, in order to lower the Doppler broadening of the capture γ-rays due to the high velocity of the recoiling nuclei in the reverse kinematics.

Special γ-ray threshold detectors have recently been developed [Seu87], made of heavy water and including ^{3}He neutron proportional counters: γ-rays with energies larger than 2.226 MeV induce the photodesintegration of the deuterium in the heavy water, the neutrons thereby produced are slowed down in the heavy water and detected in the ^{3}He counters. This apparatus is excellent for item (3), since it is completely insensitive to the 0.511 MeV radiations arising from the beam, and for item (4), for the same reason. It has however a very low sensitivity (about 10^{-4}), a high background rate (0.7 count/sec), no γ-ray energy resolution and no time resolution. Its main use lies in the study of nuclear reactions induced by stable beams on radioactive targets, for instance the ^{22}Na(p,γ)^{23}Mg reaction [Seu90].

It should be noticed that the performances of the γ-ray detectors with respect to item (2) above can be improved by registering their pulses in anticoincidence with an "umbrella" of plastic scintillators, which detect the charged cosmic rays; such a scheme reduces the "cosmic" background in the 3-10 MeV range of γ-ray energy by about a factor of 2 [Lel90a].

Some astrophysically interesting reactions produce neutrons, for example (α,n) and (d,n) reactions in inhomogeous big bang nucleosynthesis [Mal91]. In this case, a high-efficiency neutron detector with a solid angle close to 4π, such is effectively the γ-ray threshold detector just described [Seu87], would be of considerable interest, with its larger than 20 % efficiency for MeV neutrons.

The detection of light charged particles can be carried out using standard techniques [Rol88]. It can be anticipated that the radioactivity of the beam, mainly positrons, will induce supplementary problems, which still have to be investigated.

The detection of the recoil nuclei, when studying capture reactions with RIB in the reverse kinematics, represents another promising method to measure the cross sections for these reactions. These recoil nuclei have to be separated from the (much more numerous) beam particles, using magnetic analysis and/or Wien filter techniques. These methods, and their associated problems, have been discussed previously [Rol88]. The use of RIB does not introduce extra complications with respect to stable beams, except their low intensities and possible isobaric contaminations noted in Subsect. 4.1. A solution of the extra problems thereby induced could be the use of the recoil-nuclei separators in coincidence with large arrays of γ-ray detectors located around the target, with the specifications outlined above. In general, the separation of the recoil nuclei from the primary beam should be less difficult for (α,γ) than for (p,γ) reactions, due to the larger relative mass difference between the two for (α,γ). Promising developments along these lines have recently been achieved [Rol90b].

An example of the apparatus which has recently been used to study the $^{13}N(p,\gamma)^{14}O$ reaction with the $^{13}N^{1+}$ RIB of Louvain-la-Neuve is shown in Fig. 6. It has yielded the first results on a radiative proton capture reaction with a short-lived radioactive nucleus ever obtained [Dec91]. A value for the partial γ-width Γ_γ of the 5.173 MeV level in ^{14}O (Subsect. 2.1): $\Gamma_\gamma = 3.8(1.2)$ eV has been obtained by this direct method, which is in agreement with but has a lower relative experimental uncertainty than the results deduced from an undirect method, i.e. 2.7(1.3) eV ([Fer89], Subsections 2.1 and 2.2).

4.4 Off-line Detectors

Within the products of nuclear reactions induced by RIB on stable targets, some may be radioactive. The relevant cross sections can then be determined by measuring the activity of the latter collected on a suitable catcher, provided their decay characteristics, halflives and/or emitted radiations, are sufficiently different from those of the RIB nuclei. If the halflife of the products is much larger than the one of the RIB, a long enough decay time of the reaction products allows their detection free of the RIB activity; an example would be the investigation of ^{22}Na ($T_{1/2} = 2.6$ y) during a study of the $^{21}Na(p,\gamma)^{22}Mg$ reaction quickly followed by the β^+ decay of ^{22}Mg to ^{22}Na with $T_{1/2} = 3.9$ s; the ^{21}Na halflife is 22 sec.

Most reaction products and RIB nuclei are positron emitters, with however sometimes quite different β^+ maximum energies $E_{\beta+}$: when $E_{\beta+}$ for the latter is much larger than for the former, the two can be discriminated by suitable β^+ spectrometers. This is the case for ^{13}N ($E_{\beta+} = 1.2$ MeV) and ^{14}O (1.8 MeV) during the study of the $^{13}N(p,\gamma)^{14}O$ reaction, and even more so for $^{11}C(1.0$ MeV) and $^{12}N(16.4$ MeV) in the $^{11}C(p,\gamma)^{12}N$ reaction, $^{19}Ne(2.2$ MeV) and $^{20}Na(11.2$ MeV) in the $^{19}Ne(p,\gamma)^{20}Na$ reaction, whose astrophysical interest

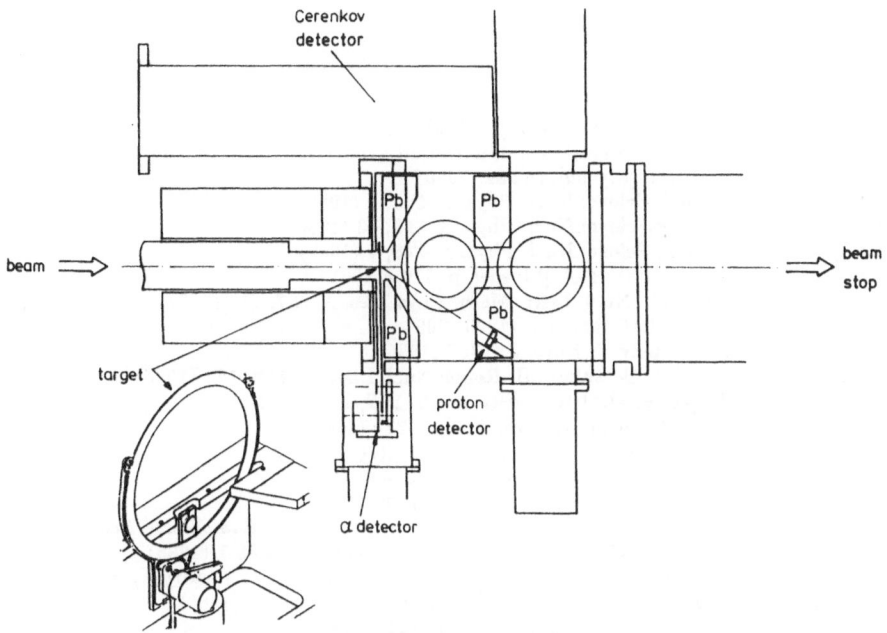

Fig. 6. Apparatus used at Louvain-la-Neuve to measure the cross section for the ^{13}N(p,γ)^{14}O reaction using a ^{13}N^{1+} Radioactive Ion Beam [Dec91]

has been outlined in Sect. 2. A new β^+ spectrometer, with special magnetic coils and positron detectors, has recently been proposed for such uses [Lel90b].

In some cases, the reaction products emit characteristic γ-ray lines, while the RIB nuclei are pure β^+ emitters i.e. only yield the 0.511 MeV radiation; an example is ^{14}O (E$_\gamma$ = 2.313 MeV) with respect to ^{13}N. Their activity can then be measured through these γ-rays, provided the γ-ray spectrometer can tolerate the high counting rate due to the 0.511 MeV radiation, which can be reduced by suitable lead shields in front of the detector.

5. Conclusions

A large range of exciting new experiments in nuclear astrophysics has recently been made possible thanks to the availability of RIB with the required energies and with high intensities, using the two-accelerator approach. Many developments still have to be carried out, both in the RIB themselfves - intensity, purity, variety -, and in the experimental techniques to measure the cross sections for astrophysically interesting reactions - new detectors, with very demanding performances with respect to their efficiency, background level, counting rate capabilities ... When these progresses in instrumentation will be achieved, a new era in Experimental Nuclear Astrophysics will be opened.

References

Agu89   P. Aguer et al.: Ref. [Kub89], (1989) 107;
         P. Aguer et al.: Private communication
Agu90   P. Aguer: Ref. [Mye90], p. 239 (1990)
Ajz85   F. Ajzenberg-Selove: Nucl. Phys. A **433**, 1 (1985)
Ajz86   F. Ajzenberg-Selove: Nucl. Phys. A **449**, 53 (1986)
Ajz87   F. Ajzenberg-Selove: Nucl. Phys. A **475**, 1 (1987)
Alo90   J.R. Alonso et al.: Ref. [Mye90], p. 112 (1990)
Ann87   R. Anne et al.: Nucl. Instr. Meth. Phys. Res. A **257**, 215 (1987)
Arn90   M. Arnould: Ref. [Mye90], (1990) 209
Bar85   F. Barker: Austr. J. Phys. **38**, 757 (1985)
Bau86   G. Baur, C.A. Bertulani, H. Rebel: Nucl. Phys. A **458**, 188 (1986)
Bec82   H.W. Becker et al.: Nucl. Instr. Meth. **198**, 277 (1982)
Bec89   F.D. Becchetti et al.: Ref. [Kub89], p. 277 (1989)
Bec90   F.D. Becchetti et al.: Phys. Rev. C **42**, R801 (1990)
Bim85   R. Bimbot et al.: Zeits. f. Phys. A **322**, 443 (1985)
Boy89   R.N. Boyd et al.: Ref. [Kub89], (1989) 39
Buc85a   L. Buchmann, J.M. d'Auria, (eds): *Proc. Accelerated Radioactive Beams Work-shop*, Parkville, Canada, Sept. 5-7, 1985, Report TRI-85-1 (1985)
Buc85b   L. Buchmann, J.M. d'Auria et al.: *The TRIUMF-ISOL facility : a proposal for an intense radioactive beams facility*, TRIUMF Report (1985)
Chu85   T.E. Chupp et al.: Phys. Rev. C **31**, 1023 (1985)
Dar90a   D. Darquennes et al.: Ref. [Mye90], (1990) 3
Dar90b   D. Darquennes, P. Decrock, Th. Delbar, W. Galster, M. Huyse, Y. Jongen, M. Lacroix, P. Leleux, I. Licot, E. Liénard, P. Lipnik, M. Loiselet, G. Ryckewaert, Sindano Wa Kitwanga, P. Van Duppen, J. Vanhorenbeeck, J. Vervier, S. Zaremba: Phys. Rev. C **42**, R804 (1990)
Dar90c   D. Darquennes, Th. Delbar, P. Lipnik: Nucl. Instr. Meth. Phys. Res. B **47**, 311 (1990)
Dec90   P. Decrock, P. Van Duppen, F. Baeten, C. Dom, Y. Jongen: Rev. Sc. Instr. **61**, 279 (1990)
Dec91   P. Decrock, Th. Delbar, P. Duhamel, W. Galster, M. Huyse, P. Leleux, I. Licot, E. Liénard, P. Lipnik, M. Loiselet, G. Ryckewaert, P. Van Duppen, J. Vanhoren-beeck, J. Vervier, M. Arnould: To be published
Des89   P. Descouvemont, D. Baye: Nucl. Phys. A **500**, 155 (1989)
Des90a   P. Descouvemont, I. Baraffe: Nucl. Phys. A **514**, 66 (1990)
Des90b   P. Descouvemont, D. Baye: Ref. [Obe90], (1990) 174
Duf86   J.P. Dufour et al.: Nucl. Instr. Meth. Phys. Res. A **248**, 267 (1986)
Fer89   P.B. Fernandez, E.G. Adelberger, A. Garcia: Phys. Rev. C **40**, 1887 (1989)
Fil83   B.W. Filippone et al.: Phys. Rev. C **28**, 2222 (1983) and references therein
Fow84   W. Fowler: Rev. Mod. Phys. **56**, 149 (1984)
Fun87   C. Funck, K. Langancke: Nucl. Phys. A **464**, 90 (1987); K. Langancke, private communication
Haa90   H. Haas et al.: Ref. [Mye90], p. 59
Huy90   M. Huyse et al.: *The ARENAS³ project*, Internal Report (1990)
Ilj90   A.S. Iljinov, V.M. Lobashev, Yu.Ts. Oganessian: Ref. [Mye90], p. 23 (1990)
Jon87   Y. Jongen, J.L. Bol, A. Chevalier, M. Lacroix, G. Ryckewaert: Nucl. Instr. Meth. Phys. Res. B **24/25**, 813 (1987)
Jon89   Y. Jongen, M. Lacroix, M. Loiselet, A. Ninane, G. Ryckewaert, S. Zaremba: IEE Trans. Nucl. Sc. 36, 1620 (1989)
Jor89   A. Jorissen, M. Arnould: Astron. Astrophys. **221**, 161 (1989)
Käp91   F. Käppeler: In *Nuclei in the Cosmos*, ed. by H. Oberhummer (Springer, Berlin 1991) this volume
Kla91   H.V. Klapdor-Kleingrothaus: In *Nuclei in the Cosmos*, ed. by H. Oberhummer (Springer, Berlin 1991) this volume
Koe88a   P.E. Koehler et al.: Phys. Rev. C **37**, 917 (1988)
Koe88b   P.E. Koehler, H.A. O'Brien: Phys. Rev. C **38**, 2019 (1988)

Kol90   J.J. Kolata et al.: Ref. [Mye90], (1990) 201
Kos90   S. Kossionides et al.: Ref. [Obe90], (1990) 49
Kub88   S. Kubono et al.: Zeits. Phys. A **331**, 359 (1988)
Kub89   S. Kubono, M. Ishihara, T. Nomura, (eds): *Proc. Int. Symp. Heavy Ion Physics and Nuclear Astrophysical Problems*, Tokyo, Japan, July 21-23, 1988 (World Scientific 1989)
Kub90   S. Kubono et al.: Ref. [Mye90], (1990) 220
Lan85   K. Langancke, O.S. Van Roosmalen, W.A. Fowler: Nucl. Phys. A **435**, 657 (1985)
Lel90a   P. Leleux et al.: private communication (1990)
Lel90b   P. Leleux, P. Lipnik: Ref. [Obe90], p. 6 (1990)
Mal91   R.M. Malaney: In *Nuclei in the Cosmos*, ed. by H. Oberhummer (Springer, Berlin 1991) this volume
Mat84   G.J. Mathews, F.S. Dietrich: Astr. J. **287**, 969 (1984)
McC90   J.B. McClelland, D.J. Vieira, (eds): *Proc. Workshop on the Science of Intense Radioactive Ion Beams*, Los Alamos, USA, April 10-12, 1990, Report LA-11964-C/UC-413 (1990)
Mue90   A.C. Mueller, R. Anne: Ref. [Mye90], p. 132 (1990)
Mun90   G. Münzenberg et al.: Ref. [Mye90] (1990) 91
Mye90   W.D. Myers, J.M. Nitschke, E.B. Norman, (eds): *Proc. First Int. Conf. Radioactive Nuclear Beams*, Berkeley, U.S.A., Oct. 16-19, 1989, (World Scientific 1990)
Nit90   J.M. Nitschke: Ref. [McC90] (1990) 260
Nom89   T. Nomura: Ref. [Kub89] (1989) 295
Nom90   T. Nomura: Ref. [Mye90] (1990) 13
Obe90   H. Oberhummer, W. Hillebrandt, (eds): *Proc. Int. Symp. Nuclear Astrophysics "Nuclei in the Cosmos"*, Baden/Vienna, Austria, June 18-22, 1990, (Max-Planck Institut für Physik und Astrophysik, Garching 1990)
Par85   P. Parker: Ref. [Buc85a] (1985) 145
Rol88   C.E. Rolfs, W.S. Rodney: *Cauldrons in the Cosmos*, The University of Chicago Press (1988)
Rol89   C. Rolfs, B. Jonson, B.W. Allardyce, H. Haas, H. Ravn: *A concept for postacceleration of radioactive ions for measurements in astrophysics*, CERN Report (1989)
Rol90a   C. Rolfs, G. Roters et al.: private communication (1990)
Rol90b   C. Rolfs, C. Barnes et al.: private communication (1990)
Seu87   S. Seuthe et al.: Nucl. Instr. Meth. Phys. Res. A **260**, 33 (1987)
Seu90   S. Seuthe et al.: Nucl. Phys. A **514**, 471 (1990)
Sin89   Sindano Wa Kitwanga, P. Leleux, P. Lipnik, J. Vanhorenbeeck: Phys. Rev. C**40**, 35 (1989)
Thi91   F.-K. Thielemann: In *Nuclei in the Cosmos*, ed. by H. Oberhummer (Springer, Berlin 1991) this volume
Top90   M. Topheide et al.: Ref. [Obe90] (1990) 16
Wag90   C. Wagemans, S. Druyts: Ref. [Obe90] (1990) 286
Wal81   R. Wallace, S.E. Woosley: Astr. J. Suppl. **45**, 389 (1981)
Wal90   G. Walker: Ref. [McC90], p. 183 (1990)
Wie89   M. Wiescher et al.: Astr. J. **343**, 352 (1989)
Woo90   S. Woosley, D. Hartmann, R. Hofmann, W. Haxton: Ap. J. **356**, 272 (1990)
Yam89   T. Yamagata et al.: Ref. [Kub89] (1989) 209

Direct Reaction Mechanism
in Astrophysically Relevant Processes

H. Oberhummer[1] and G. Staudt[2]

[1] Institut für Kernphysik, TU Wien, Wiedner Hauptstr. 8-10,
  A-1040 Wien, Austria
[2] Physikalisches Institut, Univ. Tübingen, Auf der Morgenstelle 14,
  D-7400 Tübingen, Germany

1. Introduction

Nuclear reactions in the early universe as well as in stars are among the most important driving forces for the evolution of our universe. These reactions not only power the energy production in stars, but they are also responsible for the concept of nucleosynthesis, i.e. the creation of the elements. Theoretical models for the description of these nuclear processes are essential because in most cases it is not possible to determine reaction rates directly from experiments. The reason for this is that the thermonuclear energies relevant for astrophysical processes are well below the Coulomb and/or centrifugal barrier. At these energies the reaction rates are often too small for measurements in accelerator experiments. Therefore theoretical models are essential for the determination of nuclear reaction rates at thermonuclear energies. Only with the help of such models the reaction rates can be extrapolated from experimental data at higher energies to the thermonuclear range.

This article is intended to give a simple and straightforward discussion about the importance of the direct reaction mechanism in astrophysically relevant nuclear processes. It has been realized in the last years that the direct process is important at these energies and that it can be the dominant mechanism for reactions in primordial nucleosynthesis as well as in stellar burning. Our aim is to provide an overview of the various theoretical methods used in describing direct nuclear reactions for astrophysical purposes. No attempt has been made to give a complete review of the theoretical work, instead some examples for direct reactions have been chosen to illustrate the key points.

In the next section the qualitative features of nuclear reaction mechanisms are discussed. The variety of nuclear reactions which occur in astrophysical scenarios and the relevant quantities, such as the astrophysical S-factor and the reaction rates, are introduced in Sect. 3. In Sect. 4, the progress in the treatment of optical potentials, which are an essential input for the potential model, are discussed. Sect. 5 describes the various theoretical approaches used in the theory of direct reactions. Finally, Sect. 6 shows some selected examples to demonstrate the importance of the direct model in astrophysical processes.

2. Qualitative Features of Nuclear Reaction Mechanisms

In nuclear reactions two extreme cases can be observed, compound nucleus (CN) and direct reaction (DI) processes.

1. The CN process:

   The nature of the CN mechanism was clarified more than 50 years ago by N. Bohr [Boh36, Kal37]. In this mechanism the projectile merges in the target nucleus and excites many degrees of freedom of the CN. The excitation proceeds via a multistep process and therefore has a reaction time typically of the order 10^{-16} s to 10^{-18} s. After this time the CN decays into various exit channels. The relative importance of the decay channels is determined by the branching ratios to the final states.

2. The DI process:

   The importance of the DI process was first exemplified by S.T. Butler [But50, But51, Tho51] about 40 years ago. The projectile excites only a few degrees of freedom (e.g. single-particle or collective). The excitation proceeds in one single step and has a characteristic time scale of 10^{-21} s to 10^{-22} s. This corresponds to the time the projectile needs to pass through the target nucleus; this time is much shorter than the reaction time of CN processes. However, at subCoulomb energies the reaction is now hindered by the Coulomb and centrifugal barriers. Therefore, the characteristic time scale is enhanced by a factor determined by the barrier penetration probabilities.

The question whether a given reaction favours CN or DI processes depends on the reaction considered and on the relative energies in the entrance and exit channel. In general, a given reaction involves both types of reaction mechanisms and also intermediate types (e.g. precompound reactions). However, for certain reactions and projectile energy ranges one type of reaction mechanism may dominate.

In thermonuclear scenarios the projectile energy is well below the Coulomb barrier. At these energies the competition between different reaction mechanisms is quite complicated. If the overlap between entrance- and exit-channel wave functions is large, then the DI process will be enhanced. This is the case, because subCoulomb reactions take place well outside the nucleus, so that the projectile does not have to tunnel all the way through the Coulomb barrier. On the other hand the CN process is suppressed, because the overlap with the confined bound-state type wave functions, which are responsible for the CN process, is very small due to the Coulomb and/or centrifugal barrier. At subCoulomb energies the CN formation may be further suppressed, because there exist no CN levels that can be populated, especially in light nuclei.

For projectile energies near or above the Coulomb and/or centrifugal barrier the overlap with the confined CN states increases and finally becomes dominant. Therefore, nuclear reactions above the Coulomb and/or centrifugal barrier generally proceed via the CN process. Finally, at still higher energies (greater than approximately 10 to 20 MeV) the CN process for transitions to low-lying states of the final nucleus is suppressed because of the many open exit channels. So the DI is again dominant.

Closely connected to the DI and CN mechanisms are the resonance types in the nuclear reaction cross sections. There are again two extreme cases for resonance structures. The first type (compound resonance) occurs in CN reactions and has a narrow width (order of 1 to 100 eV). This width is determined by the long reaction time corresponding to the many steps involved in the formation and decay of the CN.

The second type is called potential, optical, single-particle, potential-well or scattering resonance. This resonances type is associated with DI and is determined by the optical potential in the entrance or exit channel. Usually the widths of potential resonances are broad (order of 10 keV to a few MeV) corresponding to the short reaction times of the DI. These resonances lose their meaning for higher energies because of their broad widths, as a result of absorption into other channels. However, for energies below the Coulomb and centrifugal barriers the width of potential resonances may become very narrow due to increasing life times. Extremely narrow potential resonances can even mimick CN resonances.

Potential resonances observed in elastic scattering of ^{3}H+α, ^{3}He+α and $\alpha + \alpha$ have been discussed in [Wil77]. For inelastic scattering, such resonances have been described using coupled-channel calculations [Cha58, Tom60, Oka62, Bar67]. Most of these calculations have only model character because of the simplifying assumptions used in the optical potentials (e.g. square-well potentials). One of the observations of these papers is the fact that states with narrow widths are broadened, and the resonance energies are shifted when considering the transition to another state with a lower barrier [Lem64, McV67, Del68]. Examples for potential resonances in transfer and capture reactions will be given in this article in Sect. 6.

Many of the models used until now for the DI mechanism in transfer and capture reactions at subCoulomb energies are not well suited to describe potential resonances. One reason for this shortcoming is that the approximations used are often too crude. For instance, for sub-Coulomb energies often only the asymptotic form of the wave functions in the entrance and/or exit channel is assumed, i.e. only the Coulomb and/or centrifugal potential is considered. Such simplified calculations for subCoulomb energies have been carried out for transfer [May68, Obe89] as well as for capture reactions [Chr61, Tom63, Rol73]. In these calculations it is not possible to obtain any resonances. On the other hand, first-principle methods rarely reproduce the experimental energies and widths of potential resonances, because these methods are not flexible enough to fit the experimental data. Both the above shortcomings can be overcome by the potential models discussed later in this contribution.

In astrophysical applications the DI models cause problems when the level density of the unified nucleus near the threshold gets high (e.g. for heavy target nuclei and/or projectiles). Then the DI models may fail in reproducing all important potential resonances. In this case the DI approach is not adequate to provide useful information for astrophysics. The same is the case in hot astrophysical scenarios, because the projectile energies are close to the energy of the Coulomb barrier.

3. Nuclear Reactions in Astrophysical Scenarios

The four known interactions in nature (gravitation, weak, electromagnetic and strong interaction) determine the evolution of the universe as well as stellar evolution. The gravitational force is responsible for the structure and dynamics in the cosmos on a long-range scale. The other three forces are important in the nuclear processes of astrophysical interest. In this article we only discuss nuclear reactions and not decay processes (alpha-, beta- and gamma-decay and nuclear fission). The importance of the latter processes in astrophysical scenarios will be treated in other articles of this volume [Käp91, Kla91, Thi91, Ver91].

The determination of nuclear reaction rates for astrophysical applications requires the knowledge of cross sections at thermonuclear energies, i.e. in the region from keV to MeV. Three types of reactions can be observed: reactions induced by neutral particles (n, γ, ν), light ions (p, d, t, τ, α) and heavy ions (^{12}C, ^{16}O, ^{20}Ne, ^{28}Si). Neutron-induced reactions, mainly of the form (n,γ) are involved in primordial nucleosynthesis [Mal91] as well as in the nucleosynthesis beyond iron in the form of the s- [Käp91] and r-process [Kla91]. The photoinduced reactions (γ,p), (γ,α) and (γ,n) are responsible for the photodisintegration of nuclei in advanced burning stages, such as silicon burning in stars. The neutrino processes (ν,ν'n) and (ν,ν'p) play a role in the ν-process in supernovae [Woo90]. In primordial as well as in stellar nucleosynthesis (hydrogen and helium burning) reactions with light ions like (p,d), (p,γ), (p,α), (α,γ) and (α,n) are the dominant processes. Finally, heavy-ion induced reactions drive the advanced stages of carbon, neon, oxygen and silicon burning.

While the cross section for neutral particles has a more or less energy-independent behaviour, the cross section $\sigma(E)$ of charged-particle induced reactions below the Coulomb barrier drops sharply with decreasing energy E. Therefore, the astrophysical S-factor for charged-particle induced reactions is defined by [Bet67]

$$S(E) = E \, \sigma(E) \, \exp(2\pi\eta) \tag{3.1}$$

with the Sommerfeld parameter

$$\eta(E) = \frac{e^2 Z_a Z_A}{\hbar} \sqrt{\frac{\mu}{2E}} \quad . \tag{3.2}$$

In these equations, E and μ are the centre-of-mass (CM) energy and reduced mass in the entrance channel, and Z_a and Z_A denote the proton numbers of the projectile and target nucleus, respectively.

The approximations, under which the penetration probability through the Coulomb barrier

$$P = \exp(-2\pi\eta) \qquad (3.3)$$

in the definition of the S-factor, Eq. (3.1) was derived, are the following:

  i. Quasiclassical method (WKB technique) [Wen26, Kra26, Bri26].
 ii. Square-well potential inside and Coulomb potential of a point source outside the nuclear radius.
iii. Height of Coulomb barrier is much larger than the projectile energy.
 iv. No centrifugal barrier in the entrance channel, i.e. only s-wave projectiles.

In general, it is difficult to measure the extremely small cross sections at thermonuclear energies; so one has to rely on extrapolations from higher energies. The S-factor varies much less with projectile energy than the cross section. Therefore, the S-factor is more suited for extrapolation to thermonuclear energies than the cross section. But in order to avoid an incorrect extrapolation some aspects of quite different nature, which are responsible for an energy depence of the S-factor, have to be taken into account. First of all the approximations (i) to (iv), which are inherent in the definition of the astrophysical S-factor, generally result in a smooth energy-dependence of the S-factor. But resonances, direct reaction contributions and screening may alter this energy dependence drastically:

1. Resonances:
   A strong energy dependence of the S-factor is caused by resonances. As already discussed in Sect. 2, these resonances can be due to CN formation as well as DI processes. The widths of the resonances extend over a wide range (eV to MeV). The S-factor can vary rapidly as a function of energy in the case of very small resonance widths, or more smoothly in the case of broad resonances. Resonances near the particle threshold (just below or above the threshold) can completely dominate the reaction rate at thermonuclear energies.

2. DI mechanism:
   In the DI mechanism the reaction may take place well outside the nucleus. That means that important contributions to the cross sections also come from outside the nuclear surface resulting in an enhancement of the S-factor, because the projectile avoids tunneling all the way through the Coulomb barrier.

3. Screening:
   The cross sections may be enhanced at very low energies by screening effects of the electrons in nuclei or in plasma. These screening effects will be discussed in another article of this volume [Lan91].

Summarizing these different aspects, one can find for the energy dependence of the S-factor a nonresonant behaviour, broad resonances ($\frac{\Gamma_R}{E_R} \gtrsim 10\%$) and small resonances ($\frac{\Gamma_R}{E_R} \lesssim 10\%$), where Γ_R is the width and E_R the energy of the resonance.

The quantity of interest in the calculation of astrophysical features, such as time scales, energy production and nucleosynthesis of the elements, is the reaction rate per particle pair averaged over the velocity distribution [Rol88]

$$\langle \sigma v \rangle = \sqrt{\frac{8}{\pi \mu}} \left(\frac{1}{kT} \right)^{\frac{3}{2}} \int_0^\infty \sigma(E) \, E \, \exp\left(-\frac{E}{kT} \right) dE \qquad (3.4)$$

for a given temperature T.

The reaction rate can be parametrized in the following ways [Rol88]:

1. Neutron-induced nonresonant reactions:

$$\langle \sigma v \rangle = S(0) + \sqrt{\frac{4}{\pi}} \dot{S}(0)\sqrt{kT} + \frac{3}{4} \ddot{S}(0)kT + \dots \quad , \qquad (3.5)$$

where the dotted quantities represent derivatives with respect to E.

2. Charged-particle nonresonant reactions:

$$\langle \sigma v \rangle = \sqrt{\frac{2}{\mu}} \left(\frac{\Delta}{kT} \right)^{\frac{3}{2}} S(E_0) \exp\left(-\frac{3E_0}{kT} \right) \qquad (3.6)$$

with

$$\Delta = \frac{4}{\sqrt{3}} \sqrt{E_0 kT}$$

and

$$E_0 = (2\mu)^{\frac{1}{3}} \left(\frac{\pi e^2 Z_a Z_A kT}{2\hbar} \right)^{\frac{2}{3}} .$$

3. Reactions through a broad resonance:

$$\langle \sigma v \rangle = \alpha_1 T^m \exp\left(-\frac{\alpha_2}{kT} \right) \quad , \qquad (3.7)$$

where α_1, α_2 and m are adjustable parameters.

4. Reactions through a narrow resonance:

$$\langle \sigma v \rangle = \left(\frac{2\pi}{\mu kT} \right)^{\frac{3}{2}} \hbar^2 (\omega\gamma)_R \exp\left(\frac{-E_R}{kT} \right) \quad , \qquad (3.8)$$

where $(\omega\gamma)_R$ and E_R are the resonance strength and width, respectively.

4. Optical Potentials

In the optical model the complicated many-body problem of two colliding nuclei is reduced to the much simpler problem of two particles interacting by a potential, the optical potential. Such a replacement requires the restriction to a model space containing only one or few reaction channels.

The optical model is a one-channel model. It provides the wave functions for the relative motion of the colliding particles. On one hand, the knowledge of these wave functions allows to calculate differential cross sections as well as excitation functions including potential resonances in the elastic scattering channel. On the other hand, these wave functions are applied as "distorted waves" in DI theories. Therefore, optical potentials play an important role in cross-section calculations in direct transfer and capture reactions.

Formally the restriction to a model space containing only one or few reaction channels implies that the total wave function Ψ will be replaced by a model wave function Ψ_M. The total wave function Ψ for the colliding system obeys the complete Schrödinger equation

$$(E - H)\Psi = 0 \quad . \tag{4.1}$$

This wave function may be expanded in terms of a complete set of the internal states $\psi_a^{(\alpha)}(x_a)$ and $\psi_A^{(\alpha)}(x_A)$ of the two colliding nuclei a and A which form the partition α:

$$\Psi = \sum_\alpha \psi_a^{(\alpha)}(x_a) \, \psi_A^{(\alpha)}(x_A) \, \chi_\alpha(r_B - r_A) \equiv \sum_\alpha \psi_\alpha(x_\alpha)\chi_\alpha(r_\alpha) \quad . \tag{4.2}$$

The sum in the above expression is extended over all excited states of both nuclei. The wave function $\chi_\alpha(r_\alpha)$ describes the relative motion of the two nuclei a and A.

In practice it is impossible to use the full infinite expansion of Eq. (4.2). A practicable model can be obtained by limiting this sum and including only those channels which couple strongly to the entrance channel. Then the model wave function Ψ_M constructed in this way only represents a small part of the total wave function Ψ. As a consequence of this restriction it follows that the coupling between the model part and the rest must be represented by an effective interaction. This model is related closely on one hand to the resonating group theory of Wheeler [Whe37, Wil77] and on the other hand to the coupled channels formalism with a few inelastic channels included.

In the optical model only one channel, the entrance channel, is considered in which the nuclei a and A are in their ground states. Therefore, the model wave function Ψ_M is reduced to its simplest form

$$\Psi_M = \psi_0(x_0)\chi_0(r_0) \quad . \tag{4.3}$$

Using projection operator technique [Fes58, Fes62] the Schrödinger equation (4.1) splits into a pair of coupled equations

$$\left(E - PHP - PHQ\frac{1}{E + i\epsilon - QHQ}QHP\right) P\Psi = 0 \tag{4.4a}$$

$$\left(E - QHQ - QHP\frac{1}{E + i\epsilon - PHP}PHQ\right) Q\Psi = 0 \tag{4.4b}$$

with the projection operators

$$P = |\psi_0\rangle \langle\psi_0| \tag{4.5a}$$

$$Q = \sum_{\alpha=1}^{\infty} |\psi_\alpha\rangle \langle\psi_\alpha| \quad . \tag{4.5b}$$

The total Hamiltonian H is the sum of the internal Hamiltonians H_α of the nuclei a and A, of the kinetic energy of their relative motion T_α and of their mutual interaction V_α:

$$V_\alpha = \sum_{i=1}^{a} \sum_{j=1}^{A} v_{ij} \quad . \tag{4.6}$$

Using the fact that $H_\alpha + T_\alpha$ has no matrix elements connecting the two spaces since $P\Psi$ and $Q\Psi$ contain different internal states, Eq. (4.4a) can be rewritten as

$$\left(E - T_0(\mathbf{r}_0) - PV_\alpha P - PV_\alpha Q \frac{1}{E + i\epsilon - QHQ} QV_\alpha P \right) \chi_0(\mathbf{r}_0) = 0 \quad . \tag{4.7}$$

This expression can be interpreted as a Schrödinger equation with an effective potential

$$\begin{aligned} V_{\text{eff}} &= PV_\alpha P + PV_\alpha Q \frac{1}{E + i\epsilon - QHQ} QV_\alpha P \\ &\equiv V_{oo} + V_{o\alpha} \frac{1}{E + i\epsilon - QHQ} V_{\alpha'o} \quad . \end{aligned} \tag{4.8}$$

In this sense the optical model is a model of an effective interaction. The first term of this interaction is real whereas the second term, which arises from the coupling to all the other states, is complex, energy dependent, strongly resonant and non-local. Therefore, the effective potential (4.8) may be written in a compact form as

$$\begin{aligned} V_{\text{eff}}(r, r', E) &= V_{oo} + \Delta V(r, r', E) \\ &\equiv V_{oo} + V(r, r', E) + iW(r, r', E) \quad . \end{aligned} \tag{4.9}$$

Furthermore, there exists a dispersion relation between $\Re \Delta V = V(r, r', E)$ and $\Im \Delta V = W(r, r', E)$ [Sat83]

$$\Re \Delta V = \sum_q \frac{V_{o\alpha} |\phi_q\rangle \langle \phi_q| V_{\alpha'o}}{E - E_q} - \frac{1}{\pi} \mathcal{P} \int \frac{\Im \Delta V}{E - E_q} \, dE_q \tag{4.10}$$

with ϕ_q being the eigenstates and E_q the eigenvalues of QHQ:

$$(E_q - QHQ) \phi_q = 0 \tag{4.11}$$

and $\mathcal{P}$ denoting the Cauchy principal value.

Equation (4.7) is equivalent to the total Schrödinger equation (4.1). But now all the complications of the full problem are described by the effective potential (4.9), which can be used as a starting point for further approximations.

The first approximation deals with the strong resonances of the model wave function $\Psi_M = P\Psi$, e.g. Eq. (4.4a), which are observed when the bombarding energy is varied. In order to obtain an optical potential one is interested in an energy average of Ψ_M. It has been shown [Fes67, Aus70] that an energy average of Ψ_M is given by the energy averaged V_{eff}

$$\overline{V}_{\text{eff}}(r, r', E) = V_{\text{oo}} + \overline{V}(r, r', E) + i\overline{W}(r, r', E) \quad . \tag{4.12}$$

With the definition of the optical potential

$$U_{\text{opt}} \equiv \overline{V}_{\text{eff}} \quad , \tag{4.13}$$

Eq. (4.7) takes the form

$$(E - T(\mathbf{r}) - U_{\text{opt}}(r, r', E))\,\chi(\mathbf{r}) = 0 \quad . \tag{4.14}$$

The optical potential defined by Eq. (4.13) is nonlocal. This nonlocality is equivalent to a momentum dependence. Since most optical potentials used in the calculations are taken to be local, the nonlocal potential has to be transformed into an equivalent local potential in a further approximation. Consequently, this transformation gives rise to an additional energy dependence of the optical potential. Now Eq. (4.7) can finally be written as

$$(E - T(\mathbf{r}) - U_{\text{opt}}(\mathbf{r}, E))\,\chi(\mathbf{r}) = 0 \quad . \tag{4.15}$$

This expression is a simple one-particle Schrödinger equation in one variable $\mathbf{r}$. Summarizing the results one can say that the optical potential in Eq. (4.15) is complex, local and smoothly energy dependent. Its real term mainly consists of the potential term V_{oo} in Eq. (4.9) corrected by the real term of ΔV, its imaginary term $\overline{W}$ is given by the imaginary part of ΔV. There exists in good approximation [Mah86a] a dispersion relation (4.10) between the two parts of ΔV.

For practical applications the optical potential $U_{\text{opt}}(\mathbf{r}, E)$ has to be parametrized in such a way that observables like elastic cross sections, analyzing powers etc. are well described by the wave functions $\chi(\mathbf{r})$.

As mentioned before, in some model calculations [Tom60] square-well potentials have been applied. The most commonly used parametrization of the optical potential, however, is given as

$$U_{\text{opt}}(\mathbf{r}) = -V\,f(r, R, a) - i\,W\,f(r, R', a') - i\,W_{\text{D}}\,g(r, R', a') \quad , \tag{4.16}$$

where f is the Woods-Saxon form factor and g its derivative:

$$f(r, R, a) = (1 + \exp(x))^{-1} \quad ; \quad g(r, R', a') = 4a'\frac{\mathrm{d}}{\mathrm{d}r}f(r, R', a')$$

$$x = \frac{r - R}{a} \quad ; \quad R = r_{\text{o}}A^{1/3} \quad . \tag{4.17}$$

This means that the radial shape of the real potential (first term in Eq. (4.16)) and of the volume term of the imaginary potential (second term in Eq. (4.16)) is proportional to the nuclear density distribution (Fermi distribution) or closely related to it, whereas the surface term of the imaginary potential (third term in Eq. (4.16)) is peaked at the nuclear surface. If the colliding particles are charged, a Coulomb potential, and if the projectile has spin, a spin-orbit term has to be added. The latter has the form

$$V_{so} = \frac{1}{r}\frac{d}{dr}f(r, R_{so}, a_{so}) \quad . \tag{4.18}$$

From the fits to the elastic nucleon-nucleus scattering data it results that a depth of the real potential of about 50 MeV is required. This depth is essentially that for the shell model of the nucleus. The energy dependence is smooth, the depth decreases with increasing energy. For the radius and diffuseness parameters r_0 and a values close to 1.25 fm and 0.65 fm, respectively, have been found. More details are given in [Sat83], and optical potential parameters are summarized in [Per76].

In order to analyze elastic alpha-nucleus scattering data the simple Woods-Saxon shape is often generalized by introducing terms of higher order of the Woods-Saxon function [Mic77]. Furthermore, some analyses have been made in terms of "model-independent" parametrizations of the optical potential using either spline functions [Put77], a sum of Fourier-Bessel functions added to a Woods-Saxon form factor [Fri78], or by expressing the real part of the potential in terms of a sum of Gaussians [Gub81].

In elastic scattering of composite particles discrete ambiguities are observed. For instance, in α-nucleus scattering almost equally good fits are obtained with real potential depths of about 50, 100, 150, ... MeV. Each of the equivalent potentials reproduces the elastic scattering, but gives a different wave function in the nuclear interior. This behaviour becomes significant when the wave functions are used to calculate the cross sections of non-elastic processes such as inelastic scattering or transfer reactions [Ham83, Hoy85].

At low incident α energies a strong energy dependence of the real part of the α nucleus interaction is expected. On one hand, resonating-group method [Wad88] as well as fishbone model [Sch82] calculations predict that antisymmetrization effects induce a rapid increase of an exchange potential near the barrier radius as the energy decreases. Furthermore, this effect is predicted to become stronger at negative energies [Wad88]. On the other hand, due to dispersion relation effects, the interaction is expected to be more attractive in the energy region where the imaginary part of the potential is increasing than at higher energies, and its strength is predicted to decrease again at very low energies [Nag85, Mah86a, Mah86b]. A recent reanalysis of elastic α-^{16}O scattering data [Bus80] at energies near the Coulomb barrier agrees qualitatively with these theoretical predictions [Mic89].

A further approach to the determination of nucleus-nucleus optical potentials is the double-folding procedure. The starting point for this method is Eq. (4.8). The main part of the real optical potential is given by

$$V_{oo} = \langle \psi_o | V | \psi_o \rangle \tag{4.19}$$

with

$$\psi_o = \mathcal{A}\psi_{ao}\psi_{Ao} \quad . \tag{4.20}$$

The wave functions ψ_{ao} and ψ_{Ao} for the nuclei a and A in their ground states are antisymmetrized. The antisymmetrization operator

$$\mathcal{A} = 1 + \sum_{ij} P_{ij} \tag{4.21}$$

takes account of the exchange of two nucleons between a and A. Together with (4.6), Eq. (4.19) takes the form

$$V_{oo} = \sum_{ij} \langle \psi_{ao}\psi_{Ao}|v_{ij}^d|\psi_{ao}\psi_{Ao}\rangle + \sum_{ij} \langle ij|v_{ij}^{ex}|ij\rangle \quad . \tag{4.22}$$

where $|i\rangle$ and $|j\rangle$ refer to the single-particle wave functions of the nuclei a and A, respectively. The direct part v_{ij}^d and the exchange part v_{ij}^{ex} of the nucleon-nucleon (NN) interaction may be described by the four singlet/triplet, even/odd components of the central NN force [Sat83]

$$v^d = \frac{1}{16}\left(3v_{TE}^c + 3v_{SE}^c + 9v_{TO}^c + v_{SO}^c\right) \tag{4.23a}$$

$$v^{ex} = \frac{1}{16}\left(3v_{TE}^c + 3v_{SE}^c - 9v_{TO}^c - v_{SO}^c\right) \quad . \tag{4.23b}$$

Each component can be parametrized by a sum of three Yukawa potentials (M3Y)

$$v(r) = \sum_{k=1}^{3} a_k \frac{\exp(-s_k r)}{s_k r} \tag{4.24}$$

with parameters taken from Brueckner-Goldstone G-Matrix elements of the Reid [Ber77] or the Paris [Ana83] NN interaction. The numerical values are given in Table 1 [Cha86].

Table 1. Numerical parameters of the NN interaction

s_k	4.0 fm^{-1}	2.5 fm^{-1}	0.7072 fm^{-1}
Reid v^d	7999	-2134	0
Reid v^{ex}	4631.4	-1787.1	-7.847
Paris v^d	11061.6	-2537.5	0
Paris v^{ex}	-1524	-518.8	-7.847

The M3Y interaction has been quite successful in predicting elastic and inelastic scattering data for the collision between two nuclei. The direct part of the central potential in Eq. (4.22) is computed as a convolution of the M3Y interaction with the density distributions of the two nuclei:

$$V_d = \iint \rho_{ao}(\mathbf{r}_1)\, \rho_{Ao}(\mathbf{r}_2)\, v(s)\, d\mathbf{r}_1\, d\mathbf{r}_2 \quad , \tag{4.25}$$

where $\rho_{ao}(\mathbf{r}_1)$ and $\rho_{Ao}(\mathbf{r}_2)$ are the local density distributions of the two nuclei. These distributions may either be taken from Hartree-Fock or shell-model calculations [Bei75, Kob82], or from nuclear charge distributions measured by elastic electron scattering [Vri87]. The variable **s** in the NN interaction term $v(\mathbf{s})$ is given by

$$\mathbf{s} = \mathbf{R} + \mathbf{r}_2 - \mathbf{r}_1 \tag{4.26}$$

with **R** being the separation of the centers of mass of the two colliding nuclei.

The exchange part of Eq. (4.22) has been calculated either by correctly treating the single-nucleon knockon exchange effects arising from the Pauli principle [Cha85, Cha86], or in a microscopic approach where the finite-range exchange potential is obtained in a closed form [Kho88, Kho90].

Another result was that the exchange term can be approximated by a zero-range pseudo-potential [Sat79]. Kobos et al. [Kob84] have modified the M3Y interaction by introducing a density- and energy-dependent term. This effective interaction (DDM3Y)

$$v_{\mathrm{eff}}(E, \rho, s) = g(E, s) f(E, \rho) \tag{4.27}$$

folded with the nuclear densities gives realistic alpha-nucleus potentials

$$V = \lambda \iint \rho_{\mathrm{ao}}(\mathbf{r}_1)\, \rho_{\mathrm{Ao}}(\mathbf{r}_2)\, v_{\mathrm{eff}}(E, \rho, s)\, d\mathbf{r}_1\, d\mathbf{r}_2 \quad, \tag{4.28}$$

with a renormalization factor $\lambda \approx 1.3$ [Kob84, Abe87, Neu89a].

The first term in Eq. (4.27) is given by

$$g(E, s) = \left[7.999 \frac{\exp(-4s)}{4s} - 2134 \frac{\exp(-2.5s)}{2.5s} - 276 \left(1 - \frac{0.005E}{4}\right) \right] \quad, \tag{4.29}$$

whereas the density dependence is taken to be of the exponential form

$$f(E, \rho) = C(E) \left(1 + \alpha(E)\, e^{-\beta(E)\rho}\right) \tag{4.30}$$

with $\rho = \rho_{\mathrm{ao}} + \rho_{\mathrm{Ao}}$. The parameters $C(E)$, $\alpha(E)$ and $\beta(E)$ were determined [Pet77, Lov77] by fitting the volume integral of $v_{\mathrm{eff}}(E, \rho, s)$ to the strength of the real part of a G-matrix effective interaction obtained from Brueckner-Hartree-Fock calculations [Jeu77, Jeu81] for nuclear matter of various densities ρ and at various energies E.

Recently first results have been obtained in the description of experimental $\alpha - {}^{16}\mathrm{O}$ data at low energies by means of a double folding potential [Abe91]. Using only the real part of the potential the elastic scattering data at energies near the Coulomb barrier [Bus80] as well as the excitation functions of two potential resonances [Mac80] can be excellently reproduced. The calculated excitation energies of both the ground-state band and the $K^\pi = 0^-$ α-cluster rotational band in ${}^{20}\mathrm{Ne}$ as well as deduced $B(E2)$ values are in good agreement with experimental data. From these calculations a strong energy dependence of the potential depth due to dispersion relation effects results, which is in agreement with theoretical predictions.

5. Theoretical Descriptions of Direct Reactions

The theoretical descriptions of DI reactions at subCoulomb energies can be divided into three categories: phenomenological approaches, microscopic theories and potential models. In many cases, the cross section data from accelerator experiments can only be obtained for energies above the thermonuclear range. Therefore, theoretical methods are necessary to extend the S-factor data to the energies required in astrophysical models.

In phenomenological approaches, open parameters are adjusted to reproduce the experimental cross sections. The most popular of these methods is probably the R-matrix theory [Lan58]. In the R-matrix theory, resonance energies and widths are varied until the experimental data are reproduced.

First-principle microscopic theories, such as the resonating group method (RGM) or generator coordinate method (GCM), are based on many-nucleon wave functions of the nuclei involved and on nucleon-nucleon interactions [Wil77, Bay89]. In this approach, the explicit inclusion of the Pauli principle leads to complicated highly nonlocal potentials for the interaction between the composite nuclei in the entrance and exit channel.

The Schrödinger equation for a system with more than two nucleons cannot be solved exactly. Therefore, approximated wave functions have to be used. In the RGM [Tan81] one assumes that the A nucleons of the system can be divided in one or several partitions involving two clusters in their ground or excited states. Each partition and an excitation level define a channel. In a given channel the two clusters 1 and 2 have the nucleon numbers A_1 and A_2, proton numbers Z_1 and Z_2, spins I_1 and I_2 and parities π_1 and π_2. The cluster wave functions are denoted by $\phi_1^{I_1\pi_1}$ and $\phi_2^{I_2\pi_2}$. For the sake of simplicity we do not include other quantum numbers such as the radial or isospin quantum numbers. The RGM wave function with total spin J, magnetic quantum number M and parity π is defined as

$$\Psi^{JM\pi} = \mathcal{A} \sum_{\alpha\ell j} g_{\alpha\ell j}^{JM\pi}(r)\varphi_{\alpha\ell j}^{JM\pi}(\hat{\mathbf{r}},\xi_1^1\ldots\xi_1^{A_1},\xi_2^1\ldots\xi_2^{A_2}) \ , \qquad (5.1)$$

where $\mathcal{A}$ is the antisymmetrization operator. The partition and the excitation levels of the cluster define a channel labelled α. For simplicity we have not explicitly written the dependence of the quantum numbers and coordinates on the channel α. The angular momentum quantum number between the two clusters is ℓ, and the channel spin for the coupling of the two spins I_1 and I_2 is given by j. The channel wave function $\varphi_{\alpha\ell j}^{JM\pi}$ depends on the relative coordinate $\mathbf{r}$ as well as on the internal coordinates $(\xi_1^i, i = 1\ldots A_1)$ and $(\xi_2^j, j = 1\ldots A_2)$. It is defined by

$$\varphi_{\alpha\ell j}^{JM\pi}(\hat{\mathbf{r}},\xi_1^1\ldots\xi_1^{A_1},\xi_2^1\ldots\xi_2^{A_2}) =$$
$$[Y_\ell(\hat{\mathbf{r}}) \otimes [\phi_1^{I_1\pi_1}(\xi_1^1\ldots\xi_1^{A_1}) \otimes \phi_2^{I_2\pi_2}(\xi_2^1\ldots\xi_2^{A_2})]^I]^{JM} \ . \qquad (5.2)$$

In many applications only one channel (i.e. $\alpha = 1$) is taken into account. However, in order to calculate transfer or inelastic reactions additional channels have to be introduced.

The main drawback of the RGM is that it requires extensive analytical calculations without systematic character when going from one reaction to another. Consequently, the application of the RGM is essentially restricted to reactions involving only a small number of nucleons. This problem can be overcome by the GCM [Hil53]. The GCM is similar to the RGM, but it allows systematic calculations, well adopted to a numerical approach. Nowadays, most microscopic calculations involving p-shell nuclei are performed in the GCM framework [Bay89].

In the GCM the relative wave functions $g_{\alpha\ell j}^{JM\pi}(r)$ are expanded in a Gaussian basis, yielding

$$g_{\alpha\ell j}^{JM\pi}(r) = \int f_{\alpha\ell j}^{J\pi}(R)\, \Gamma_\ell(r, R)\, dR \quad , \tag{5.3}$$

where R is the generator coordinate, and $\Gamma_\ell(r, R)$ is a projected Gaussian function centered at R [Bay77]. In the GCM the calculation of the radial wave functions $g_{\alpha\ell j}^{JM\pi}(r)$ is replaced by the calculation of the generator functions $f_{\alpha\ell j}^{J\pi}(R)$. Inserting eq. (5.3) in (5.1) we obtain

$$\Psi^{JM\pi} = \sum_{\alpha\ell j} \int f_{\alpha\ell j}^{J\pi}(R)\, \phi_{\alpha\ell j}^{JM\pi}(R)\, dR \quad , \tag{5.4}$$

where $\phi_{\alpha\ell j}^{JM\pi}(R)$ is defined by

$$\phi_{\alpha\ell j}^{JM\pi}(R) = \mathcal{A}\, \Gamma_\ell(r, R)\, \varphi_{\alpha\ell j}^{JM\pi} \quad . \tag{5.5}$$

If the oscillator parameter of the clusters are identical, $\phi_{\alpha\ell j}^{JM\pi}(R)$ is a projected Slater determinant involving A_1 orbitals at a distance of $-RA_2/A$ and A_2 orbitals at RA_1/A. The Slater determinants are easy to handle numerically. Matrix elements involving Slater determinants are computed by using Brink's formula [Bri66].

A practical problem with the GMC in actual calculations is that the integral occuring in eq. (5.4) is replaced by a finite sum over a set of generator coordinates. Accordingly, the asymptotic behaviour of the radial wave functions for large distances between the clusters is Gaussian and hence unphysical for scattering as well as for bound states. This drawback can be solved with the Microscopic R-Matrix Method (MRM). Details concerning this method are given in [Bay77].

It is obvious that a fully microscopic approach like RGM and GCM is more satisfying, since it is a first-principle approach. Such a microscopic model starts from the nucleon-nucleon interaction and does not contain any free parameters. It is therefore possible to predict physical properties of the system independently of experimental data. However, in most cases the fully microscopic approach rarely reproduces physical quantities which are fundamental for the calculation of astrophysical cross sections, such as thresholds or resonance energies and their widths. Nevertheless, there has been considerable progress in the description of astrophysically relevant processes by allowing adjustments of the nucleon-nucleon interaction parameters. Review articles on microscopic theories including examples of astrophysical processes are found in [Lan86, Lan88, Des90].

Potential models are based on the description of the dynamics of the reaction by a Schrödinger equation with local optical potentials in the entrance and/or exit channels. Such models are the "Distorted Wave Born Approximation" (DWBA) [Aus70, Sat83, Gle83] for transfer or the "Direct Capture" model (DC) [Chr61, Tom63, Rol73] for capture reactions.

The differential cross section for the transfer reaction $a + A \rightarrow b + B$ with $a - x = b$, $A + x = B$ (stripping) using light projectiles and ejectiles (for $a \leq 4$ and $x = 1$ or $x = 3$) is given in zero-range DWBA by [Sat83, Gle83]

$$\frac{d\sigma}{d\Omega} = \frac{\mu_\alpha \mu_\beta}{(2\pi\hbar^2)^2} \frac{k_\beta}{k_\alpha} \frac{2I_B + 1}{2I_A + 1} \sum_{\ell s j} C^2 S_{\ell j} N \frac{\sigma_{\ell s j}(\vartheta)}{2s + 1} \tag{5.6}$$

with the zero-range normalisation constant

$$N = \frac{1}{2} a D_0^2 \quad . \tag{5.7}$$

The reduced cross section without spin-orbit coupling is

$$\sigma_{\ell s j}(\vartheta) = \sum_m |t_{\ell s j}^m|^2 \tag{5.8}$$

with the reduced transition amplitude

$$t_{lsj}^m = \frac{1}{\sqrt{2\ell + 1}} \int \chi_\beta^{(-)*} \left(k_\beta, \frac{A}{B} r \right) u_{\ell j}(r) [i^\ell Y_\ell^m(\hat{r})]^* \chi_\alpha^{(+)}(k_\alpha, r) \, dr \quad . \tag{5.9}$$

With spin-orbit coupling and neglecting interference terms between different l the following expressions are valid:

$$\sigma_{\ell s j}(\vartheta) = \sum_{m M_a M_b} |t_{\ell s j}^{m M_a M_b}|^2 \tag{5.10}$$

with

$$t_{lsj}^{m M_a M_b} = \frac{1}{\sqrt{2\ell + 1}} \sum_{M_a' M_b'} (-1)^{I_a - M_a'}$$

$$\times \left\langle I_b \quad I_a \quad M_b' \quad -M_a' \, | \, s \quad M_b' - M_a' \right\rangle$$

$$\times \left\langle s \quad j \quad M_b' - M_a' \quad m + M_a - M_b \, | \, \ell \quad M_b' - M_a' + m + M_a - M_b \right\rangle$$

$$\times \int \chi_{M_b' M_b}^{(-)*} \left(k_\beta, \frac{A}{B} r \right) u_{\ell j}(r) [i^\ell Y_\ell^m(\hat{r})]^* \chi_{M_a' M_a}^{(+)}(k_\alpha, r) \, dr \quad . \tag{5.11}$$

The quantities μ_α, μ_β and k_α, k_β are the reduced masses and wave numbers in the entrance channel α and exit channel β, respectively. The spin and magnetic spin quantum numbers of the projectile, ejectile, target and residual nucleus are given by (I_a, I_b, I_A, I_B) and (M_a, M_b, M_A, M_B), respectively. The orbital angular momentum quantum number ℓ, the spin quantum number s and the total angular momentum quantum number j refer to the cluster x bound in the residual nucleus B. The optical wave functions in the entrance and exit channels are characterized by $\chi^{(+)}$ and the time-reversed solution $\chi^{(-)}$. The spectroscopic factor and the isospin Clebsch-Gordan coefficient for the partition $B = A + x$ are given by C and $\mathcal{S}_{\ell j}$, respectively. Expressions similar to the above equations will be obtained, if the finite range of the interaction potential is taken into account [Sat83].

The above expressions are written for a stripping reaction, where the nucleon cluster x is stripped from the projectile a. The corresponding formulae for a pick-up reaction can be obtained easily with the help of the reciprocity theorem for the reduced cross sections [Sat83].

The DWBA is based on the premise that elastic scattering in the entrance and exit channel is dominant compared to the flux into other channels. The DWBA is well established for higher projectile energies ($\gtrsim 10 - 20\,\mathrm{MeV}$) and transitions to low lying states of the residual nucleus. For lower energies precompound (e.g. exciton) or compound (e.g. Hauser-Feshbach) models have to be used. However, for subCoulomb energies pure Coulomb scattering (Rutherford scattering) is the dominant process. Therefore, compared to this direct process the coupling to other channels is small. So the description with the DWBA may again be adequate at these energies. Furthermore, at subCoulomb energies there are only a few channels open (somtimes only two or three). Therefor only a small imaginary part of the optical potentials, which describes absorption into other channels, is necessary. Examples of DWBA calculations will be shown later in this article (Sect. 6).

The direct capture process $a + A \rightarrow \gamma + B$, which is entirely electromagnetic, is treated in first-order perturbation theory. As examples, we quote the expressions for the electric dipole E1 and electric quadrupole E2 capture [Chr61, Rol73]:

$$\sigma_{E1} = \frac{16\pi}{9} \left(\frac{E_\gamma}{\hbar c}\right)^3 \frac{e^2 \mu_\alpha^3}{\hbar^2 k_\alpha} \frac{3}{(2I_a + 1)(2I_A + 1)} \left(\frac{Z_a}{m_a} - \frac{Z_A}{m_A}\right)^2 C^2 \mathcal{S}_{\ell_\beta J_\beta}$$

$$\times \sum_{\ell_\alpha J_\alpha I} (2J_\beta + 1)(2J_\alpha + 1) \, \max(\ell_\alpha, \ell_\beta)$$

$$\times \left\{\begin{matrix} 1 & \ell_\beta & \ell_\alpha \\ I & J_\alpha & J_\beta \end{matrix}\right\}^2 a_I^2 |R_{1\beta\alpha}|^2 \tag{5.12}$$

with the radial integral

$$R_{1\beta\alpha} = \frac{1}{k_\alpha} \int u_\beta^*(r) \, \mathcal{O}_{E1}(r) \, \chi_\alpha(r) \, \mathrm{d}r \tag{5.13}$$

and

$$\sigma_{E2} = \frac{4\pi}{75} \left(\frac{E_\gamma}{\hbar c} \right)^5 \frac{e^2 \mu_\alpha^5}{\hbar^2 k_\alpha} \frac{5}{(2I_a + 1)(2I_A + 1)} \left(\frac{Z_a}{m_a^2} + \frac{Z_A}{m_A^2} \right)^2 C^2 \mathcal{S}_{\ell_\beta J_\beta}$$

$$\times \sum_{\ell_\alpha J_\alpha I} (2J_\beta + 1)(2J_\alpha + 1)(2\ell_\beta + 1) \langle \ell_\beta 020 | \ell_\alpha 0 \rangle^2$$

$$\times \left\{ \begin{matrix} 2 & \ell_\beta & \ell_\alpha \\ I & J_\alpha & J_\beta \end{matrix} \right\}^2 a_I^2 |R_{2\beta\alpha}|^2 \tag{5.14}$$

with the radial integral

$$R_{2\beta\alpha} = \frac{1}{k_\alpha} \int u_\beta^*(r) \, \mathcal{O}_{E2}(r) \, \chi_\alpha(r) \, \mathrm{d}r \quad . \tag{5.15}$$

The coefficients a_I^2 are calculated in LS coupling to

$$a_I^2 = (2I + 1)(2I_A + 1)(2L_B + 1)(2S_B + 1)$$

$$\times \left\{ \begin{matrix} I & L_A & S_B \\ L_B & I_B & \ell_\beta \end{matrix} \right\}^2 \left\{ \begin{matrix} I & L_A & S_B \\ I_A & I_a & I_A \end{matrix} \right\}^2 \quad . \tag{5.16}$$

In the above expressions, the energy of the emitted photon is E_γ. The charge and mass of the projectile and target nucleus are Z_a, m_a, Z_A and m_A, respectively. The orbital angular momentum and total angular momentum quantum numbers of the nuclei in the entrance and exit channels are ℓ_α, J_α, ℓ_β and J_β, respectively. The spin quantum number, orbital angular momentum and total angular momentum quantum numbers are characterized by S, L and I, respectively, with indices a, A and B corresponding to the projectile, target and residual nucleus, respectively. The symbol $\{ \ldots \}$ is the $6j$ symbol. The radial wave functions in the entrance and exit channel are given by χ_α and u_β, respectively. The spectroscopic factor and the isospin Clebsch-Gordan coefficient for the partition $B = A + a$ are given by C and $\mathcal{S}_{\ell_\beta J_\beta}$, respectively. The $\mathcal{O}_{E\ell}$ are the multipole operators.

For the validity of the DC the same arguments hold as already given above for the DWBA. Again, also for projectile energies below the Coulomb and/or centrifugal barrier the description using the DC can be adequate. Examples for DC calculations will be shown later in this article (Sect. 6).

The potential models can be derived from microscopic theories essentially by allowing approximations in the antisymmetrization procedure [Wil77]. In principle, the potential models neglect the antisymmetrization of the optical potentials in the entrance and exit channel as well as exchange processes. However, the effects of the Pauli principle can be taken into account phenomenologically by fitting the parameters of the optical potentials to elastic scattering data or to phase shifts obtained from microscopic models. Furthermore, also exchange processes such as knock out and heavy particle pickup and stripping have to be considered in some cases. This may be necessary e.g. to describe differential reaction cross sections on light target nuclei.

Important for the success of the potential models is the fact that the input data for the optical potentials can be taken from realistic models, i.e. from

semimicroscopic or microscopic formalisms such as the folding-potential model
or RGM and GCM discussed above. In this respect the potential models combine the first-principle approach of a microscopic theory with the flexibility of
a phenomenological method.

6. Examples

In this section applications of the potential model to a few selected transfer
and capture reactions which occur in primordial or stellar nucleosynthesis are
presented. Available data for these reactions (see e.g. [Rol88] and references
therein) suggest the presence of a DI mechanism at subCoulomb energies. Although only a few examples for the application of the DI to nuclear reactions
in astrophysical scenarios are given, we believe that many light-ion reactions
at thermonuclear energies can proceed via this mechanism.

The numerical calculations for the examples discussed in this section were
performed using the DWBA code TETRA [Bac90] for transfer reactions and
the DC code TEDCA [Kra90] for direct capture. Both codes are especially
designed for the subCoulomb (thermonuclear) energy range.

6.1 Transfer Reactions with Non-resonant Behaviour

First the reactions ^{7}Li(p,α)^{4}He and ^{19}F(p,α)^{16}O are considered. The DI may
be dominant for these processes, because pick-up reactions with a high Q-value
have a low separation energy of the transferred particle group in the target
nucleus and a high energy in the exit channel. The low separation energy tends
to enhance the wave function in the nuclear exterior, and the energy of the
ejectile is well above the Coulomb barrier in the exit channel. Both effects
increase the direct contribution to the cross section in the thermonuclear region
[Obe89].

The reaction ^{7}Li(p,α)^{4}He ($Q = 17.35\,$MeV) is involved in primordial and
stellar nucleosyntheses at subCoulomb energies. The available data for this
reaction include angular distributions and cross sections (or, equivalently, astrophysical S-factors) over a wide range of energies ($E_{CM} = 13\,$keV to 1 MeV)
[Spi71, Rol86, Eng90, Eng91] . The direct contribution to the nuclear S-factor
is calculated in zero range DWBA using a triton cluster formfactor [Rai90a,
Rai90b].

In the entrance channel, an optical potential of the Woods-Saxon type was
employed. For energies below 200 keV only a real potential was used. These
parameters were assumed to be constant in the whole energy range between
10 keV and 1 MeV ($V = -50.5\,$MeV, $r_0 = 1.25\,$fm, $a = 0.6\,$fm). Above 200 keV,
a surface absorption potential was added. The depth of this imaginary part
slowly increases with energy up to $W_s = -2.0\,$MeV ($r = 1.5\,$fm, $a = 0.3\,$fm). In

practice, both the angular distributions and the $S(E)$-data are rather insensitive to the shape and well depth of the proton-nucleus potential.

In the exit channel, a double folding potential was used for the real part. The depth was determined by fitting the elastic $\alpha\alpha$ scattering data at $E_\alpha = 34.2\,\mathrm{MeV}$ [Kuk75]. The phase shifts are reproduced well. The folding potential also agrees well with the RGM-potential obtained by Friedrich [Fri84] and with the Woods-Saxon potential given by Marquez [Mar83]. The angular distributions and the $S(E)$-data of $^7\mathrm{Li}(\mathrm{p},\alpha)^4\mathrm{He}$ proved very sensitive to slight variations of the α-nucleus potential.

The cluster form factor for the three-nucleon transfer was calculated using a Woods-Saxon potential with $r_0 = 1.1\,\mathrm{fm}$ and $a = 0.986\,\mathrm{fm}$. The depth of the potential well was chosen to reproduce the correct separation energy. The SU(3) cluster spectroscopic amplitude is given by $S^{1/2} = 1.084$ [Kur75] and the zero-range normalization constant is $D_0^2 = 2.56 \cdot 10^5\,\mathrm{MeV}^2\,\mathrm{fm}^3$ [Hoy85].

Parity conservation for $^7\mathrm{Li}(\mathrm{p},\alpha)^4\mathrm{He}$ requires the orbital angular momentum L_i in the incident channel to be odd and L_f in the exit channel to be even. Thus, the main contribution is given by the partial waves $L_\mathrm{i} = 1$ and $L_\mathrm{f} = 0$ and 2. Although higher partial waves are suppressed strongly by the centrifugal barriers, their contributions were calculated up to $L_\mathrm{i} = 5$ and $L_\mathrm{f} = 6$. Important contributions to the DWBA-integral came from the nuclear exterior which leads to an enhancement of the direct contribution compared to the compound nucleus process.

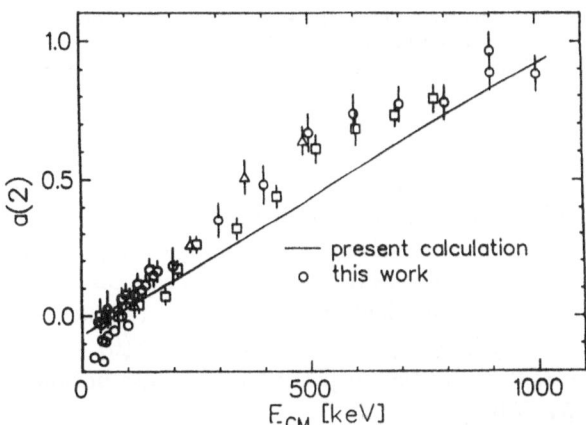

Fig. 1. Angular distributions of the reaction $^7\mathrm{Li}(\mathrm{p},\alpha)^4\mathrm{He}$. The a_2-coefficient of the angular distribution is plotted as a function of CM energy. Results are shown in the figure for new data [Eng91] and for older data [Spi71, Rol86]. The solid line represents the DWBA-calculation of [Rai90a]

In the energy range of $26\,\mathrm{keV}$ to $1\,\mathrm{MeV}$ the experimental angular distributions show a monotonic increase of the a_2 coefficient in the Legendre polynomial expansion, $W(\theta) = 1 + a_2 P_2(\cos\theta)$. The DWBA calculation gives the correct energy dependence of the a_2 coefficient (Fig. 1). The a_2 coefficient changes sign at approximately $60\,\mathrm{keV}$. This behaviour is also reproduced by the DWBA calculation. The calculated a_4 coefficient is about a factor of ten smaller than the a_2 value, consistent with the experimental data [Eng91]. Simple penetrability arguments would not predict the correct experimental angular distribution. Because of the high Q-value of the reaction the optical wave function in the exit channel is almost energy independent for the considered proton energy range. Since both the s- and d-wave in the exit channel are determined almost exclusively by the same p-wave in the entrance channel, one would expect an almost energy independent a_2 coefficient, which is clearly in contradiction to the experimental data.

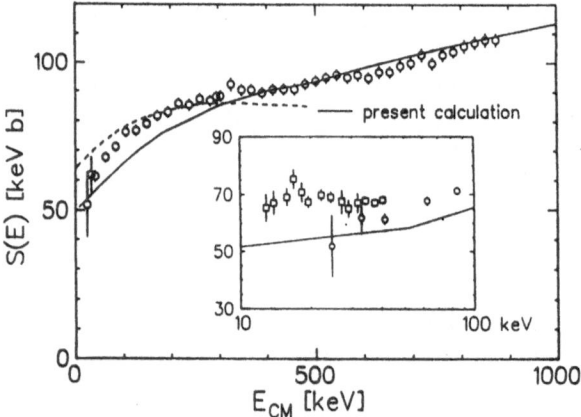

Fig. 2. The astrophysical S-factor of the reaction $^7\mathrm{Li}(p,\alpha)^4\mathrm{He}$ as a function of CM-energy. The solid line represents the DWBA-calculation of [Rai90a]. Also plotted is a previous R-matrix fit by [Bar72] (dashed line). The rise of the low-energy points is due to electron screening

The absolute values of the astrophysical S-factor were calculated in the energy range of 10 keV to 1 MeV and compared with the experimental data (Fig. 2). Note that the comparison makes sense only for energies higher than about 50 keV because here the effects of electron screening are expected to be negligible. The theoretical calculation yields an astrophysical S-factor of $S(0) = 50\,\mathrm{keV\,b}$ compared to the previously extrapolated value $S(0) = 52 \pm 8\,\mathrm{keV\,b}$ [Rol86]. An R-matrix analysis by Barker [Bar72] of the data of Spinka et al. [Spi71] led to $S(0) = 65\,\mathrm{keV\,b}$.

Until now, DI has only been applied for three-nucleon transfer to energies higher than approximately 20 MeV. It is interesting that both the bound-state formfactor and the spectroscopic factor used in the calculation at thermonuclear energies also reproduce the data at higher energies fairly well [Rai90b].

As a second example the reaction $^{19}\mathrm{F}(p,\alpha)^{16}\mathrm{O}(\mathrm{g.s.})$ ($Q = 8.115\,\mathrm{MeV}$), which is responsible for the burning of $^{19}\mathrm{F}$ in the fourth branch of the CNO cycle, is discussed. The direct contribution to this reaction in zero-range DWBA for thermonuclear energies is calculated in [Her91].

While the transitions to the excited states of $^{16}\mathrm{O}(\mathrm{g.s.})$ show marked resonance peaks, no resonant behaviour is observed for the ground-state transition

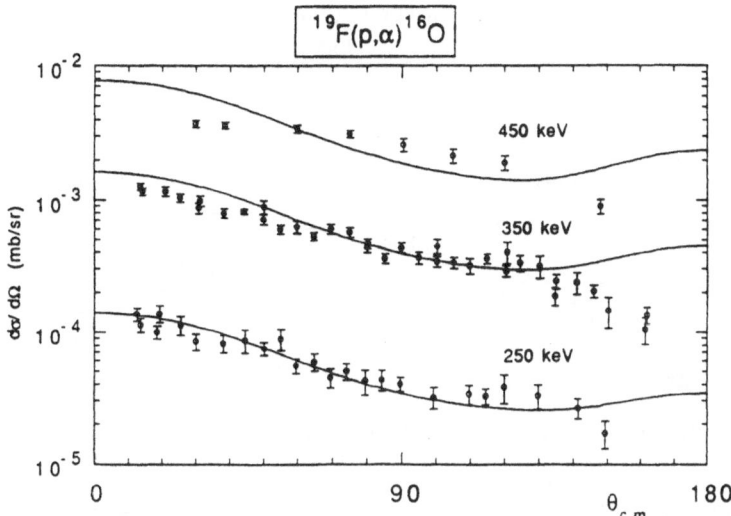

Fig. 3. Differential cross sections of the reaction $^{19}F(p,\alpha)^{16}O(g.s.)$ at projectile energies of 250, 350 and 450 keV. The experimental data are taken from [Lor78]. The solid lines are the DWBA calculation of [Her91]

up to about 700 keV [Lor78]. This indicates that the excited ^{20}Ne states in this energy range have only a small $\alpha + {}^{16}O(g.s.)$ component, meaning that the decay of the compound nucleus into the ground state of the exit channel is suppressed. As a consequence, the ground-state transition may be dominated by a direct mechanism.

The reaction $^{19}F(p,\alpha)^{16}O(g.s.)$ was already investigated at projectile energies greater than 18 MeV. The angular distributions and the absolute magnitude of the cross sections could be well reproduced by microscopic finite-range DWBA [Wal88b, Neu89b]. An essential improvement in the calculation of the absolute cross sections for (p,α) reactions was obtained for the first time by using the double-folding procedure for the α-optical potential [Wal88a].

In the entrance channel an energy-dependent optical potential of the Woods-Saxon form with $V = V_0 + V_1 \cdot E_{LAB}$ ($V_0 = -56.5\,\text{MeV}$, $V_1 = 2.0$, $r_0 = 1.12\,\text{fm}$, $a = 0.48\,\text{fm}$, $r_c = 1.20\,\text{fm}$) and $W = W_0 + W_1 \cdot E_{LAB}$ ($W_0 = -0.625\,\text{MeV}$, $W_1 = -1.5$, $r_0 = 1.25\,\text{fm}$, $a = 0.55\,\text{fm}$) was used. No spin-orbit term was included. In the exit channel, a real double-folding potential [Abe87] was employed. With a normalization factor of $\lambda = 1.312$ and a Coulomb radius $r_c = 1.25\,\text{fm}$, the elastic α-^{16}O data [Bus80] were reproduced near 5 MeV. Also for the exit channel, a volume absorption term had to be introduced ($W = -2.0\,\text{MeV}$, $r_0 = 1.25\,\text{fm}$, $a = 0.6\,\text{fm}$). The volume integral for this imaginary part fits well into an energy dependent analysis of the α-^{16}O optical potential [Abe91]. The formfactor, identical to the one used for energies greater than 18 MeV [Wal88b, Neu89b], was calculated in the microscopic model [Hoy85].

In Fig. 3, the experimental differential cross sections at 250, 350 and 450 keV projectile energy are shown together with the results of the calculations. Good

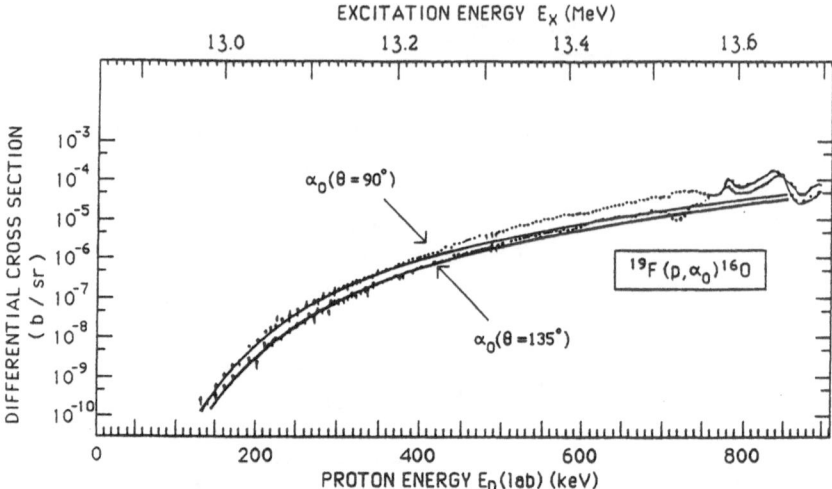

Fig. 4. Excitation functions of the reaction ^{19}F(p,α)^{16}O(g.s.) at $\theta_{LAB} = 90°$ and 135° for projectile energies in the range up to 900 keV. The experimental data are taken from [Lor78]. The solid lines are the DWBA-calculation of [Her91]

agreement was found not only for the angular distributions, but also for the absolute values of the cross section without any normalization factor.

In Fig. 4, the experimental excitation functions for $\theta_{LAB} = 90°$ and 135° are compared with the calculated results. The experimentally observed energy dependence below 450 keV is nicely reproduced. The calculated astrophysical S-factors at zero energy are $S(0) = 8.755\,\text{MeV}\,\text{b}$, $\dot{S}(0) = -3.48\,\text{b}$ and $\ddot{S}(0) = 20.1\,\text{MeV}^{-1}\,\text{b}$, where the dotted quantities represent derivatives with respect to E.

Further examples for calculations with the potential model at thermonuclear energies for astrophysically relevant non-resonant transfer reactions are D(d,n)^{3}He and D(d,p)^{3}H [Grü90], ^{3}He(τ,2p)^{4}He [May68] and ^{7}Li(d,n)^{8}Be [Rau90b].

6.2 Transfer Reactions with Resonant Behaviour

Transfer reactions with resonant behaviour and their explanation by potential resonances are discussed. As an example, the ground-state transitions of ^{9}Be(p,d)^{8}Be ($Q = 0.56\,\text{MeV}$) and ^{9}Be(p,α)^{6}Li ($Q = 2.13\,\text{MeV}$) at subCoulomb energies is given [Rau91]. These reactions are of importance in determining light element abundances in the inhomogeneous big bang [Mal88] and in power generating fusion reactors [McN71a, McN71b, Elw71].

The level scheme shows a strong $J^\pi = 1^-$ resonance at 287 keV and a 1^+ resonance at 416 keV [Ajz88]. The transferred neutron and triton are relatively loosely bound in the target nucleus (separation energies $S_n = 1.67\,\text{MeV}$, $S_t = 17.69\,\text{MeV}$). Therefore the DI process may contribute significantly.

Both reactions were already analyzed at subCoulomb energies using the R-matrix formalism [Sie73]. However, such an analysis has the disadvantage that a large number of open parameters (resonance energies and widths, interaction radii) has to be adjusted to the experimental data. In the R-matrix analysis [Sie73] 21 such open parameters were necessary. A preliminary account for the nonresonant cross section of the reaction ^{9}Be(p,d)^{8}Be was given in [Rau90a].

The optical potentials in the entrance and exit channels were given by folding potentials with a normalization factor λ. In the entrance channel a parity-dependent potential was used. The λ were adjusted to reproduce the resonance energies of the $J^\pi = 1^-$ resonance at 287 keV and the 1^+ resonance at 416 keV. Values of $\lambda = 1.310$ and $\lambda' = 0.801$ were obtained, respectively. The unprimed or primed quantities refer to even or odd partial waves, respectively. A small imaginary part had to be introduced in the entrance channel (surface term, $W_s = 1.5$ MeV, $W'_s = 0.5$ MeV $r_s = 1.2$ fm and $a_s = 0.3$ fm) in order to describe the absorption into other channels. For the optical potentials in the exit channels again the folding procedure with $\lambda_d = 0.47$, $\lambda'_d = 0.56$, $\lambda_\alpha = 2.28$ and $\lambda'_\alpha = 2.10$ was used. The ^{6}Li-α potential was adjusted to reproduce experimental phase shifts [Mey67].

The bound states were described by Woods-Saxon potentials with the geometry parameters $r_0 = 1.4$ fm and $a = 0.30$ fm for the (p,d) and $r_0 = 1.30$ fm and $a = 0.84$ fm for the (p,α) reaction. The depths of these potentials were adjusted to reproduce the separation energies of the neutron and the triton from the target nucleus, respectively. The Coulomb radius parameters were set to $r_p^c = 1.33$ fm, $r_d^c = 1.30$ fm, $r_\alpha^c = 2.0$ fm and $r_t^c = 1.40$ fm [Per76]. The spectroscopic factor for the (p,d) reaction is $S = 0.5804$ [Coh67]. For the (p,α) reaction the spectroscopic factors are given by $S_{P_{1/2}} = 0.0347$, $S_{P_{3/2}} = 0.0224$ and $S_{F_{5/2}} = 0.0009$ [Kur75]. For the zero- range normalization constant a value of $D_0^2 = 1.56 \cdot 10^4$ MeV^2 fm^3 for (p,d) [Knu77] and of $D_0^2 = 2.56 \cdot 10^5$ MeV^2 fm^3 for (p,α) [Hoy85] was employed.

In Fig. 5 (upper part) the experimental differential cross sections [Sie73] at 110 keV are compared to the results of the DWBA calculation. The total cross section in the energy range up to 700 keV is also shown in Fig. 5 (lower part). The DWBA calculation reproduces both the angular distribution and the energy behaviour, including both resonances. For the astrophysical S-factor at zero energy values of $S(0) = 21$ MeV b and $\dot{S}(0) = 82$ b for the (p,d) reaction and $S(0) = 15$ MeV b and $\dot{S}(0) = 30$ b for (p,α) are obtained.

Contrary to previous claims that the DI cannot explain the cross sections of the reactions ^{9}Be(p,d)^{8}Be and ^{9}Be(p,α)^{6}Li below the Coulomb barrier [Sie73] the present results show that the DWBA reproduces the experimental data. Additionally, the resonances in these reactions can be interpreted as potential resonances and thus be consistently described within the potential model.

6.3 Direct Capture

Some examples of calculations with the potential model at thermonuclear energies for astrophysically relevant direct capture reactions are D(p,γ)^{3}He [Don67], ^{3}He(α,γ)^{7}Be [Tom63], ^{7}Be(p,γ)^{8}B [Tom65], ^{12}C(p,γ)^{13}N [Rol74a], ^{12}C(α,γ)^{16}O [Lan83a, Lan85, Fun85] and ^{15}N(p,γ)^{16}O [Rol74b]. Direct capture calculations

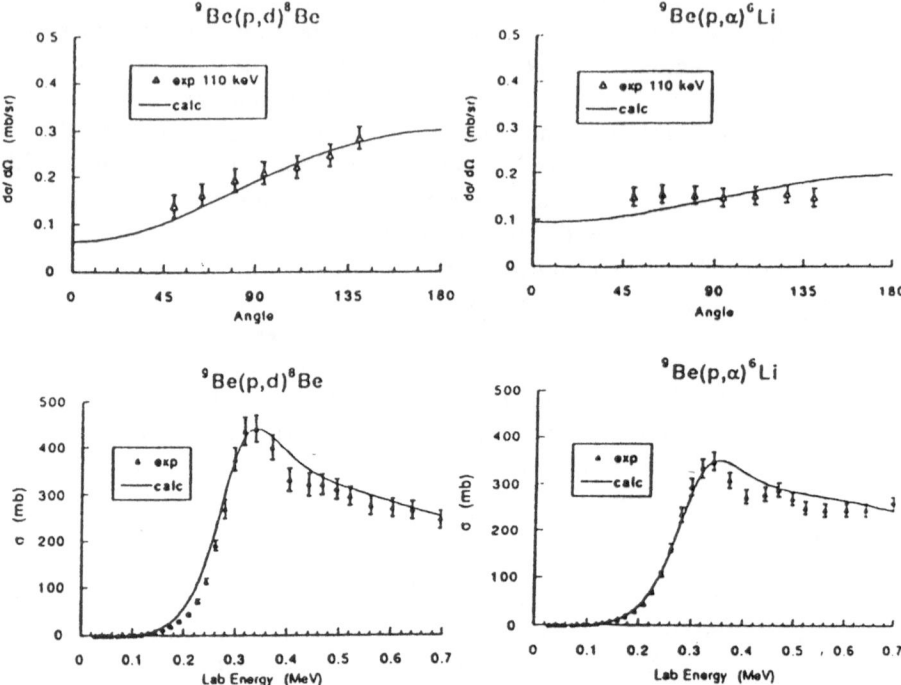

Fig. 5. Differential cross sections for a projectile energy of 110 keV (upper part) and total cross section for projectile energies below 700 keV (lower part) of the reactions ^{9}Be(p,d)^{8}Be and ^{9}Be(p,α)^{6}Li. The experimental data are taken from [Sie73], the full line is the DWBA-calculation of [Rau91]

for (n,γ) processes have been performed for lighter nuclei [Mug81, Lyn90] as well as for r- process nuclei [Mat83]. However, in the above calculations sometimes only the Coulomb and centrifugal potential for the calculation of the wave functions were considered correctly. The nuclear potential was either neglected or taken into account using simplifying assumptions like a hard sphere.

In the following the reaction ^{16}O(α,γ)^{20}Ne ($Q = 4.73$ MeV) at subCoulomb energies is, considered [Kra91]. Besides the triple-alpha process and the capture reaction ^{12}C(α,γ)^{16}O, this reaction is of interest in helium burning of red giants.

The reaction ^{16}O(α,γ)^{20}Ne was already analysed using R-matrix theory [Hah87]. However, besides the many open parameters in such a phenomenological approach it is not possible to determine the nonresonant part of the cross section, because interference with the resonant amplitude can occur. Therefore it is very difficult with such a model to extrapolate the astrophysical S-factor to stellar energies. Microscopic (GCM) [Des83, Bay85, Des86, Bay88, Bay89] and semimicroscopic (orthogonality condition model, OCM) [Lan83b, Lan84] calculations of the reaction ^{16}O(α,γ)^{20}Ne cannot reproduce the energy behaviour of the astrophysical S-factor [Bay85]. Also both of these methods give contra-

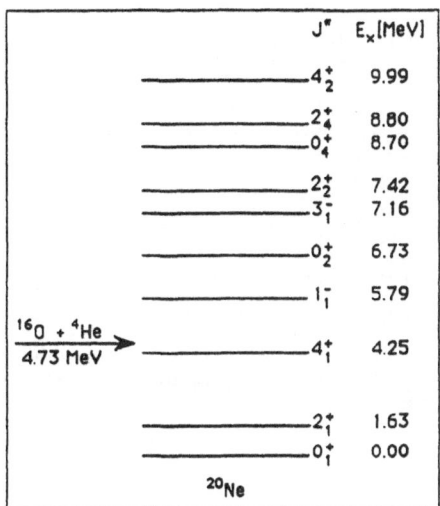

Fig. 6. Level diagram of ^{20}Ne showing the natural parity states relevant for ^{16}O$(\alpha, \gamma)^{20}$Ne

dictory results for the energy dependence of the S-factor [Bay85]. The level diagram showing the natural parity states of ^{20}Ne relevant for ^{16}O$(\alpha, \gamma)^{20}$Ne is given in Fig. 6. As mentioned at the end of Sect. 4, the excitation energies of these states as well as B(E2) transition probabilities were calculated using the double-folded ^{16}O$-\alpha$ potential, and good agreement with experimental data was found [Abe91]. For the ground-state band ($K^{\pi} = 0_1^+$) a harmonic-oscillator quantum-number of $Q = 8$ corresponding to a (sd)^4 configuation in ^{20}Ne was taken. For the $K^{\pi} = 0^-$ and $K^{\pi} = 0_4^+$ bands the next possible higher value of $Q = 9$ and $Q = 10$ were applied, respectively. In order to reproduce the excitation energies of these states different strength factors λ for the ^{16}O$-\alpha$ potential were used (Table 2). The members of the $K^{\pi} = 0^-$ and $K^{\pi} = 0_4^+$ bands are assumed to have a marked ^{16}O$+\alpha$-structure. For the ground-state band with $K^{\pi} = 0_1^+$ this structure is supposed to be somewhat weaker. On the other hand the $K^{\pi} = 0_2^+$ band is assumed to have a dominant ^{8}Be$+^{12}$C-structure [Hor78, Kaz84]. In previous microscopic ^{16}O$(\alpha, \gamma)^{20}$Ne calculations [Des83, Bay85, Lan84] this clustering was not included. Therefore in these calculations resonances due to the $K^{\pi} = 0_2^+$ band could not be reproduced. These resonances, however, are observed in the ^{16}O$(\alpha, \gamma)^{20}$Ne reaction [Hah87] as well as in the ^{16}O$+\alpha$ scattering [Cam53] and in the α-transfer reaction ^{16}O$(^7$Li, t$)^{20}$Ne [Mid70]. This means that there must also be a weak ^{16}O(g.s.)$+\alpha$ cluster component for the members of the $K^{\pi} = 0_2^+$ band. In order to describe this observation in the potential model different potential strengths for the $\alpha+^{16}$O (Q=8) configuration of the ground-state band and the $K^{\pi} = 0_2^+$ band were employed.

Table 2. Natural parity states in ^{20}Ne and their reproduction by the potential model

J^π	K^π	$E_x^{a)}$ [MeV]	E_{CM}^α [MeV]	λ	$\Gamma_{exp}^{a)}$ [keV]	Γ_{calc} [keV]
0^+	0_1^+	0.00		1.237		
2^+	0_1^+	1.63		1.220		
1^-	0^-	5.79	1.05	1.267	0.028 ± 0.003	0.038
0^+	0_2^+	6.73	1.99	1.012	19.0 ± 0.9	49
3^-	0^-	7.16	2.43	1.277	8.2 ± 0.3	11.5
2^+	0_2^+	7.42	2.69	1.025	15.1 ± 0.7	94
0^+	0_4^+	≈8.7	≈3.97	1.298	>800	>1200
4^+	0_2^+	9.99	5.26	1.012	155 ± 30	450

a): [Ajz87]

In the theoretical treatment of the α-capture reaction to the ground and first excited state of ^{20}Ne in the DC model, first the bound-state wave functions of the 0^+ and 2^+ ($K^\pi = 0_1^+$) final states were calculated using a double-folding potential. The normalization factors λ are determined by the bound-state energies (Table 2). The spectroscopic factors for the ground-state transition and the transition to the first excited state are $S = 0.312$ and $S = 0.311$, respectively [Fuj78].

The relevant contributions to the cross section in the energy range up to 2.5 MeV are given by positive and negative parity contributions. The E1 transitions stem from the negative-parity partial waves, the E2 transitions from the positive-parity partial waves in the entrance channel. Therefore the strength factors λ of the $\alpha+^{16}$O optical potential were adjusted to the 1^-, 3^-, 0^+ ($K^\pi = 0_2^+$) and 0^+ ($K^\pi = 0_4^+$) resonance energies, respectively (Table 2).

The calculation in the potential model reproduces the experimental data of [Hah87] and the recently observed resonance structure [Kne91] in the considered energy range quite well (Table 2). This was not possible in the microscopic GCM and OCM calculations [Des83, Bay85, Lan84], which show no resonances in that energy range. For the astrophysical S-factor at zero energy values of $S(0) = 0.0024$ MeV b, $\dot{S}(0) = 0.026$ b for the ground-state transition and $S(0) = 0.77$ MeV b, $\dot{S}(0) = 1.26$ b for the total transition were obtained, respectively.

The calculated widths of the resonances are somewhat higher than the experimental values (Table 2). This is probably due to the Pauli barrier [Sai82] in the entrance channel which is not included in our model. The inclusion of such a barrier can indeed reduce the calculated widths to the experimental values.

These results show that it is possible to obtain the resonant and nonresonant part of the cross section consistently. Therefore this method seems to be suitable for extrapolating the astrophysical S-factor to thermonuclear energies. Furthermore, it was shown that the observed resonances can be interpreted as potential resonances.

7. Summary

- There is strong evidence that light-ion induced nuclear reactions at sub-Coulomb energies may be dominated by a direct reaction mechanism.

- The assumption of this mechanism has implications on the energy dependence of reaction cross sections (astrophysical S-factors) and therefore on nuclear reaction rates in astrophysical scenarios (primordial and stellar nucleosynthesis).

- Direct reaction models such as DWBA for transfer and DC for capture reactions are based on the description of the dynamics of the reaction by a Schrödinger equation. In the entrance and/or exit channel local optical potentials are used. These are calculated from a semimicroscopic model, such as the folding potential approach. New computer codes (TETRA and TEDCA) are especially designed for reactions in the subCoulomb energy range.

- Potential resonances – well known from elastic and inelastic scattering processes – are also observed in transfer and capture reactions and are described by DWBA and DC calculations.

- We have shown that the direct reaction approach can explain the differential cross sections as well as the excitation functions of the transfer reactions $^7\mathrm{Li}(\mathrm{p},\alpha)^4\mathrm{He}$ and $^{19}\mathrm{F}(\mathrm{p},\alpha)^{16}\mathrm{O}$ in the non-resonant range below the Coulomb barrier. DWBA calculations can reproduce the experimental data including the resonant behaviour of the reaction $^9\mathrm{Be}(\mathrm{p},\mathrm{d})^8\mathrm{Be}$ and $^9\mathrm{Be}(\mathrm{p},\alpha)^6\mathrm{Li}$ at low energies, which means that these resonances can be interpreted as potential resonances. Finally it is possible to obtain the resonant and nonresonant part of the $^{16}\mathrm{O}(\alpha,\gamma)^{20}\mathrm{Ne}$ cross section consistently. The study of further reactions is in progress.

Acknowledgements

We wish to thank P. Descouvemont for fruitful discussions and the help in writing this article. We are indebted to G. Raimann for many suggestions and comments on the manuscript. Furthermore, we wish to thank our coworkers H. Abele, B. Bach, K. Grün, H. Herndl, H. Krauss, G. Raimann and T. Rauscher for their contributions. We want to thank the Fonds zur Förderung der wissenschaftlichen Forschung in Österreich (projects P6760T, P7838-TEC) for its support. One of us (H.O.) is indebted to the Alexander von Humboldt-Stiftung for a research fellowship at the University of Tübingen.

References

Abe87   H. Abele, H.J. Hauser, A. Körber, W. Leitner, R. Neu H. Plappert, T. Rohwer, G. Staudt,M. Strasser, S. Welte, M. Walz, P.D. Eversheim, F. Hinterberger: Z. Phys. **A326**, 373 (1987)
Abe91   H. Abele et al.: to be published
Ajz87   F. Ajzenberg-Selove: Nucl. Phys. **A475**, 1 (1987)
Ajz88   F. Ajzenberg-Selove: Nucl. Phys. **A490**, 1 (1988)
Ana83   N. Anantaraman, H. Toki, G.F. Bertsch: Nucl. Phys. **A398**, 269 (1983)
Aus70   N. Austern: *Direct Nuclear Reaction Theories*, (John Wiley & Sons, New York 1970)
Bac90   B. Bach, K. Grün, G. Raimann: code TETRA, unpublished
Bar67   A.C.L Barnard: Phys. Rev. **67**, 1135 (1967)
Bar72   F.C. Barker: Ap. J. **173**, 477 (1972)
Bay77   D. Baye, P.H. Heenen, M. Libert-Heinemann: Nucl. Phys. **A291**, 230 (1977)
Bay83   D. Baye, P. Descouvemont: Ann. Phys. (N.Y.) **165**, 115 (1985)
Bay88   D. Baye, P. Descouvemont: Phys. Rev. **C38**, 2463 (1988)
Bay89   D. Baye, P. Descouvemont: In *Proc. 5th Int. Conf. on Clustering Aspects in Nuclear and Subnuclear Systems, Kyoto, Japan 1988*, J. Phys. Soc. Jpn. **58**, Suppl. 103 (1989)
Bei75   M. Beiner, H. Flocard, N. Van Giai, P. Quentin: Nucl. Phys. **A238**, 29 (1975)
Ber77   G.F.Bertsch, J. Borysowicz, H.McManus, W.G. Love: Nucl. Phys. **A284**, 399 (1977)
Bet67   H.A. Bethe: In *Les Prix Nobel*, (Almquist and Wiksells, Stockholm 1967)
Boh36   N. Bohr: Nature **137**, 344 (1936)
Bri26   L. Brilloin: J. Phys. Rad. **7**, 321, 353 (1926)
Bri66   D. Brink: In *Proc. Int. School Enrico Fermi 36*, (Academic Press, New York 1966) p. 247
Bus80   M. Buser: *PhD thesis*, (Universität Basel 1980)
But50   S.T. Butler: Phys. Rev. **80**, 1095 (1950)
But51   S.T. Butler: Proc. Roy. Soc. (London) **A208**, 36 (1951)
Cam53   J.R. Cameron: Phys. Rev. **90**, 839 (1953)
Cha58   D.M. Chase, L. Wilets: Phys. Rev. **110**, 1080 (1958)
Cha85   A.K. Chaudhuri, D.N. Basu, B. Sinha: Nucl. Phys. **A439**, 415 (1985)
Cha86   A.K. Chaudhuri, B. Sinha: Nucl. Phys. **A455**, 169 (1986)
Chr61   R.F. Christy, I. Duck: Nucl. Phys. **A24**, 89 (1961)
Coh67   S. Cohen, D. Kurath: Nucl. Phys. **A101**, 1 (1967)
Del68   P.P. Delsanto, M.F. Roetter, H.G. Wahsweiler: Phys. Lett. **28B**, 246 (1968)
Des83   P. Descouvemont, D. Baye: Phys. Lett. **127**, 286 (1983)
Des86   P. Descouvemont, D. Baye: Nucl. Phys. **A259**, 374 (1986)
Des90   P. Descouvemont: In *Lectures of the Joliot-Curie school 1990*, in press
Don67   J.M. Donhowe, J.A. Ferry, W.G. Monrad, R.G. Herb: Nucl. Phys. **A102**, 383 (1967)
Elw71   A.J. Elwyn, J.E. Monanhan, J.P. Schiffer: Nucl. Fusion **11**, 551 (1971)
Eng90   S. Engstler, C. Angulo, U. Greife, J.F. Harmon, K. Neldner, G. Raimann, C. Rolfs, U. Schröder, E. Somoraj: In *Proceedings of the International Symposium on Nuclear Astrophysics, Baden/Vienna 1990*, ed. by H. Oberhummer, W. Hillebrandt (Max-Planck Institut for Physics and Astrophysics, Garching 1990) p. 185
Eng91   S. Engstler et al.: to be published
Fes58   H. Feshbach: Ann. Phys. **5**, 357 (1958)
Fes62   H. Feshbach: Ann. Phys. **19**, 287 (1962)
Fes67   H. Feshbach, A. Kerman, R.H. Lemmer: Ann. Phys. **41**, 230 (1967)
Fri78   E. Friedman, C.J. Batty: Phys. Rev. **C17**, 34 (1978)
Fri84   H. Friedrich: Phys. Rev. **C30**, 1102 (1984)
Fuj78   Y. Fujiwara, H. Horiuchi, R. Tamagaki: In *International Conference on Clustering Aspects of Nuclear Structure and Nuclear Reactions*, (AIP Conference Proceedings 47, Winnipeg 1978) p. 570
Fun85   C. Funck, K. Langanke, A. Weiguny: Phys. Lett. **152**, 11 (1985)

Gle83   N.K. Glendenning: *Direct Nuclear Reactions*, (Academic Press, New York 1983)
Grü90   K. Grün, H. Krauss, T. Rauscher, H. Oberhummer, G. Raimann, K. Langanke, T. Warmann: In *Proceedings of the International Conference on Muon Catalyzed Fusion, Vienna 1990*, in press
Gub81   H.P. Gubler, U. Kiebele, H.O. Meyer, G.R. Plattner, I. Sick: Nucl. Phys. **A351**, 29 (1981)
Hah87   K.H. Hahn, K.H.Chang, T.R. Donughue, B.W. Filippone: Phys. Rev. **C36**, 892 (1987)
Ham83   J.J. Hamill, P.D. Kunz: Phys. Lett. **129B**, 5 (1983)
Her91   H. Herndl, H. Abele, G. Staudt, B. Bach, K. Grün, H. Oberhummer, G. Raimann: submitted to Phys. Rev. C
Hil53   D.L. Hill, J.A. Wheeler: Phys. Rev. **89**, 1102 (1953)
Hor78   H. Horiuchi: In *International Conference on Clustering Aspects of Nuclear Structure and Nuclear Reactions*, (AIP Conference Proceedings 47, Winnipeg 1978) p. 144
Hoy85   F. Hoyler, H. Oberhummer, T. Rohwer, G. Staudt, H.V. Klapdor: Phys. Rev. **C31**, 17 (1985)
Jeu77   J.P. Jeukenne, A. Lejeunne, C. Mahaux: Phys. Rev. **C16**, 80 (1977)
Jeu81   J.P. Jeukenne, C. Mahaux: Z. Phys. **A302**, 233 (1981)
Käp91   F. Käpeller: In *Nuclei in the Cosmos*, ed. by H. Oberhummer (Springer-Verlag, Berlin 1991) this volume
Kal37   F. Kalckar, N. Bohr: K. Danske Vidensk. Selsk. Mat.-fys. Medd. **14**, 10 (1937)
Kaz84   H. Kazama, K. Kato, H. Tanaka: Prog. Theor. Phys. **71**, 215 (1984)
Kho88   D.T. Khoa: Nucl. Phys. **A484**, 376 (1988)
Kho90   D.T. Khoa, A. Faessler, N. Ohtsuka: J. Phys. **G16**, 1253 (1990)
Kla91   H.V. Klapdor-Kleingrothaus: In *Nuclei in the Cosmos*, ed. by H. Oberhummer (Springer-Verlag, Berlin 1991) this volume
Kne91   H. Knee, A. Denker, H.W. Drotleff, J.W. Hammer, S. Küchler, D. Streit: Verhandl. DPG **26**, 437 (1991)
Knu77   L.D. Knutson: Ann. Phys. **106**, 1 (1977)
Kob82   A.M. Kobos, B.A. Brown, P.E. Hodgson, G.R. Satchler, A. Budzanowski: Nucl. Phys. **A384**, 65 (1982)
Kob84   A.M. Kobos, B.A. Brown, R. Lindsay, G.R. Satchler: Nucl. Phys. **A425**, 205 (1984)
Kra26   H.A. Kramers: Z. Phys. **39**, 828 (1926)
Kra90   H. Krauss: code TEDCA, unpublished
Kra91   H. Krauss et al: to be published
Kuk75   V.I. Kukulin, G. Neudatchin, Y.F. Smirnov: Nucl. Phys. **A245**, 429 (1975)
Kur75   D. Kurath, D.J. Millener: Nucl. Phys. **A238**, 269 (1975)
Lan58   A.M. Lane, R.G. Thomas: Rev. Mod. Phys. **30**, 257 (1958)
Lan83a   K. Langanke, S. Koonin: Nucl. Phys. **A410**, 334 (1983)
Lan83b   K. Langanke: Phys. Lett. **B131**, 21 (1983)
Lan84   K. Langanke: Z. Phys. **317**, 325 (1984)
Lan85   K. Langanke, S. Koonin: Nucl. Phys. **A439**, 384 (1985)
Lan86   K. Langanke: In *Nuclear Astrophysics*, ed. by J.W.Negele, E. Vogt (Plenum Press, New York 1986) p.223
Lan88   K. Langanke, H. Friedrich: In *Advances in Nuclear Physics*, Vol. 17, ed. by M. Lozano, M.I. Gallardo, J.M. Arias (Springer-Verlag, Berlin 1988) p.241
Lan91   K. Langanke: In *Nuclei in the Cosmos*, ed. by H. Oberhummer (Springer-Verlag, Berlin 1991) this volume
Lem64   R.H. Lemmer, C.M. Shakin: Annals of Physics **27**, 13 (1964)
Lor78   H. Lorenz-Wirzba: *PhD thesis*, (Universität Münster 1978)
Lov77   W.G. Love: Phys. Lett. **72B**, 4 (1977)
Lyn90   J.E. Lynn, S. Raman: In *Proceedings of the 7. international Symposium on capture gamma ray spectroscopy and related problems*, (Pacific Grove 1990) in press
Mac80   J.D. MacArthur, H.C. Evans, J.R. Leslie, H.B. Mak: Phys. Rev. **C22**, 356 (1980)
Mah86a   C. Mahaux, H. Ngo, G.R. Satchler: Nucl. Phys. **A449**, 354 (1986)
Mah86b   C. Mahaux, H. Ngo, G.R. Satchler: Nucl. Phys. **A456**, 134 (1986)
Mal88   R.A. Malaney, W. Fowler: In *Origin and Distribution of the Elements*, ed. by G. Mathews (World Scientific, Singapore 1988) p. 76

Mal91   R.A. Malaney: In *Nuclei in the Cosmos*, ed. by H. Oberhummer (Springer-Verlag, Berlin 1991) this volume
Mar83   L. Marquez: Phys. Rev. **C28**, 2525 (1983)
Mat83   G.J. Mathews, A. Mengoni, F.-K. Thielemann, W.A. Fowler: Ap.J. **270**, 740 (1983)
May68   R. May, D.D. Clayton: Ap.J. **153**, 855 (1968)
McN71a   J.R. McNally Jr.: Nucl. Fusion **11**, 187 (1971)
McN71b   J.R. McNally Jr.: Nucl. Fusion **11**, 554 (1971)
McV67   K.W.McVoy, L. Heller, M. Bolsteri: Rev. Mod. Phys. **39**, 245 (1967)
Mey67   V. Meyer, R.E. Pixley, P. Truöl: Nucl. Phys. **A101**, 321 (1967)
Mic77   F. Michel, R. Vanderpoorten: Phys. Rev. **C16**, 142 (1977)
Mic89   F. Michel,Y. Kondo, G. Reidemeister: Phys. Lett. **B220**, 479 (1989)
Mid70   R. Middleton: In *Proceedings of the International Conference on Nuclear Reactions Induced by Heavy Ions, Heidelberg 1969*, ed. by R. Bock, W.R. Hering (North-Holland, Amsterdam 1970) p.263
Mug81   S.F. Mughabghab, M. Divadeenam, N.E. Holden: In *Neutron cross sections series, Vol. 1*, ed. by S.F. Mughabghab, R.R. Kinsey, C.L. Dunford (Academic Press, New York 1981)
Nag85   M.A. Nagarajan, C. Mahaux, G.R. Satchler: Phys. Rev. Lett. **54**, 1136 (1985)
Neu89a   R. Neu, S. Welte, H. Clement, H.J. Hauser, G.Staudt, H. Müther: Phys. Rev **C39**, 2145 (1989)
Neu89b   R. Neu, H. Abele, P.D.Eversheim, F. Hinterberger, H. Oberhummer, H. Jäntsch, E. Striebel, G.Staudt, M. Walz: J. Phys. Soc. Jpn. **58**, Suppl. 574 (1989)
Obe89   H. Oberhummer, H. Herndl, H. Leeb, G. Staudt: Kerntechnik **53**, 211 (1989)
Oka62   S. Okai, T. Tamura: Nucl. Phys. **31**, 185 (1962)
Per76   C.M. Perey, F.G. Perey: At. Data and Nucl. Data Tables **17**, 1 (1976)
Pet77   F. Petrovich, D. Stanley, J.J. Bebelacqua: Phys. Lett. **71B**, 259 (1977)
Put77   W.L. Put, A.M.J. Paans: Nucl. Phys. **A291**, 93 (1977)
Rai90a   G. Raimann, B. Bach, K. Grün, H. Herndl, H. Oberhummer, S. Engstler, C. Rolfs, H. Abele, R. Neu, G. Staudt: Phys. Lett **249B**, 191 (1990)
Rai90b   G. Raimann, H. Abele, R. Neu, G.Staudt, B. Bach, K. Grün, H. Herndl, H. Oberhummer, S. Engstler, C. Rolfs: In *Proceedings of the International Symposium on Nuclear Astrophysics, Baden/Vienna 1990*, ed. by H. Oberhummer, W. Hillebrandt (Max-Planck Institut for Physics and Astrophysics, Garching 1990) p. 216
Rau90a   T. Rauscher, H. Krauss, K. Grün, H. Oberhummer: In *Proceedings of the International Symposium "Nuclei in the Cosmos", Baden/Vienna 1990*, ed. by H. Oberhummer, W. Hillebrandt (Max-Planck-Institut für Physik und Astrophysik, Garching 1990) p. 220
Rau90b   T. Rauscher, H. Krauss, K. Grün, H. Oberhummer: In *Proceedings of the International Conference on Primordial Nucleosynthesis and Evolution of the Universe, Tokyo 1990*, in press
Rau91   T. Rauscher et al.: to be published
Rol73   C. Rolfs: Nucl. Phys. **A217**, 29 (1973)
Rol74a   C. Rolfs, R.E. Azuma: Nucl. Phys. **A227**, 291 (1974)
Rol74b   C. Rolfs, W.S. Rodney: Nucl. Phys. **A235**, 450 (1974)
Rol86   C. Rolfs, R.W. Kavanagh: Nucl. Phys.**A455**, 179 (1986)
Rol88   C.R. Rolfs, W.S. Rodney: *Cauldrons in the Cosmos*, (University of Chicago Press, Chicago 1988)
Sai82   S. Saito, E.W. Schmid: Z. Phys. **306**, 37 (1982)
Sat79   G.R. Satchler, W.G. Love: Phys. Rep. **55**, 183 (1979)
Sat83   G.R. Satchler: *Direct Nuclear Reactions*, (Clarendon Press, Oxford 1983)
Sch82   E.W. Schmid, S. Saito, H. Fiedeldey: Z. Phys. **A306**, 37 (1982)
Sie73   A.J. Sierk, T.A. Tombrello: Nucl. Phys. **A210**, 29 (1973)
Spi71   H. Spinka, T. Tombrello, H. Winkler: Nucl. Phys. **A164**, 1 (1971)
Tan81   Y.C. Tang: In *Topics in Nuclear Physics II*, Lecture Notes in Physics, Vol. 145 (Springer, Berlin 1981) p. 572
Thi91   F.-K. Thielemann: In *Nuclei in the Cosmos*, ed. by H. Oberhummer (Springer, Berlin 1991) this volume
Tho51   R.G. Thomas: Phys. Rev. **84**, 1061 (1951)

Tom60	T.A Tombrello, G.C. Phillips: Nucl. Phys. **20**, 648 (1960)
Tom63	T.A Tombrello, P.D. Parker: Phys. Rev. **131**, 6 (1963)
Tom65	T.A Tombrello: Nucl. Phys. **71**, 459 (1965)
Ver91	J. Vervier: In *Nuclei in the Cosmos*, ed. by H. Oberhummer (Springer, Berlin 1991) this volume
Vri87	H. De Vries, C.W. De Jager, C. De Vries: At. Data and Nucl. Tables **36**, 495 (1987)
Wad88	T. Wada, H. Horiuchi: In *Proc. 5th Int. Conf. on Clustering Aspects in Nuclear and Subnuclear Systems*, Kyoto, 1988 ed. by Y. Sakuragi, T.Wada, Y. Fugiwara, p. 36
Wal88a	M. Walz, R. Neu, G. Staudt, H. Oberhummer, H. Cech: J. Phys. **G14**, L1 (1988)
Wal88b	M. Walz: *PhD thesis*, (Universität Tübingen 1988)
Wen26	G. Wentzel: Z. Phys. **38**, 518 (1926)
Whe37	J.A. Wheeler: Phys. Rev. **52**, 1107 (1937)
Wil77	K. Wildermuth, Y.C. Tang: *A Unified Theory of the Nucleus, Clustering Phenomena in Nuclei, Vol. 3* (Vieweg & Sohn, Braunschweig 1977)
Woo90	S. Woosley, D. Hartmann, R. Hofmann, W. Haxton: Ap. J. **356**, 272 (1990)

Nuclear Reaction Rates – from Laboratory Experiments and in the Stellar Plasma

Karlheinz Langanke

Institut für Theoretische Physik I, Westfällische Wilhelms-Universität, Wilhelm-Klemm-Straße 9, D-4400 Münster, Germany

1. Introduction

The two goals in nuclear astrophysics are the attempts to understand the energy generation of stars at all stages of stellar evolution and to explain the abundancies of the elements and their isotopes as we observe them in nature. Both goals are closely related as nuclear processes have been identified as the enormous energy source which stabilizes the stars and governs their evolution by transmuting nuclear species into other nuclear species with lower total mass, thus simultaneously creating new elements. It is therefore not surprising that studies of stellar evolution require as input the reaction rates for those nuclear processes which are part of the network of many nuclear reactions taking place at each stage of stellar evolution.

For a nuclear process with two nuclear species in the entrance channel this reaction rate is defined as

$$R = \frac{n_1 n_2}{1 + \delta_{12}} < \sigma v > \quad , \tag{1}$$

where n_1, n_2 are the number densities of the colliding nuclei and σ is the pure nuclear reaction cross section. The denominator takes care of the special case of two identical nuclei in the entrance channel. The quantity

$$< \sigma v >= \sqrt{\frac{8}{\pi \mu}} \frac{1}{(kT)^{3/2}} \int_0^\infty dE \, \sigma(E) \, E \exp\left(-\frac{E}{kT}\right) \tag{2}$$

is the average of the product σv over the relative velocities v between the two colliding nuclei which, for hydrostatic stellar burning stages with temperature T, is well described by a Maxwellian distribution. Here, k is Boltzmann's constant, E and μ are the relative collision energy and the reduced mass in the entrance channel. Considering that for charged particle reactions, on which we will focus in this work, the nuclear process in general takes most effectively place at energies well below the Coulomb barrier, at which the reaction cross section is strongly suppressed by the barrier penetration probability, one conveniently defines the cross section in terms of the astrophysical S-factor

$$S(E) = E \, \sigma(E) \, \exp\left\{\frac{2\pi \, Z_1 Z_2 \, e^2}{\hbar v}\right\} \tag{3}$$

where

$$P(E) = \exp\left\{-\frac{2\pi \, Z_1 Z_2 \, e^2}{\hbar v}\right\} \tag{4}$$

is the penetration factor through an s-wave Coulomb barrier between point charges Z_1, Z_2. For the further evaluation of Eq. (2) one usually distinguishes between non-resonant processes and reactions proceeding via narrow or broad resonances.

By definition $S(E)$ is expected to be a slowly varying function in energy for non-resonant nuclear reactions. In this case $S(E)$ might be expanded in a McLaurin series

$$S(E) = S(0) + \dot{S}(0)E + \frac{1}{2}\ddot{S}(0)E^2 + \dots \tag{5}$$

With this expansion and a Gaussian approximation for the energy-dependent exponential, Eq. (2) can be evaluated as

$$<\sigma v> = \sqrt{\frac{2}{\mu}}\frac{\Delta}{(kT)^{3/2}}S_{\text{eff}}(E_0)\exp\left\{-\frac{3E_0}{kT}\right\} \tag{6}$$

with

$$S_{\text{eff}}(E_0) = S(0)\left[1 + \frac{5}{12\tau} + \frac{\dot{S}(0)}{S(0)}\left(E_0 + \frac{35E_0}{12\tau}\right) + \frac{\ddot{S}(0)}{2S(0)}\left(E_0^2 + \frac{89E_0^2}{12\tau}\right)\right] \quad . \tag{7}$$

The quantity E_0 defines the effective mean energy for thermonuclear fusion reactions at a given temperature T,

$$E_0 = 1.22\left(Z_1^2 \, Z_2^2 \, \mu \, T_6^2\right)^{1/3} \text{keV} \quad , \tag{8}$$

where T_6 measures the temperature in 10^6 K. The quantities τ and Δ are given by

$$\tau = \frac{3E_0}{kT}; \quad \Delta = \frac{4}{\sqrt{3}}(E_0 kT)^{1/2} \quad . \tag{9}$$

This formalism is not appropriate for resonant reactions, as in this case $S(E)$ is not well approximated by a McLaurin expansion. For narrow resonances with width $\Gamma \ll E_R$, where E_R is the resonance energy, the Maxwell-Boltzmann function in (2) changes very little over the resonance region. Using this and by approximating $\sigma(E)$ by a Breit-Wigner resonance formula, one finds

$$<\sigma v> = \left(\frac{2\pi}{\mu kT}\right)^{3/2}\hbar^2 \, (w\gamma)_R \exp\left\{-\frac{E_R}{kT}\right\} \quad , \tag{10}$$

where we have introduced the resonance strength

$$(w\gamma)_R = \frac{(2J+1)}{(2J_p+1)(2J_T+1)}(1+\delta_{12})\frac{\Gamma_1\Gamma_2}{\Gamma_{\text{tot}}} \tag{11}$$

in terms of the partial and total resonance widths (Γ_1, Γ_2 and Γ_{tot}), the angular momentum of the state J and the angular momenta of the target J_T and of the projectile J_p.

For broad resonances, Eq. (2) is usually evaluated numerically. Again, $\sigma(E)$ might be approximated by a Breit-Wigner resonance formula, where, however, the energy dependences of the various width parameters have to be taken into account (for a detailed derivation of the respective formulae see Ref. [Rol88]). In practical applications it is often advantageous to consider various resonant and non-resonant contributions to the $S(E)$-factor separately. Then, however, possible interference terms between these contributions might arise, for which Eq. (2) usually is also evaluated numerically.

In stars charged-particle nuclear reactions proceed at such low energies ($E \approx E_0$) that a direct experimental determination of the cross section is often not possible with existing techniques. Hence an extrapolation down to stellar energies of cross sections measured at energies accessible in the laboratory is usual and necessary. In order not to be unreliable such an extrapolation should have a strong theoretical foundation and should be based on as much experimental information as possible. In particular, resonant states in the stellar energy regime (whose properties can often be determined indirectly, e.g. by transfer reactions) have to be described very accurately. Often, however, also the reproduction of bound states just below the threshold, or of broad resonances at higher energies is essential for a reasonable extrapolation, as these states can contribute to the cross section at stellar energies via their high-energy or low-energy wings, respectively. This situation is realized in the $^{12}C(\alpha, \gamma)^{16}O$ reaction which we will use in Sect. 2 to discuss several different theoretical approaches often used to extrapolate measured cross section into the astrophysically relevant energy regime.

To reduce the uncertainty in the determination of the relevant cross sections required in (2), the obvious strategy seems to be an improvement of the experimental effort in order to measure the cross section at even lower energies, ultimately in the regime of astrophysical interest. However, this aim is plagued by an experimental problem, which becomes increasingly important with decreasing energy: In laboratory measurements the target is usually in the form of an atomic or molecular gas or a solid. Thus, in these experiments the measured cross section is influenced by the presence of electrons in the target and does not represent the pure nuclear cross section as required in (2). For astrophysical use, it is therefore necessary to disentangle nuclear and electronic contributions to the measured data. In Sect. 3 we will discuss first experimental and theoretical attempts to study this novel field at the edge between nuclear and atomic physics.

As in laboratory scattering experiments at very low collision energies, nuclear reaction cross sections in stellar plasmas are also enhanced; this time, however, due to plasma shielding effects. As in the plasma electrons approximately form a free electron gas rather than are being bound to atomic nuclei, the physical description of plasma screening is quite different from the situation observed in the collision experiments. In Sect. 4 we will very briefly discuss plasma shielding in terms of two important and illuminating examples: the fate of ^{7}Be in the sun and the triple-alpha reaction under the conditions expected to exist on neutron star surfaces.

2. Theoretical Models for Extrapolating Measured Data

2.1 Example: The ^{12}C$(\alpha, \gamma)^{16}$O E1 Capture Reaction

In this section the ^{12}C$(\alpha, \gamma)^{16}$O reaction will serve as an illustrative example to present several theoretical approaches frequently used to extrapolate measured cross sections into the energy regime of astrophysical interest.

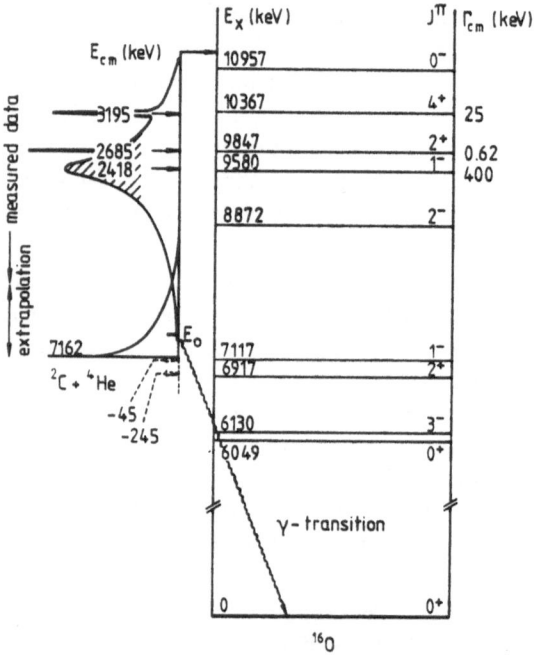

Fig. 1. The ^{16}O level scheme and a sketch of the low-energy ^{12}C$(\alpha, \gamma)^{16}$O cross section

The triple-alpha reaction $(3\alpha \rightarrow {}^{12}$C) and the subsequent ^{12}C$(\alpha, \gamma)^{16}$O reaction are the main fusion processes in stellar helium burning, where the cross section of the latter determines the relative abundancies of ^{12}C and ^{16}O at the end of this stellar burning stage [Fow84]. Besides its obvious importance for studies of elemental abundancies, the magnitude of the ^{12}C$(\alpha, \gamma)^{16}$O cross section also has a strong impact on the evolution of massive stars and, possibly, even on the supernova mechanism [Woo85]. Despite its importance, the ^{12}C$(\alpha, \gamma)^{16}$O cross section at astrophysically important energies ($E_0 = 300$ keV for $T = 10^8$K) is rather uncertain. Experimentally the cross section has been measured down to about 1 MeV [Dye74, Kre88, Red87]. The data are dominated by E1 capture from a broad $J = 1^-$ resonance at $E_R \approx 2.45$ MeV into the ^{16}O ground state (see Fig. 1). Cascade transitions in which the capture occurs first into another bound state of ^{16}O which then decays into the ^{16}O ground state are also observed, but are probably not too important. At astrophysical energies, however, one expects that the ^{12}C$(\alpha, \gamma)^{16}$O cross section will strongly be influenced by a bound 1^- state just below the α-threshold ($E_R \approx -45\,$keV). On the contrary, in the measured data the contribution of this state can only by deduced from its interference with the broad resonance. Moreover, non-resonant dipole capture processes, which mainly should contribute via their interferences with the two low-energy 1^- states, may not be negligible. The situation is further complicated by the fact that, due to isospin selection rules, a dipole capture in the ^{12}C$(\alpha, \gamma)^{16}$O reaction is strongly suppressed. It is therefore conceivable that quadrupole capture might be quite important at astrophysical energies. If so, the E2 part of the ^{12}C$(\alpha, \gamma)^{16}$O cross section at $E = 300\,$keV will be dominated by the high-energy wing of a bound 2^+ state ($E_R \approx -245\,$keV), which is known to have a large α-spectroscopic factor. Again, non-resonant E2 capture processes and their interference with the 2^+ bound state might not be negligible.

Important additional informations about the ^{12}C$+\alpha$ scattering states in the entrance channel can be obtained from precision measurements of the elastic α-scattering of ^{12}C [Cla68, Pla87]. It is therefore reasonable to extrapolate the measured ^{12}C$(\alpha, \gamma)^{16}$O E1 capture cross sections by simultaneously fitting the elastic ^{12}C$ + \alpha$ phase shifts deduced from the experimental ^{12}C$(\alpha, \alpha)^{12}$C cross sections.

In the following we will mainly restrict our discussion to the ^{12}C$(\alpha, \gamma)^{16}$O E1 capture cross sections. These can either be deduced from measured angular distributions (these experiments are extraordinary difficult at low energies due to the very tiny cross sections) or by adopting special geometries in the experimental setup which suppress the unwanted E2 contribution. In principal, the E2 cross sections can also be determined from the angular distribution measurements. But the presently available data are probably not accurate enough to allow for a safe determination of E2 cross sections.

2.1.1 Parameter Fits: K-Matrix and R-Matrix Approaches: In nuclear astrophysics measured data which are dominated by the contributions from a few resonant or bound states are often extrapolated in terms of R-matrix or K-matrix fits. The appeal of these methods rests on the fact that in both approaches analytical expressions, which allow for a rather simple parametrization of the data in terms of theses states, can be derived from underlying formal reaction theories. The parameters can be deduced from their formal relation to experimentally measured properties of these levels or, if not known, are determined by best fits to the cross section data. In their strategies the K-matrix and R-matrix approaches are rather similar. Detailed presentations of these methods can be found in Refs. [Lan58, Hum90]. The differences between the K-matrix and R-matrix approaches as well as possible advantages and disadvantages of the two methods in actual applications are discussed in Refs. [Bar87, Fil89, Bar91]. Both methods have in common that no rigorous description exists to parametrize possible non-resonant (background) contributions.

As the strategies adopted in K-matrix and R-matrix applications are rather similar, we will in the following restrict ourselves to a short presentation of the K-matrix approach to the $^{12}C(\alpha, \gamma)^{16}O$ reaction.

Applying the K-matrix formalism to the $^{12}C(\alpha, \gamma)^{16}O$ reaction [Fil89], the observed E1 capture cross section into the ^{16}O ground state ($J = 0^+$) is given by

$$\sigma_{E1}(E) = \frac{3\pi}{k_\alpha^2} |T_{\gamma\alpha}|^2 .$$ (12)

Here, the indices α, γ refer to the $\alpha + ^{12}C$ and $^{16}O + \gamma$ channels, respectively, and k_α is the wave number. Considering that the flux in the γ-channel is several orders of magnitude smaller than the one in the α-channel, one has in perturbation theory

$$T_{\gamma\alpha} = \frac{2ip_\gamma p_\alpha K_{\gamma\alpha}}{1 + i\mu_\alpha K_{\alpha\alpha}} .$$ (13)

Analogously, the elastic phase shifts can be expressed in terms of the K-matrix elements in the α-channel

$$\delta = \arctan(-\mu_\alpha K_{\alpha\alpha}) .$$ (14)

In Eqs. (13,14) the quantities p_γ, p_α and μ_α are known functions of the wave number k_α, the photon wave number k_γ and the Sommerfeld parameter η [Fil89].

For the low-energy $^{12}C(\alpha, \gamma)^{16}O$ E1 capture the K-matrix can be approximated by a 2-level plus background parametrization where the levels correspond to the physical states at $E_R = -45\,keV$ and at ≈ 2.45 MeV:

$$K_{\alpha\gamma} = \frac{g_{\gamma 1}g_{\alpha 1}}{E - E_1} + \frac{g_{\gamma 2}g_{\alpha 2}}{E - E_2} + B_{\gamma\alpha} ;$$ (15)

$$K_{\alpha\alpha} = \frac{g_{\alpha 1}^2}{E - E_1} + \frac{g_{\alpha 2}^2}{E - E_2} + B_{\alpha\alpha} .$$ (16)

While there exist well-defined prescriptions how to parametrize physical levels within the K-matrix, no such prescription exists for the background term. (Note that the situation is similar within R-matrix approaches where the background is often simulated by an *unphysical* level at high energies). To account for the roughly linear decrease of the p-wave phase shifts at $E > 3.3$ MeV, the background term $B_{\alpha\alpha}$ has to be energy-dependent. In the K-matrix fits of Ref. [Fil89], the energy-dependence of the background terms $B_{\alpha\alpha}, B_{\alpha\gamma}$ were either described by introduction of an „echo-pole" [McV67], which plays a similar role to the hardsphere scattering phase shifts in the R-matrix, or by a linear polynomial.

The experimental knowledge of the parameters E_1 and $g_{\gamma 1}$ seems to be precise enough to replace these quantities by their experimental values. In

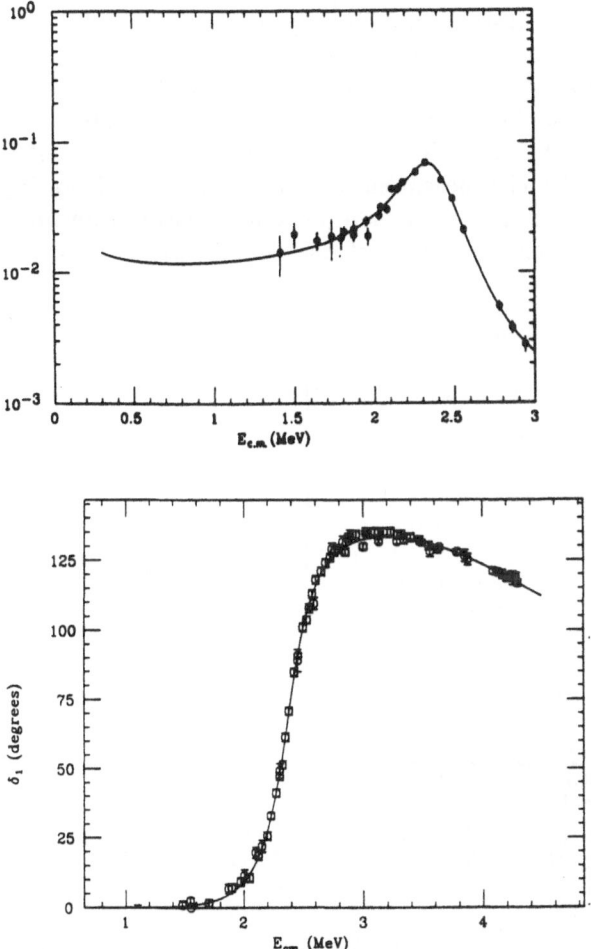

Fig. 2. Simultaneous K-matrix fit to the E1 data of Dyer and Barnes [Dye75] (above) and the elastic δ_1 phase shifts for $^{12}C(\alpha, \alpha)^{12}C$ [Cla68, Pla87] (below)

the K-matrix analysis of Ref. [Fil89], the $^{12}C(\alpha, \gamma)^{16}O$ $E1$ data and the p-wave phase shifts have been *simultaneously* fitted with eight free parameters: $g_{\alpha1}, E_2, g_{\alpha2}, g_{\gamma2}$ to describe the two 1^- levels and four parameters to approximate the background terms $B_{\alpha\alpha}$ and $B_{\alpha\gamma}$. The obtained K-matrix fits reproduce the E1 data and the p-wave phase shifts very well (see Fig. 2).

As has already been pointed out by Barnes and Dyer [Dye74], the fitted reduced α-width of the bound state (given by the parameter $g_{\alpha1}$) is strongly correlated to the background term. In order to determine the uncertainty in the astrophysical S-factor at $E_0 = 300\,$keV without concern of this correlation, the parameter $g_{\alpha1}$ should be eliminated from the set of free parameters and might be replaced by fixed values of $S_{E1}(E_0)$ within the K-matrix analyses. As is evident from Fig. 3, the data can be fitted with equal quality for a wide range of $S_{E1}(E_0)$-values. It seems therefore appropriate to quote a range of acceptable values for $S_{E_1}(E_0)$ rather than a "best value". Ref. [Fil89] recommends $S_{E_1}(E_0) = 0.03 \pm^{0.14}_{0.03}$ MeVb which corresponds to the average of the three analysed data sets. To reduce the uncertainty in $S_{E_1}(E_0)$ one might attempt to measure precisely the E1 capture cross section at energies above the 1^- resonance (i.e. $E > 3\,\mathrm{MeV}$) which might determine the background contribution with better accuracy. Precise measurements of the α-spectrum following delayed β-decay of ^{16}N can also be used as additional information allowing a better determination of the level parameters of the two relevant 1^- states in ^{16}O [Ji90].

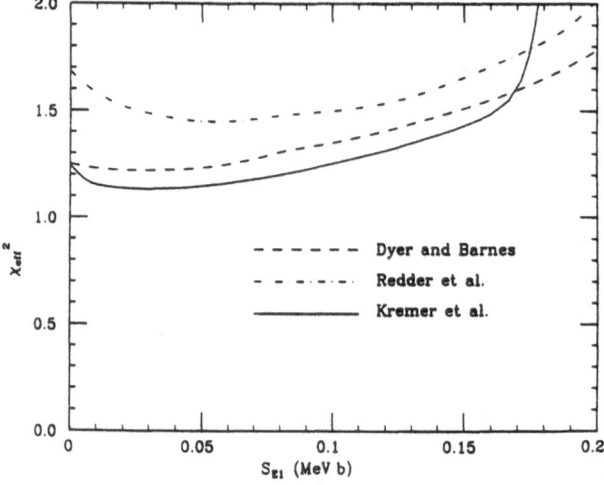

Fig. 3. The quality of the K-matrix fits to the combined E1 capture data and the elastic δ_1 phase shifts. The fits have been performed for each of the three E1 data sets separately

The $^{12}\mathrm{C}(\alpha,\gamma)^{16}\mathrm{O}$ $E1$ cross sections have been fitted within several R-matrix approaches. These studies traditionally apply a 3-level R-matrix model, where two levels correspond to the two lowest physical 1^- states in ^{16}O and the third level, placed at a rather high energy, is introduced to account for possible background contributions. A review of these R-matrix fits can be found in [Bar91]. It is observed that the obtained $S_{E_1}(E_0)$-values depend rather sensitively on the adopted channel radius as well as on model assumptions about the γ-width of the third level (background term). Barker gives some physical arguments why this γ-width can be set to zero in these fits and finds, applying this assumption, noticeably larger $S_{E_1}(E_0)$-values than for example Kremer et al. who treated this quantity as a fit parameter [Kre88] (see Table 1 below).

It seems to be that there are appreciable discrepancies between the E1 data of Redder et al. and Kremer et al. [Bar91, Fil89]. New measurements should be made to resolve these.

2.1.2 Parameter Fits: Potential Models: Potential models are also widely used to extrapolate measured data to astrophysical energies. In theses approaches, usually local potentials of standard parametrizations (e.g. Woods-Saxon or

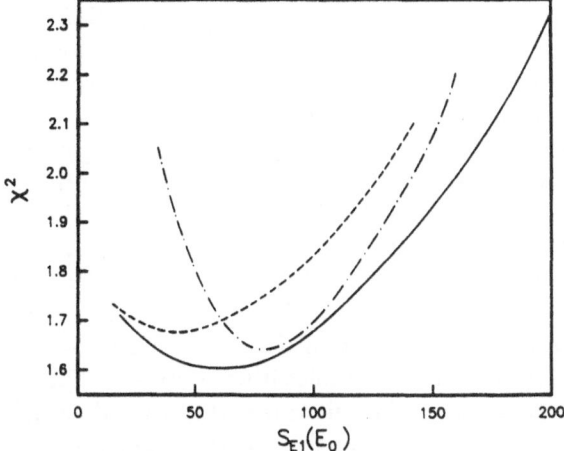

Fig. 4. The quality of the Hybrid R-matrix fits to the combined E1 capture data and the elastic δ_1 phase shifts. The fits have been performed for each of the three E1 data sets (solid, Ref. [Red87]; dashed, Ref. [Dye74]; dashed-dotted, Ref. [Kre88]) separately

Gaussian form factors) are adjusted to relevant experimental data. The derived cross sections at astrophysical energies are then obtained by using the scattering informations (e.g. phase shifts, wave functions) supplied by these potentials. Potential models are useful in physical situations in which only a few channels are open (at best one for capture reactions and two for particle reactions).

2.1.3 Hybrid Models: Combination of R-Matrix and Potential Model: Generally it is not possible to describe two closely located states with the same quantum numbers by the same *local* potential. For this reason, simple potential models cannot be applied to the extrapolation of the $^{12}C(\alpha, \gamma)^{16}O$ E1 data. To overcome this problem, Koonin et al. have proposed the Hybrid R-matrix formalism as a combined potential and R-matrix model [Koo74]. It is motivated by the observation that the broad 1^- resonance at $E_R = 2.44$ MeV can be described most simply and very accurately as an $l = 1$ state in a central $\alpha + ^{12}C$ potential which in the Hybrid R-matrix model is assumed to account additionally for resonant contributions from *all* states at higher energies as well as for non-resonant background contributions. The potential model results for the elastic scattering and the capture cross sections (taking only the radial matrix elements) are then converted into equivalent R-functions to which an additional single level pole, corresponding to the bound 1^- state at $E_R = -45$ keV, is added. The latter is necessary as the bound state has a dominantly shell model like structure and is not appropriately described by an $\alpha + ^{12}$ C cluster ansatz with structureless fragments. (This is different, if *antisymmetrized* $\alpha + ^{12}$ C product wave functions are used, see below).

Assuming a Woods-Saxon form factor for the $\alpha + ^{12}$ C potential, the free parameters of the Hybrid R-matrix model are the three potential parameters, the reduced α-width of the bound state (introduced via a single level pole) and a scaling parameter in the relative part of the electric dipole operator. (Note that the exact fragment masses must be used to ensure that the dipole operator does not vanish.) The ^{16}O ground state, as required for the calculation of the capture cross section in the potential model part, is derived from an $\alpha + ^{12}$C potential of Gaussian form with the constraints to reproduce the correct binding energy and the rms-radius of this state. Adopting the same data sets as in the K-matrix analysis of Ref. [Fil89], the Hybrid R-matrix model yields the following ranges of acceptable values for $S_{E_1}(E_0)$: $0.04 \pm_{0.04}^{0.11}$ MeVb, if the E1 data of Dyer and Barnes are fitted, $0.06 \pm_{0.06}^{0.12}$ MeVb for the Redder et al.-data and $0.08 \pm_{0.05}^{0.08}$ MeV $\cdot b$ for the Kremer et al.-data; see Fig. 4. The quality of the Hybrid R-matrix fits is comparable to those obtained in the R-matrix and K-matrix analyses. In previous Hybrid R-matrix studies it was observed that this model usually gave smaller χ^2-values for the best fit to the E1 capture data than the R-matrix or K-matrix analyses. Obviously in these studies the parameters were more constrained to the capture data than to the elastic phase shifts, as it turned out that the parametrization determined in these fits reproduce the elastic p-wave phase shifts at the higher energies rather badly.

For comparison Table 1 summarizes a few results for $S_{E_1}(E_0)$ as obtained in recent K-matrix, R-matrix and Hybrid R-matrix approaches. This list is far from being complete; further results can be found in Refs. [Bar87, Bar91, Dye74, Fil89, Kre88, Red87]. However, Table 1 shows that our knowledge of $S_{E_1}(E_0)$ for the ^{12}C$(\alpha, \gamma)^{16}$O reaction is rather uncertain. Clearly, further experimental and theoretical effort is necessary to determine this important cross section with the accuracy ($\approx 20\%$) required in astrophysical applications.

Table 1. $S_{E1}(E_0)$-factors obtained by extrapolating the E1 data sets of Barnes and Dyer (Ref. [Dye74]), Redder et al. (Ref. [Red87]) and Kremer et al. (Ref. [Kre88]) adopting various theoretical models. The S-factors are given in MeV$\cdot b$. In the K-matrix fits a linear background was assumed. In the model-dependent R-matrix [Bar91] approach a zero γ-width was assumed for the background, while in the model-independent K-matrix [Fil89] and R-matrix [Kre88] the γ-strength of the background was treated as a free parameter

extrapolation method	Ref. [Dye74]	Ref. [Red87]	Ref. [Kre88]
model-independent K-matrix	$0.02\pm_{0.02}^{0.23}$	$0.08\pm_{0.08}^{0.21}$	$0.05\pm_{0.05}^{0.13}$
model-independent R-matrix	—	—	$0.01\pm_{0.01}^{0.13}$
model-dependent R-matrix	—	$0.26\pm_{0.16}^{0.14}$	$0.15\pm_{0.07}^{0.17}$
Hybrid R-matrix	$0.04\pm_{0.04}^{0.11}$	$0.06\pm_{0.06}^{0.12}$	$0.08\pm_{0.05}^{0.08}$

2.1.4 Microscopic Models: The pure phenomenological models discussed so far are able to reproduce data very nicely. However, they harbor the danger of introducing too many parameters so that the predictive power of the model is lost. Microscopic cluster models, which have been proven in recent years to be a reliable tool for studies of light nuclear reactions, clearly have a greater demand [Wil77, Tan78, Lan86]. These microscopic models, which are based on many-nucleon wave functions for the internal states of the reacting nuclei and effective nucleon-nucleon interactions and treat the antisymmetrization of all nucleons involved in the reaction exactly, are intellectually satisfying and give insight in the physics governing the nuclear process. Often, especially for light nuclear reactions at low energies, the microscopic models give a qualitatively good description of the physical nature of the reaction process. Thus, they seem to be very useful for determining reaction cross sections at astrophysical energies. However, first-principle microscopic approaches will in general not be as accurate as is required for applications to astrophysically important reactions where any useful description requires a careful and accurate account of physically relevant data, especially of energy positions and widths of resonances and bound states which happen to fall into the stellar energy range or to dominate the reaction cross section in this very regime. Thus, *a physically accurate description* of the nuclear reactions becomes more important than *a microscopically pure derivation* of cross sections from first principles. Consequently a quantitative correct prediction of astrophysically relevant reaction cross sections requires some kind of fitting or manipulation of the input (usually adjustment of parameter(s) in the effective nucleon-nucleon interaction). Note, however, that the microscopic models describe light nuclear reactions usually well enough so that the fitting procedure represents only some minor correction and does not influence the fundamental statements about the physics of the reaction process. Detailed descriptions of microscopic models such as the Resonating Group Method (RGM) and the Generator Coordinate Method (GCM) can be found in Refs. [Wil77, Tan78, Lan86]. Thus a brief description of these models suffices here.

In the microscopic cluster models the many-nucleon wave function Ψ is conveniently decomposed into contributions from various channels, where in general each channel ν is defined by a given number N_ν of clusters and internal wave functions $\phi_i(\xi_i)$ $(i = 1, 2, \ldots, N_\nu)$ depending on the translationally invariant internal coordinates ξ_i in the respective clusters. For simplicity we will restrict ourselves in the following to 2-cluster wave functions, noting that a generalization is formally straightforward, but practically rather tedious. Then,

$$\Psi = \sum_\nu \mathcal{A}_\nu \left\{ \phi_{1\nu}(\xi_{1\nu}) \phi_{2\nu}(\xi_{2\nu}) g_\nu(\mathbf{r}_\nu) \right\}. \tag{17}$$

The internal wave functions $\phi_{i\nu}$ are antisymmetrized with respect to exchange of nucleons in the various clusters and $\mathcal{A}_\nu$ is the rest antisymmetrizer which accounts for the appropriate Pauli exchange of nucleons between the different clusters. The relative motion functions g_ν in the various channels depend on the relative coordinate. They can be determined by solving the many-body Schrödinger equation $(H - E)\psi = 0$ for the full microscopic Hamiltonian H in the subspace of Hilbert space spanned by the internal channel functions

$$\Psi_\nu(\mathbf{r}) = \mathcal{A}_\nu \left\{ \phi_{1\nu}(\xi_{1\nu}) \phi_{2\nu}(\xi_{2\nu}) \delta(\mathbf{r}_\nu - \mathbf{r}) \right\}. \tag{18}$$

Then we have

$$< \Psi_\nu | H - E | \Psi > = 0 \tag{19}$$

which has to be solved for all ν. Noting that Ψ can be expanded in terms of the channel functions

$$\Psi = \sum_\nu \int d^3\mathbf{r}\, g_\nu(\mathbf{r})\Psi_\nu(\mathbf{r}), \qquad (20)$$

where g_ν plays the role of an expansion coefficient, and using this expansion in (19) leads to a set of coupled-channel integro-differential equations (RGM equations) from which the functions g_ν can be determined after imposing the appropriate boundary conditions [Wil77]. The model space should include all physically relevant elastic and inelastic fragmentations. In practical applications, the internal cluster degrees of freedom are usually described by harmonic oscillator shell model functions with quantum numbers (angular momentum, spin, isospin, parity etc.) chosen in accordance with those of the physical clusters although the choice of more sophisticated spatial wave functions is in principle possible. For s-shell nuclei the spatial dependence is often approximated by a sum of Gaussians with different width parameters. The parameters in the spatial wave functions are chosen to ensure a correct reproduction of physical properties of the free nuclear clusters (e.g. binding energy, rms-radius etc.), where the various cluster functions might have different parameters.

A construction of many-body functions with good quantum numbers is then straightforward. Often the description of a nuclear reaction is improved by considering "cluster pseudo-states" in the sense of Tang et al. [Tho73]. The pseudo-states correspond to (unphysical) excited cluster states obtained by diagonalization of the cluster Hamiltonian in a multidimensional cluster model space (e.g. sum of Gaussians). The pseudo-state method is a tractable way of enlarging the numbers of channel functions in order to take account of nuclear distortion effects or to qualitatively simulate 3-body break-up channels.

A practical drawback of the Resonating Group Method formulation is that the many-body matrix elements are defined in terms of the translationally invariant coordinates $\xi_{1\nu}$, $\xi_{2\nu}$ and thus their evaluation is very tedious. This disadvantage is overcome within the Generator Coordinate Method [Wil77, Bay85, Lan86] in which the many-body function Ψ is written as a superposition of two-center harmonic oscillator shell model (2CSM) functions ϕ_ν,

$$\Psi = \sum_\nu \int d^3\mathbf{R}\, f_\nu(\mathbf{R})\,\phi_\nu(\mathbf{R}), \qquad (21)$$

where ϕ_ν could represent the antisymmetrized product of two Slater determinants in potential wells separated by the distance $\mathbf{R}$. In general, $\phi_\nu(\mathbf{R})$ will contain spurious center-of-mass components. This can be avoided in the 2CSM by using identical oscillator parameters for the potential wells. This condition, however, represents a severe restriction of the allowed classes of GCM basis functions.

The only microscopic study of the $^{12}C(\alpha,\gamma)^{16}O$ dipole capture to date has been performed by Baye and Descouvemont within the framework of the Generator Coordinate Method [Des87b]. Their model space was spanned by several antisymmetrized product states of possible fragmentations of the $A = 16$-system [Des87a] : elastic and inelastic $\alpha + {}^{12}C$ fragmentations with the ${}^{12}C$ nucleus in either its ground state or in its first excited 2^+ state and p+${}^{15}N$ and n+${}^{15}O$ fragmentations containing both $T = 0$ and $T = 1$ admixtures. Owing to the inclusion of the $T = 1$ components the model space allowed for electric dipole transitions between the calculated many-body states.

The microscopic Hamiltonian contained the Coulomb force, a zero-range spin-orbit interaction as well as an effective central part which is based on the frequently used Volkov $V2$ force. Some of the interaction parameters have been adjusted to reproduce experimental data important for the low energy ${}^{12}C(\alpha, \gamma){}^{16}O$ reaction; in particular, Bartlett and Heisenberg terms not present in the original version of the force have been introduced into the $V2$ interaction [Des87a]. Furthermore, the potential parameters were allowed to be parity-dependent. To reproduce the properties of the ${}^{16}O$ ground state the potential parameters had to be chosen differently for this state than for the other positive parity levels. With this adjustment of the potential parametrization Baye and Descouvemont were able to reproduce the energy positions of the ${}^{16}O$ ground state as well as of the 1_1^- and 2_1^+ states in ${}^{16}O$ which are all of key-importance for the ${}^{12}C(\alpha, \gamma){}^{16}O$ reaction. Furthermore the calculated ${}^{16}O$ level scheme agrees well with the experimentally observed one. The electromagnetic transition strengths between the various ${}^{16}O$ levels at excitation energies below 16 MeV are reproduced sufficiently well.

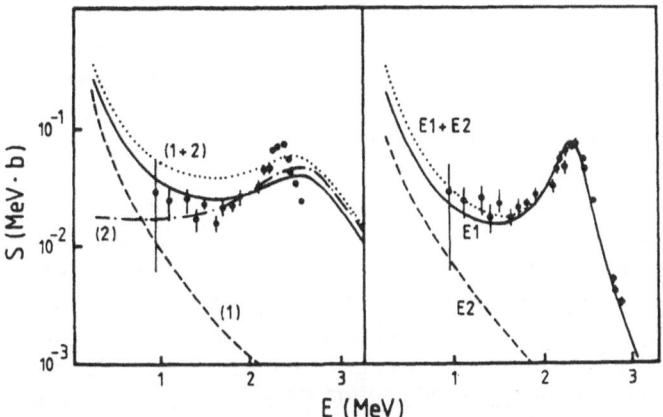

Fig. 5. Comparison of the microscopically calculated ${}^{12}C(\alpha, \gamma){}^{16}O$ S-factor with the data of Ref. [Red87]. The left side shows the uncorrected results for the E1 capture (solid), the contributions from the bound (1) and resonant (2) 1^- states and their combined effect (dotted). The differences between the total E1 capture data and the combined effects are due to background contributions. The right side shows the corrected E1 and E2 S-factors (from Ref. [Des87b])

As pointed out above, an accurate reproduction of important experimental input data is an undispensable ingredient in any meaningful estimate of the ${}^{12}C(\alpha, \gamma){}^{16}O$ reaction rate at stellar energies. Although the multichannel

GCM study yields a valuable approximation to the ^{16}O level scheme [Des87a] as well as to the ^{12}C$(\alpha, \gamma)^{16}$O reaction [Des87b] obtained results are still not accurate enough for astrophysical purposes (see Fig. 5). Descouvemont and Baye have therefore consistently corrected some of these results by replacing calculated values by the experimentally observed ones. For example, the GCM study overestimates the $B(E1)$ transition strengths for the decay of the 1_1^- level into the ^{16}O ground state ($5.7 \cdot 10^{-4}$ W.u. compared to the experimental value of $3.6 \pm 0.4 \cdot 10^{-4}$ W.u.). This slight discrepancy, however, could be dissolved by introducing a small effective charge. A similar correction has been made for the 2_1^+ state. Finally Descouvemont and Baye had to correct their results for the input data of the broad 1_2^- resonance. The corrected E1 contribution to the ^{12}C$(\alpha, \gamma)^{16}$O S-factors at energies below $E \approx 3$ MeV are also shown in Fig. 5 where they are compared to the experimental data of Redder et al. [Red87]. The agreement with the data is fair. The authors emphasize [Des87b] "that the theoretical curve is the result of a microscopic calculation in which a few inaccurate theoretical quantities are replaced by their experimental counterparts" rather than a fit to the data. The analysis of Descouvemont and Baye predict an E1 contribution to the astrophysical S-factor at $E = 300$ keV of $S_{E1}(E_0) = 0.16$ MeVb.

2.2 The ^{12}C$(\alpha, \gamma)^{16}$O E2 Capture at Astrophysical Energies

For completeness, we will briefly discuss the status of the E2 contribution to the ^{12}C$(\alpha, \gamma)^{16}$O cross section at low energies. Angular distributions have been measured by Dyer and Barnes [Dye74] and, for a larger energy range, by Redder et al. [Red85, Red87] From the Redder-data two E2 capture data sets have been analysed owing to slightly different strategies of error handling in the data [Fil89, Red87]. However, considering the still large uncertainties the two data sets essentially agree with each other.

Extrapolations of the data sets have been performed within the R-matrix and K-matrix models leading to a rather large range of possible $S_{E2}(E_0)$-values; e.g. $0.06 - 0.34$ MeVb [Red87], $0.06 - 0.12$ MeV $\cdot b$ [Pla87], $0.05 - 0.18$ MeV $\cdot b$ [Bar91] for R-matrix fits and $0.0 - 0.16$MeV $\cdot b$ [Fil89] for the K-matrix fit, where the latter range combines the results obtained for various treatments of the background, which, however, showed strong scatter among each other.

Data-independent studies of the ^{12}C$(\alpha, \gamma)^{16}$O E2 cross section, in which the results are not fitted to the E2 capture data, have been performed within the multichannel [Des87b] and single-channel [Des84] Generator Coordinate Method and within multichannel and single-channel microscopic potential models [Lan85, Fun85]. (As in the microscopic cluster model, the microscopic potential model derives the capture cross section from antisymmetrized many-body wave functions. The relative wave functions between the clusters, however, are determined from phenomenological potentials which are adjusted to reproduce relevant experimental data within the adopted model space. In contrast, conventional potential models treat the fragment nuclei as structureless particles.) The obtained results agree reasonably well with the data and suggest $S_{E2}(E_0)$

to be in the range $0.05 - 0.10\,\mathrm{MeV} \cdot b$. A conventional potential model study of the low-energy $^{12}\mathrm{C}(\alpha,\gamma)^{16}\mathrm{O}$ E2 capture process has been reported by Oberhummer et al. [Kra90].

3. Electron Screening in Laboratory Experiments

For astrophysical applications one is interested in the cross sections of charged-particle-induced fusion reactions for bare nuclei, $\sigma_{\mathrm{nucl}}(E)$, at very low energies (in practice near zero energy). Depending on the astrophysical scenario to be studied these cross sections have then to be corrected for possible screening effects caused by plasma electrons (using models, see Sect. 4). Since the astrophysically most important energy region lies far below the Coulomb barrier, where the cross sections are extremely small, a direct measurement of these cross sections in the laboratory is difficult if not impossible with presently available techniques. Therefore, one measures the cross sections to energies as low as possible and then extrapolates the data to the relevant energy region. It was generally believed that the uncertainty of this procedure is reduced by steadily lowering the energies at which data can be taken in the laboratory.

Assenbaum et al. [Ass87] pointed out that this procedure, however, might be problematic as at very low energies the experimentally measured cross section does not represent the required case for bare nuclei: the cross section is increased due to screening effects connected with the electrons present in the target (and in the projectile). To understand the argumentation of these authors, let us consider the $^{3}\mathrm{He}(\mathrm{d},\mathrm{p})^{4}\mathrm{He}$ fusion reaction. One can study experimentally this reaction, for example, by shooting an ion beam of deuterons with the c.m. energy E on an atomic $^{3}\mathrm{He}$ gas target [Kra87, Eng88]. At large separations of target and projectile the scattering system consists of a deuteron (bare nucleus) and a $^{3}\mathrm{He}$ atom (with two electrons in the ground state). This description, however, is no longer valid at those separations at which the deuteron is tunneling through the joint Coulomb barrier and is certainly inadequate at radii of a few Fermis at which the nuclear reaction takes place. If we approximate the radius at which the tunnel process starts by the classical turning point,

$$R(E) = \frac{Z_1 Z_2 e^2}{E}, \tag{22}$$

one finds radii of order $10^{-2}\,\mathring{A}$ at energies of a few keV. Under these conditions, the atomic scattering system corresponds in a good approximation to an atom with the combined charges $Z_1 + Z_2$ in the center (a $^{5}\mathrm{Li}$ atom in our example), in which the two electrons are stronger bound than in the $^{3}\mathrm{He}$ atom. This gain of electronic binding energy might be transferred to the relative motion of the two nuclei which then tunnel through the barrier with an effective energy E_{eff} being slightly higher than the asymptotic energy E:

$$E_{\text{eff}} = E + E_{\text{el}}(A + B) - E_{\text{el}}(A) - E_{\text{el}}(B) = E + \Delta E. \qquad (23)$$

Here, $E_{\text{el}}(A+B)$ is the electronic binding energy in the combined atom $A+B$, while $E_{\text{el}}(A)+E_{\text{el}}(B)$ is the sum of the electronic binding energies in the asymptotically separated reaction partners. As $E_{\text{eff}} > E$ (the Coulomb barrier becomes thinner due to screening effects) the experimental cross section $\sigma_{\text{exp}}(E)$ should be larger than that of bare nuclei, $\sigma_{\text{nucl}}(E)$. One might define this enhancement as

$$f(E) = \frac{\sigma_{\text{exp}}(E)}{\sigma_{\text{nucl}}(E)}. \qquad (24)$$

To derive an estimate for $f(E)$, we assume that the astrophysical S-factor is roughly constant over the small energy interval ΔE. Then, we find

$$f(E) = \frac{E E_{\text{eff}}}{\exp}[-2\pi(\eta(E_{\text{eff}}) - \eta(E)] \approx \exp[\pi\eta(E)\frac{\Delta E}{E}]. \qquad (25)$$

Thus, the enhancement f(E) should increase exponentially with decreasing energy E. For our example ($\Delta E \approx 120$ eV), one expects that the measured cross section exceeds the pure nuclear cross section at E = 5 and 10 keV by about 45% and 14%, respectively. Note that the $^3\text{He}(d, p)^4\text{He}$ reaction has been measured in the laboratory down to an energy of 5.88 keV [Eng88, Kra87].

The above discussion has been confirmed in more realistic descriptions of the electronic screening effects [Ass87, Blü89]. These studies are based on a Born-Oppenheimer approximation assuming that, at the energies involved, electronic and nucleonic degrees of freedom are well separated and that the elimination of the electronic degrees of freedom leads to an effective screening potential V_{sc} between the colliding nuclei. Thus, if electrons are present the two nuclei have to penetrate a shielded Coulomb barrier

$$V_{\text{eff}}(r) = V_{\text{coul}}(r) - V_{\text{sc}}(r) = \frac{Z_1 Z_2 e^2}{r} - V_{\text{sc}}(r) \qquad (26)$$

rather than $V_{\text{coul}}(r)$. As the cross sections at energies far below the Coulomb barrier are dominated by the barrier penetration, the f(E) factor in these models can be approximated by

$$f(E) = \frac{P_{\text{eff}}(E)}{P_{\text{coul}}(E)}, \qquad (27)$$

where P_{eff} and P_{coul} are the WKB-penetrabilities calculated with the shielded and unshielded Coulomb potentials, respectively. It turns out that the enhancement factors calculated with various screening potentials (Thomas-Fermi, Hartree-Fock, phenomenological universal screening potential) support the argumentation of Ref. [Ass87] that measured cross sections at low energies noticeably exceed those for bare nuclei. Furthermore, these model calculations also find that $f(E)$ should increase exponentially with decreasing energy. However, the various models predict a somewhat different strength of the screening effects: the $f(E)$ factor calculated within the Thomas-Fermi model or using the universal potential of Biersack et al. [Bie80] is noticeably larger than that derived from the spherical Hartree-Fock approach.

To test the validity of the various models by comparison with data, one has to take into account that the above model predictions are relative statements with respect to the pure nuclear cross sections. As the latter cannot be measured yet in the laboratory (such experiments would require a procedure involving crossed beams of bare nuclei, which is yet not feasible due to extremely low luminosities), one has to rely on model predictions for $\sigma_{\mathrm{nucl}}(E)$ at low energies. Fortunately, there are reactions induced by light nuclides, where the experimental cross sections are well reproduced by theoretical calculations in the energy regime which should be uneffected by electron screening effects (i.e., at the higher energies where, say, $f(E) < 1.01$). These theoretical studies do not represent fits to the data but have been performed within the microscopic multichannel cluster model (see Sect. 3). Thus, the low-energy cross sections calculated in these models might be viewed as a physical extrapolation of the measured nuclear cross sections into the energy regime in which one expects electron screening effects to occur. In the following we will identify the cross sections for bare nuclei, $\sigma_{\mathrm{nucl}}(E)$ in Eq. (24), with the microscopically calculated cross sections.

Until now the most extensive studies of electron screening effects on fusion reactions have been performed for the $^3\mathrm{He}(\mathrm{d,p})^4\mathrm{He}$ reaction. Here detailed experimental measurements have been carried out down to an energy of 5.88 keV (for experimental details, see Refs. [Eng88, Kra87]). Theoretically, this reaction has been studied successfully within a multichannel cluster approach [Blü90a] adopting a model space spanned by p + α and d + $^3\mathrm{He}$ configurations. As the deuteron can be easily distorted, additional d + $^3\mathrm{He}$ pseudo-states have been included. The cluster calculation was able to simultaneously reproduce: (i) the $^3\mathrm{He}(\mathrm{d,p})^4\mathrm{He}$ cross sections at E = 40 keV to about 1 MeV, (ii) the tensor analyzing powers measured with polarized deuterons on top of the $\frac{3}{2}^+$ resonance (E = 254 keV), which dominates the cross sections at lower energies [Blü90b], and (iii) all experimentally known levels in $^5\mathrm{Li}$ at excitation energies below about 20 MeV. At the very low energies at which we expect electronic screening effects to be present in the data ($E \leq 40$ keV), the calculated astrophysical S-Factor is well approximated by

$$S(E) \approx 5.54 + 47.27E - 393.4E^2(\mathrm{MeV} \cdot b) \qquad (28)$$

(here: E in units of MeV). Comparing the calculated cross sections with data, the expected enhancement due to screening effects is apparent (Fig. 6). One also observes the exponential increase of the $f(E)$ factor with decreasing energy, as expected from Eq. (25). In Fig. 6 we also compare the data with predictions, where we have used Eqs. (3) and (28) for $\sigma_{\mathrm{nucl}}(E)$ and have derived the $f(E)$ factors both using a spherical Hartree-Fock approach and adopting the universal screening potential [Bie80]. The latter model calculations reproduce the data well. (Note that the data of Engstler et al. [Eng88] are only a *relative* measurement normalized to the *absolute* cross sections of Krauss et al. [Kra87]. It has been pointed out [Blü89] that the adopted normalization might be too high by about 5%.) The Hartree-Fock approach underestimates the data. The success of the universal screening potential is somewhat surprising as, in contrast to the Hartree-Fock approach, it does not fulfill the united atom test: the difference $V_{\text{eff}}(r = 0) - V_{\text{eff}}(r \to \infty)$ is larger than the difference in electronic binding energy ΔE as defined in Eq. (23). Results obtained by describing the screening effects within the statistical Thomas-Fermi model are rather similar to those of the universal screening potential. Again, the united atom test is not fulfilled in this model.

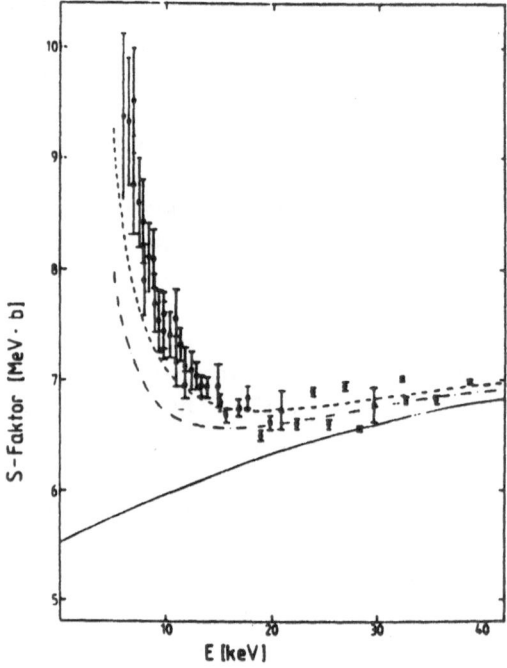

Fig. 6. Comparison of the experimental ^{3}He(d, p)^{4}He data [Eng88, Kra87] with theoretical predictions based on the microscopically calculated nuclear S-factor from Ref. [Blü90a] (solid line) and describing electron screening effects in the Hartree-Fock model (dashed-dotted) and by the universal screening potential of Ref. [Bie80] (dashed line)

An important nuclear reaction, for which screening effects have been discussed to be probably present in the low-energy data, is the ^{3}H(d, n)^{4}He reaction, which is considered to be the fuel in future fusion reactors. Again, the observed cross sections [Jar84] are well reproduced within the microscopic cluster model [Blü91], which predicts that the cross sections are strongly dominated by the resonant contributions of the $\frac{3}{2}^+$ resonance at E = 64 keV. In fact, it has been shown that *all* Los Alamos data [Jar84] at $E > 16$ keV (being here virtually free of screening effects) can be reproduced by a single Breit-Wigner resonance [Lan89]. This fit, however, predicts cross sections which are noticeably lower than the data at $E < 16$ keV. This discrepancy can be removed if one proposes [Lan89] that these low-energy data are enhanced due to electronic screening effects. In Fig. 7 we compare the data with the best fit to the data at $E > 16$ keV and with calculated S-factors in which screening effects are included. The latter calculations have been carried out within the Thomas-Fermi and Hartree-Fock models and considered that in the experiments the tritium target was a *molecular* gas. Thus, in the energy balance one has to account for the dissociation energy of the molecule and the kinetic energy which the second atom in the molecule might carry away after the molecular bond is broken. The latter quantity has been estimated within a Coulomb explosion model (see Ref. [Eng88]). From Fig. 7 one observes that the best fit to the Los Alamos data at $E > 16$ keV is consistent with the low-energy data at $E < 16$ keV, if the latter data are corrected for electron screening effects.

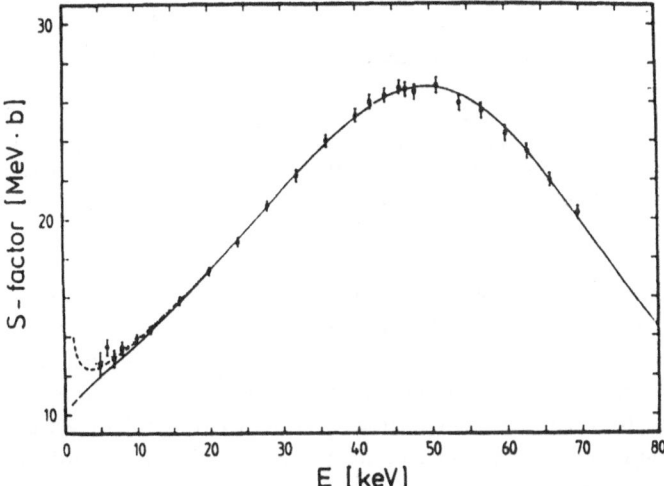

Fig. 7. Comparison of the Los Alamos ^{3}H(d, n)^{4}He data [Jar84] with theoretical predictions based on the Breit-Wigner best fit [Lan89] to the $\frac{3}{2}^+$ resonance at 64 keV (solid line) and describing electron screening effects in the Hartree-Fock (dashed line) and Thomas-Fermi model (dotted line)

The presence of electron screening effects has also been discussed for the data of the ^{3}He(^{3}He, 2p)^{4}He reaction [Typ91], which is the major source for ^{4}He production in the sun. Here, the microscopically calculated cross sections, which agree well with the data at $E \geq 80$ keV, are somewhat lower than the trend in the low-energy data at $E \leq 80$ keV, perhaps indicating screening effects. Definite conclusions, however, cannot be drawn due to the relatively large uncertainties of the data.

Experimentally, further indications for electron screening effects have been observed in reactions involving Li isotopes [Eng90]. However, quantitative statements concerning these reactions cannot yet be made due to the lack of reliable estimates for $\sigma_{nucl}(E)$. Recently, more sophisticated models have been proposed to study screening effects [Ben90]: the quantitative results agree well with the predictions of the models presented here.

To conclude this section, electron screening effects on low-energy fusion reactions can be investigated with presently available experimental techniques. The

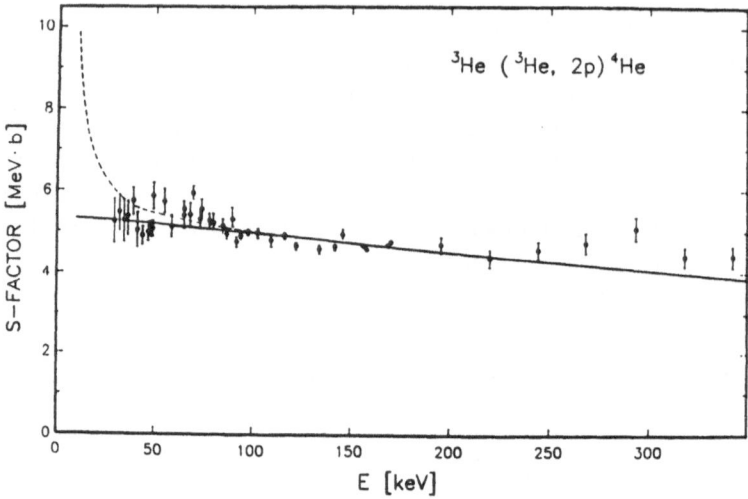

Fig. 8. Comparison of the calculated ^{3}He(^{3}He, 2p)^{4}He S-factors at low energies (solid line) with the data of Ref. [Kra87]. The dashed line indicates the expected experimental S-factor due to the enhancement by electron screening effects [Typ91]

theoretical studies suggest that the observed enhancements can be described in first order in models based on the Born-Oppenheimer approximation and that the coupling of nuclear and electronic degrees of freedom can be neglected. Definite conclusions, however, have to await the results of improved measurements and calculations. For the future, the experimental and theoretical investigations of electron screening effects appear to offer the challenging possibility to study the interplay of atomic and nuclear physics.

4. Screening Effects in the Plasma

In a plasma shielding effects might also be important as the Coulomb interaction of reacting nuclei with surrounding plasma particles can enhance the thermonuclear reaction rates. Here the physical situation is quite different than is encountered in the nuclear physics laboratory experiments discussed in the previous section. In the plasma, the electrons (mainly) occupy free continuum states rather than being bound to an atomic nucleus. The theoretical description of plasma screening strongly depends on the plasma conditions. Rather than reviewing models relevant to various astrophysical scenarios (here the reader is referred to Refs. [Sal54, Sal69, Ich84, Yak89, Kra86]), we will in the following briefly discuss two nuclear physics problems subject to two extreme screening conditions: the fate of ^{7}Be in the sun (weak screening limit) and the pycnonuclear triple-alpha reaction under conditions expected to exist on neutron star surfaces (screening in very dense plasma).

In stellar plasma during hydrostatical burning stages (e.g. in our sun) one usually has a situation in which the average Coulomb energy between neighbouring ions $\overline{E}_{coul}$ is much less than the thermal energy kT of the plasma. The nearly perfect stellar gas is well desribed in this "weak screening" [Sal54] limit by the Debye-Hückel theory (see e.g. Ref. [Cla83]), in which, due to shielding effects of all other charges in the plasma, the Coulomb potential around an ion charge Z_1, has to be replaced by the shielded Coulomb potential (Debye-Hückel potential)

$$V_1(r) = \frac{Z_1 e}{r} \exp\left\{-\frac{r}{R_D}\right\}, \qquad (29)$$

where R_D is the Debye-radius

$$R_D = \left(\frac{kT}{4\pi e^2 \sum_i Z_i^2 \overline{n}_{Z,i}}\right)^{1/2}. \qquad (30)$$

Here $\overline{n}_{Z,i}$ represents the average number of a plasma species with charge Z_i. In the range of applicability of the Debye-Hückel theory ($\overline{E}_{coul} \ll kT$), the radius R_D is much larger than the average distance between neighbouring particles. An ion with charge Z_2 passing close to Z_1 will then feel the shielded Coulomb interaction

$$V(r) = \frac{Z_1 Z_2 e^2}{r} \exp\left\{-\frac{r}{R_D}\right\}. \qquad (31)$$

The fact that the ions have to penetrate a shielded Coulomb barrier means that the corresponding nuclear reaction rate in the plasma is enhanced compared to that of bare nuclei as (ideally) determined in the laboratory,

$$< \sigma v >_{plasma} = f_{plasma} < \sigma v >_{bare\ nuclei}. \qquad (31)$$

Here we have introduced the enhancement factor f_{plasma} which can be estimated by the ratios of penetrabilities calculated for the shielded (Eq. (32)) and unshielded Coulomb barriers. Noting that at the astrophysically most effective energies E_0 the classical turning point for these barriers is usually much less than R_D one might expand the exponential in (31) finding that the Debye-Hückel shielding essentially results in a small increase of the fusion energy of the colliding ions by the amount $-U_0 = Z_1 Z_2 e^2 / R_D$. Thus, in a good approximation the enhancement factor is given by

$$f_{plasma} = e^{-U_0/kT} \approx 1 - \frac{U_0}{kT}. \qquad (33)$$

If this weak screening formula, which is originally due to Salpeter [Sal54], is applied to nuclear reactions under solar conditions, the enhancement is found to be of the order of a few percent. Note that, similarly to Eq. (32), under most astrophysical conditions (however, with non-vanishing temperature) the generalized reaction rate integral can be separated into the conventional expression without screening (2) and a screening factor, e.g. [Ich84, Sal69], which generally might be dependent on the charges of the reacting nuclei and on the density and the temperature. As we will discuss in the next section, expression (32) is not longer valid under extreme conditions (very high densities and/or very low temperatures) as it has been derived under the assumption of a Maxwell-Boltzmann distribution for the relative velocities of the reacting nuclei.

Formally weak screening in the plasma and in the laboratory scattering experiments have rather similar effects on the nuclear reaction rates: In both cases the screening effects are well accounted for by a slight shift of the collision energy of the reacting nuclei. The different physical nature of the shielding processes, however, shows up in the derivation and estimates of these energy shifts.

4.1.1 Example for Weak Screening: The Fate of ^{7}Be in the Sun:
Screening effects play an interesting role in the fate of ^{7}Be in the sun. Thus their correct treatment is also relevant for the predictions of the neutrino flux to be observed in the Davis experiment which owing to threshold effects is mainly sensitive to neutrinos originating from the ^{8}B-decay following a synthesis of this nucleus by proton-capture on ^{7}Be. The amount of ^{8}B produced in the center of the sun is determined by the rates of proton- and electron-capture on ^{7}Be, where the latter is the dominant mechanism to destroy ^{7}Be. The fate of ^{7}Be is considered the most uncertain nuclear physics input to the solar neutrino problem [Bah86a].

At very low energies ($E \leq 500\,\text{keV}$) there exist essentially two experimental ^{7}Be(p, γ)^{8}B data sets measured by Kavanagh et al. [Kav69] and, more recently, by Filippone et al. [Fil83]. These data are strongly dominated by non-resonant dipole capture from ^{7}Be + p $s-$ und $d-$wave scattering states into the only bound ^{8}B level, the ground state at $E = -138\,\text{keV}$ with $J^\pi = 2^+$. To determine the reaction rate in the solar core ($E \approx 0 - 20\,\text{keV}$), the measured cross sections, which extend to energies as low as $E = 117\,\text{keV}$, have very recently been extrapolated adopting the energy dependence of the S-factor as calculated in a microscopic 3-cluster RGM approach [Kol88, Joh90]. In this study, the model space was spanned by antisymmetrized p$+^3$He$+^4$He 3-cluster basis wave functions, where the dynamics between the three clusters were determined form the many-body Schrödinger equation using two different effective NN-interactions. One parameter in these interactions has been adjusted to reproduce the binding energy of the ^{8}B ground state. Note that this microscopic approach describes the main properties of the individual cluster (^{3}He,^{4}He,^{7}Be) very well.

Both effective interactions predict a similar energy-dependence of the $S(E)$-factor at $E \leq 500\,\text{keV}$. In fact, it also agrees reasonably well with energy-dependences derived in different approaches; e.g. from a microscopic 3-cluster study within the Generator Coordinate Method [Des88] and from direct potential model analyses [Bar86, Kim87]. Fig. 9 shows extrapolations of the measured cross sections using the energy dependence of the 3-cluster RGM approach. A non-weighted average of these extrapolations gives $S(0) = 0.0225 \pm 0.0018\,\text{keV}b$, where the dominating part of the uncertainty arises from the obvious differences in the absolute magnitude of the two data sets. Barker has pointed out that

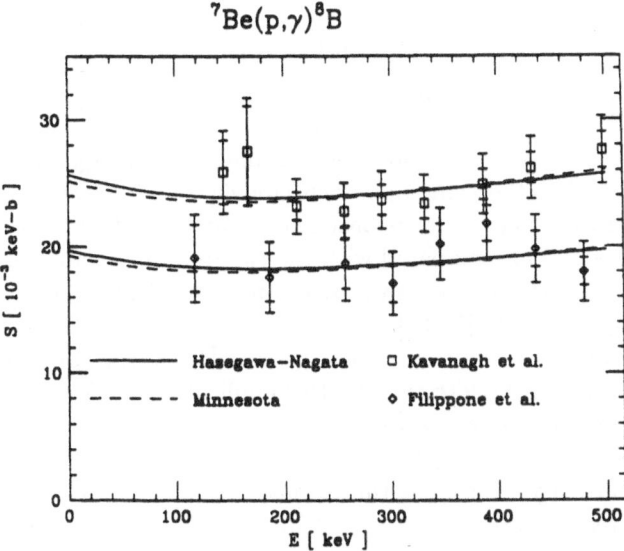

$^7\mathrm{Be}(\mathrm{p},\gamma)^8\mathrm{B}$

Fig. 9. Extrapolation of the $^7\mathrm{Be}(\mathrm{p},\gamma)^8\mathrm{B}$ data (Refs. [Fil83, Kav69]) using the energy-dependence of the microscopic 3-cluster calculation [Joh90]. The RGM study has been performed for two different NN-interactions (Hasegawa-Nagata force [Fur80]; Minnesota force [Chw73])

the accepted value for the $^7\mathrm{Li}(\mathrm{d},\mathrm{p})^8\mathrm{Li}$ cross section, to which the $^7\mathrm{Be}(\mathrm{p},\gamma)^8\mathrm{B}$ data have been normalized, might be too large [Bar86] which would lead to a reduction of the recommended $^7\mathrm{Be}(\mathrm{p},\gamma)^8\mathrm{B}$ S-factor at solar energies. From Eq. (33) one finds that in the solar core the $^7\mathrm{Be}(\mathrm{p},\gamma)^8\mathrm{B}$ reaction is enhanced by about 20% due to plasma shielding effects.

The atomic nucleus $^7\mathrm{Be}$ decays via capture of one of its bound electrons with a lifetime of $\tau_{\mathrm{LAB}} = 77$ days. As the lifetime of a nucleus against electron capture is proportional to the square of the electron wave function at the nucleus, one expects the lifetime to be longer in a plasma of free electrons than for the atomic nucleus. In fact, one finds $\tau \approx 1.2\,\tau_{\mathrm{LAB}}$ if the $^7\mathrm{Be}$-nucleus decays by capture of continuum electrons under solar conditions [Bah86b]. Iben et al., however, have pointed out that there is a non-negligible probability for $^7\mathrm{Be}$-nuclei in the sun to have a bound K-shell electron [Ibe67]. This occupation probability is proportional to the Boltzmann factor $\exp\left\{-E_{\mathrm{K}}/kT\right\}$, where E_{K} is the difference between the binding energy of the K-shell electron and the continuum threshold which in the solar plasma is strongly reduced owing to screening effects compared with the atomic value ($E_{\mathrm{K}} \cong 216.6$ eV). Similarly plasma screening reduces the probability to find a K-shell electron at the nucleus. Iben et al. and subsequently Bahcall and Moeller [Bah86b] determined these quantities within the Debye-Hückel approximation by solving the Schrödinger equation for the shielded Coulomb potential (31) with $Z_1 = 4, Z_2 = -1$. Very recently Johnson et al. [Joh90] have pointed out that this treatment is not quite appropriate as for a nucleus Z_1 with a bound electron the plasma "sees" a point charge $(Z_1 - 1)$ rather than Z_1. In this case, the Ecker-Weizel potential

$$V_{EW}(r) = -Z_1 e^2 \left\{ \frac{1}{r} \exp\left\{ -\frac{r}{R_D} \right\} + \frac{1}{R_D} \right\}, \tag{34}$$

which is found to give a qualitativly correct description of the discrete and continuous spectra [Kra86], might be used to determine E_K. The authors of Ref. [Joh90], which adopted a somewhat more sophisticated screening potential, obtained $E_K \approx 56$ eV. Within the same approach [Joh90], which treats the electron capture of ^{7}Be in the solar core self-consistently, one finds the capture rate to be increased by about 17% compared with the case of pure continuum electron capture. However, this amounts only to an increase of roughly 2% compared to the presently used rate [Ibe67].

All studies, which estimate the electron capture rate of ^{7}Be in the sun, are based on the Debye-Hückel theory. Note that for this application the assumptions, on which this model is based, are weakly violated. It seems therefore to be desirable to calculate the lifetime of ^{7}Be in the sun with models which go beyond the Debye-Hückel theory. A study within a finite-temperature Hartree-Fock approach is in progress [Joh91].

4.1.2 Example for Strong Screening: The Pycnonuclear Triple-alpha Fusion

Rate: At high densities and low temperatures (i.e. $\overline{E}_{coul} \gg kT$) the electron gas is degenerate and nearly perfect [Cla83], while the ions are located in a lattice that maximizes the inter-ion separation. This physical situation is usually described in the Wigner-Seitz approximation, where the plasma is divided into neutral spheres around each nucleus with charge Z such that the sphere contains the Z electrons closest to the nucleus. The radius r_0 of this cell is connected with the number density of nuclei $\overline{n}_n$ by $\frac{4\pi}{3} r_0^3 \cdot \overline{n}_n = 1$. Under theses physical conditions, temperature is insufficient to induce thermonuclear reactions. Rather, nuclear fusion takes place through pycnonuclear reactions; that is the ions can penetrate the Coulomb barrier due to zero-point motion and the plasma screening energy. Discussion of pycnonuclear reactions can be found in Refs. [Cam59, Sal69, Sch91b].

If for a given density ρ the temperature exceeds a critical value, the melting temperature [Yak89]

$$T_m \approx 1.3 \cdot 10^3 \cdot Z^2 \left(\frac{\rho}{A} \right)^{1/3} K, \tag{35}$$

the ion lattice will melt as the thermally induced fluctuations of the ions will destroy the long-range order. The plasma is then best described as a strongly coupled quantum fluid. The gaseous regime occurs at [Yak89]

$$T \gg 2.3 \cdot 10^5 Z^2 \left(\frac{\rho}{A} \right)^{1/3} K. \tag{36}$$

Note that for a one-component plasma, even if $T = 0$, crystallization does not occur beyond a critical density ρ_m [Yak89]. For $\rho > \rho_m$ the ions rather form a strongly coupled quantum liquid where the crystallization is suppressed by large zero-point motion. For a hydrogen plasma the critical density is calculated as $\rho_m \approx 1.7 \times 10^4 \, \text{g/cm}^3$, while it is about $6.8 \times 10^7 \, \text{g/cm}^3$ for a plasma made of ^{4}He nuclei and their associated electrons [Cep80].

In a scenario recently proposed to explain γ-ray bursts of neutron stars [Bla90], one of the basic nuclear ingredients is the triple-alpha fusion rate $3\alpha \rightarrow ^{12}C$, which has to be known in a density/temperature regime ($T \approx 10^6 K$, $\rho \approx 10^9 \, \text{g/cm}^3$) where a helium plasma is best described as a quantum liquid. The neutralizing electron gas can be considered as an homogeneous background.

The triple-alpha fusion rate can then be defined as

$$r_{3\alpha} = \frac{1}{6} n_\alpha^3 \, g^{(3)}(r_{12} = 0, r_{23} = 0) \, V_r^2 \, \Gamma_\gamma (E_{3\alpha}). \tag{37}$$

Here, V_r is the reaction volume (expected to be of order of the nuclear volume), n_α is the α-particle number density and $\Gamma_\gamma(E)$ is the radiative width of the carbon-like 3α-state at energy E, which can be related to the experimentally known decay width Γ_γ of the first excited 0^+ state in ^{12}C via $\Gamma_\gamma(E_{3\alpha}) \approx 0.54 \, \Gamma_\gamma$ [Sch91b]. The 3-body correlation function $g^{(3)}$ defines the probability to find three α-particles (i = 1,2,3) at specified distances (r_{12}, r_{23}). In the Kirkwood approximation, $g^{(3)}$ can be expressed in terms of the 2-body correlation function $g^{(2)}$ as

$$g^{(3)}(r_{12}, r_{23}) = g^{(2)}(r_{12}) g^{(2)}(r_{13}) g^{(2)}(r_{23}). \tag{38}$$

The quantity $g^{(2)}(r)$ can then be obtained within the hypernetted chain (HNC) formalism which has been successfully applied to charged boson systems.

For the many-body wave function of the helium liquid Ψ a product of Jastrow-type two-particle wave functions has been used [Sch91a, Mül91]

$$\Psi(\{\mathbf{r}_i\}) = \prod_{i<j} f(\mathbf{r}_i - \mathbf{r}_j) = \prod_{i<j} f(|\mathbf{r}_i - \mathbf{r}_j|), \tag{39}$$

where the latter identity follows from the isotropy and homogeneity of the system. The optimal wave function $f(r)$ can be obtained by minimizing the energy of the system, where the internuclear interaction is described by a phenomenological potential [Ali66] which reproduces the low-energy $\alpha + \alpha$ phase shifts including the ^{8}Be ground state resonance at $E = 92$ keV with a width of 6.7 eV. Asymptotically (for r larger than a radius r_A), $f(r)$ behaves like $e^{-\Gamma/r}$, where the constant Γ can be derived from the ion plasma frequency. For $r < r_A$, the optimal $f(r)$ within the HNC approximation can be determined by a iterative solution of a set of coupled non-linear equations [Usm82, Sch91b, Lan91].

In Fig. 10 the calculated fusion rate Eq. (37) is shown as a function of density. In the regime $\rho \leq 10^9 \, \text{g/cm}^3$, as it is relevant for the study of γ-ray bursts, this rate is slightly higher than obtained by Fushiki and Lamb [Fus87] in an S-matrix approach and by Schramm et al. in a non-optimized hypernetted chain approach [Sch91b].

Fig. 10 shows a pronounced peak at $\rho \approx 3 \cdot 10^9 \, \text{g/cm}^3$. This is an indication that the ^{4}He liquid undergoes a phase transitions to ^{8}Be matter owing to the fact that at these densities the plasma screening suffices to induce the strong

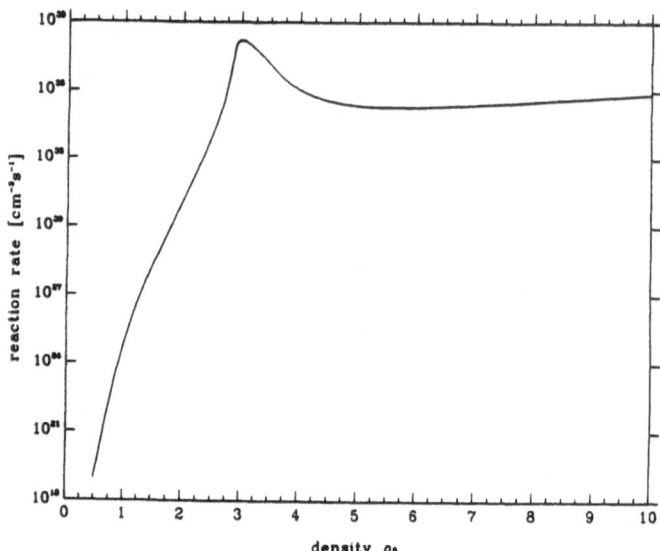

Fig. 10. The triple-alpha reaction rate (37) as calculated in an optimal hypernetted chain approach [Mül91]

resonant short-ranged correlations between two α-particles corresponding to the ^{8}Be ground state. However, for a quantitative determination of the critical density, at which this transition occurs, the HNC approximation might not be accurate enough and should be followed by an essentially exact, but more involved Monte Carlo study of the helium liquid. Furthermore, the HNC calculation investigates the transition from a helium liquid to a ^{8}Be liquid. However, at the densities involved ^{8}Be matter will exist as a Coulomb lattice rather than a quantum liquid. Consequently the phase transition will occur at a somewhat lower density. Following Baym, Pethick and Sutherland [Bay71] the equilibrium density of a helium liquid and a ^{8}Be lattice of fcc-structure can be estimated as $\rho \approx 2.3 \cdot 10^9 \mathrm{g/cm}^3$.

It is not clear whether a physical scenario exists where this phase transition might be realized. A possible site is the crust of a slowly accreting neutron star [Bla90]. However, a recent study of the pycnonuclear triple-alpha reaction [Sch91b] indicates that the α-particles in the helium liquid are probably fused to ^{12}C before the system reaches the critical density.

Parts of the work presented here has greatly benefitted from collaborations with H.J. Assenbaum, G. Blüge, B.W. Filippone, J. Humblet, C. Johnson, E. Kolbe, S.E. Koonin, D. Lukas, H.-M. Müller, C. Rolfs, S. Schramm, F.K. Thielemann and S.Typel.

References

Ali66	S. Ali, A.R. Bodmer: Nucl. Phys. **80**, 99 (1966)
Ass87	H.J. Assenbaum, K. Langanke, C. Rolfs: Z. Phys. **A327**, 461 (1987)
Bah86a	J.N. Bahcall, R.K. Ulrich: Rev. Mod. Phys. **60**, 297 (1986)
Bah86b	J.N. Bahcall, R.K. Ulrich: Rev. Mod. Phys. **60**, 297 (1986)
Bar86	F.C. Barker, R.H. Spear: Austr. J. Phys. **307**, 847 (1986)
Bar87	F.C. Barker: Aust. J. Phys. **24**, 777 (1971); **40**, 25 (1987)
Bar91	F.C. Barker, T. Kajino: to be published in Aust. J. Phys.
Bay71	G. Baym, C. Pethick, P. Sutherland: Astrophys. J. **170**, 299 (1971)
Bay85	D. Baye, P. Descouvemont: Ann. Phys. **165**, 115 (1985)
Ben90	Gy. Bencze: Nucl. Phys. **A492**, 459 (1989);
	L. Bracci, G. Fiorentini, V.S. Melezhik, G. Mezzorani, P. Quarati: Nucl. Phys. **A513**, 316 (1990);
	Z. Azrak et al.: preprint (1990)
Bie80	J.P. Biersack, L.G. Haggmark: Nucl. Instr. Meth. **174**, 257 (1980)
Bla90	O. Blaes, R. Blandford, P. Madau, S.E. Koonin, Astrophys. J. **363**, 612 (1990)
Blü89	G. Blüge, K. Langanke, H.-G. Reusch, C. Rolfs: Z. Phys. **A333**, 219 (1989)
Blü90a	G. Blüge, K. Langanke: Phys. Rev. **C41**, 1191 (1990);
	G. Blüge: doctoral thesis, Münster (1990)
Blü90b	G. Blüge, K. Langanke, M. Plagge, K.R. Nyga, H. Paetz gen.Schieck: Phys. Lett. **B238**, 137 (1990)
Blü91	G. Blüge, K. Langanke: submitted to Few-Body Systems
Cam59	A.G.W. Cameron: Ap.J. **130**, 916 (1959)
Cep80	D.M. Ceperley, B.J. Alder: Phys. Rev. Lett. **45**, 566 (1980)
Chw73	F.S. Chwieroth, R.E. Brown, Y.C. Tang, D.R. Thompson, Phys. Rev. **C8**, 938 (1973)
Cla68	G.J. Clark, D.J. Sullivan, P.B. Treacy: Nucl. Phys. **A110**, 481 (1968);
	C.M. Jones, G.C. Phillips, R.W. Harris, E.H. Beckner: Nucl. Phys. **37**, 1 (1962)
Cla83	D.D. Clayton: *Principles of Stellar Evolution and Nucleosynthesis* (Chicago-Press, Chicago, 1983).
Des84	P. Descouvemont, D. Baye, P.-H. Heenen: Nucl. Phys. **A430**, 426 (1984)
Des87a	P. Descouvemont: Nucl. Phys. **A470**, 309 (1987)
Des87b	P. Descouvemont, D. Baye: Phys. Rev. **C36**, 1249 (1987)
Des88	P. Descouvemont, D. Baye: Nucl. Phys. **A487**, 420 (1988)
Dye74	P. Dyer, C.A. Barnes: Nucl. Phys. **A233**, 495 (1974)
Eng88	S. Engstler, A. Krauss, K. Neldner, C. Rolfs, U. Schröder, K. Langanke: Phys. Lett. **B202**, 179 (1988)
Eng90	S. Engstler, C. Angulo, U. Greife, J.F. Harmon, K. Neldner, G. Raimann, C. Rolfs, U. Schröder, E. Somorjai: Proceedings of the International Symposium on Nuclear Astrophysics *Nuclei in the Cosmos*, eds. H. Oberhummer and W. Hillebrandt (MPI für Physik und Astrophysik, München, 1990), p. 185
Fil83	B.W. Filippone, A.J. Elwyn, C.N. Davids, D.D. Koetke: Phys. Rev. Lett. **50**, 412 (1983);
	Phys. Rev. **C28**, 2222 (1983)
Fil89	B.W. Filippone, J. Humblet, K. Langanke: Phys. Rev. **C40**, 515 (1989)
Fow84	W.A. Fowler: nobel lecture, Rev. Mod. Phys. **56**, 149 (1984)
Fun85	C. Funck, K. Langanke, A. Weiguny: Phys. Lett. **B152**, 11 (1985)
Fur80	H. Furutani, H. Kanada, T. Kaneko, S. Nagata: Progr. Theor. Phys. Suppl. **68**, 215 (1980)
Fus87	I. Fushiki, D.Q. Lamb: Ap. J. **317**, 368 (1987)
Hum90	J. Humblet, P. Dyer, B.A. Zimmerman: Nucl. Phys. **A271**, 210 (1976);
	J. Humblet: Phys. Rev. **C42**, 1582 (1990)
Ibe67	I. Iben, K. Kalata, J. Schwartz, Ap. J. **150**, 1001 (1967)
Ich84	S. Ichimaru: Rev. Mod. Phys. **54**, 1017 (1982);
	S. Ichimaru, K. Utsumi: Ap. J. **278**, 382 (1984)
Jar84	N. Jarmie, R.E. Brown, R.A. Hardekopf: Phys. Rev. **C29**, 2031 (1984);
	R. E. Brown, N. Jarmie, G.M. Hale: Phys. Rev. **C35**, 1999 (1987)

Ji90   X. Ji, B.W. Filippone, J. Humblet, S.E. Koonin: Phys. Rev. **C41**, 1736 (1990)
Joh90   C. Johnson, E. Kolbe, S.E. Koonin, K. Langanke: Caltech preprint MAP-134 (1990)
Joh91   C. Johnson, E. Kolbe, S.E. Koonin, K. Langanke: to be submitted to Astrophys. J.
Kav69   R.W. Kavanagh, T.A. Tombrello, J.M. Mosher, D.R. Goosman: Bull. Am. Phys. Soc. **14**, 1209 (1969)
Kim87   K.H. Kim, M.H. Park, B.T. Kim: Phys. Rev. **C35**, 363 (1987)
Kol88   E. Kolbe, K. Langanke, H.J. Assenbaum: Phys. Lett. **B214**, 169 (1988)
Koo74   S.E. Koonin, T.A. Tombrello, G. Fox: Nucl. Phys. **A220**, 221 (1974)
Kra86   W.-D. Kraeft, D. Kremp, W. Ebeling, G. Röpke: *Quantum Statistics of Charged Particle Systems* (Akademie-Verlag, Berlin 1986)
Kra87   A. Krauss, H.W. Becker, H.P. Trautvetter, C. Rolfs, K. Brand: Nucl. Phys. **A465**, 150 (1987)
Kra90   H. Krauss, K. Grün, T. Rauscher, H. Oberhummer, H. Abele, G. Staudt: Proceedings of the International Symposium on Nuclear Astrophysics *Nuclei in the Cosmos*, eds. by H. Oberhummer, W. Hillebrandt (MPI für Physik und Astrophysik, München 1990), p. 200
Kre88   R.M. Kremer, C.A. Barnes, K.H. Chang, H.C. Evans, B.W. Filippone, K.H. Hahn, L.W. Mitchell: Phys. Rev. Lett. **60**, 1475 (1988)
Lan58   A.M. Lane, R.G. Thomas: Rev. Mod. Phys. **30**, 257 (1958)
Lan85   K. Langanke, S.E. Koonin: Nucl. Phys. **A410** (1983) 334; **A439**, 384 (1985)
Lan86   K. Langanke, H. Friedrich, Advances in Nuclear Physics **17**, eds. by J.W. Negele, E. Vogt (Plenum, New York 1986)
Lan89   K. Langanke, C. Rolfs: Mod. Phys. Lett. **A4**, 2101 (1989)
Lan91   K. Langanke, D. Lukas, H.-M. Müller, S. Schramm, S.E. Koonin: submitted to Z. Phys. A
McV67   K.W. McVoy: Phys. Lett. **17**, 42 (1965); Ann. Phys. (N.Y.) **43**, 91 (1967)
Mül91   H.-M. Müller, K. Langanke: to be published
Pla87   R. Plaga, H.W. Becker, A. Redder, C. Rolfs, H.P. Trautvetter, K. Langanke: Nucl. Phys. **A465**, 291 (1987)
Red85   A. Redder, H.W. Becker, J. Görres, M. Hilgemeier, A. Krauss, C. Rolfs, U. Schröder, H.P. Trautvetter, K. Wolke: Phys. Rev. Lett. **55**, 1262 (1985)
Red87   A. Redder, H.W. Becker, C. Rolfs, H.P. Trautvetter, T.R. Donoghue, T.C. Rinckel, J.W. Hammer, K. Langanke: Nucl. Phys. **A462**, 385 (1987)
Rol88   C.E. Rolfs, W. Rodney: *Cauldrons in the Cosmos* (Chicago-Press, 1988)
Sal54   E.E. Salpeter: Austr. J. Phys. **7**, 373 (1954)
Sal69   E.E. Salpeter, H.M. van Horn: Ap. J. **155**, 183 (1969)
Sch91a   S. Schramm, S.E. Koonin: Astrophys. J., in press
Sch91b   S. Schramm, K. Langanke, S.E. Koonin: submitted to Astrophys. J.
Tan78   Y.C. Tang, M. LeMere, D.R. Thompson: Phys. Rep. **C47**, 167 (1978)
Tho73   D.R. Thompson, Y.C. Tang: Phys. Rev. **C8**, 1649 (1973)
Typ91   S. Typel, G. Blüge, K. Langanke, W.A. Fowler: to be published in Z. Phys. A
Usm82   Q.N. Usmani, B. Friedman, V.R. Pandharipande: Phys. Rev. **B25**, 4502 (1982)
Wil77   K. Wildermuth, Y.C. Tang: *A Unified Theory of the Nucleus* (Vieweg, Braunschweig 1977).
Woo85   S.E. Woosley, *Proceedings of the Workshop on Accelerated Radioactive Beams*, eds. by J.M. d'Auria, L. Buchmann (Parksville, Canada 1985)
Yak89   D.G. Yakovlev, D.A. Shalybkov: Sov. Sci. Rev. Astrophys. Space Phys. **7**, 331 (1989)

Abundances in Galaxies

B.E.J. Pagel

NORDITA, Blegdamsvej 17, DK-2100 Copenhagen Ø, Denmark

1. Introduction

The observed distribution of abundances of elements in the Solar System, stellar populations and the interstellar medium in our own and other galaxies is affected by nucleosynthesis in the Big Bang and in former generations of stars and by many aspects of the evolution of galaxies. A rough scheme for the resulting "chemical evolution" of the Galaxy (GCE) is shown in Fig. 1, and some observational data on H II regions and relatively unevolved stars, revealing clues about these processes, will be described in what follows. Evolved stars and planetary nebulae often show modified abundances in their atmospheres (particularly for Li, C, N, O and s-process elements) resulting from nuclear reactions in their interiors followed by dredge-up due to convective mixing [Joh89]; these important effects are beyond the scope of the present article.

2. Big Bang Nucleosynthesis and Light Element Abundances

Fig. 2 shows the outcome of standard (homogenous) Big Bang nucleosynthesis (SBBN) calculations [Yan84, Del90, Oli90] giving predicted primordial abundances as a function of the baryon:photon number ratio η, or equivalently (given the present-day temperature of the microwave background) the average baryonic mass-density. This is expressed in the form of $\Omega_{b0} h_{100}^2$ where Ω_{b0} is the fraction of the present-day closure density contributed by baryons and h_{100} the Hubble constant in units of 100 km s^{-1} Mpc^{-1} or $(10^{10}$ yrs$)^{-1}$ which lies somewhere in the range 0.7 ± 0.3. Horizontal lines represent limits based on observations [Pag82, Sha83a, Yan84, Boe85, Pag87, Pag91] which lead within the framework of SBBN theory to the limits on η shown by the two tall vertical lines. The lower limit to η is based on primordial $(D + {}^3He)/H$, which

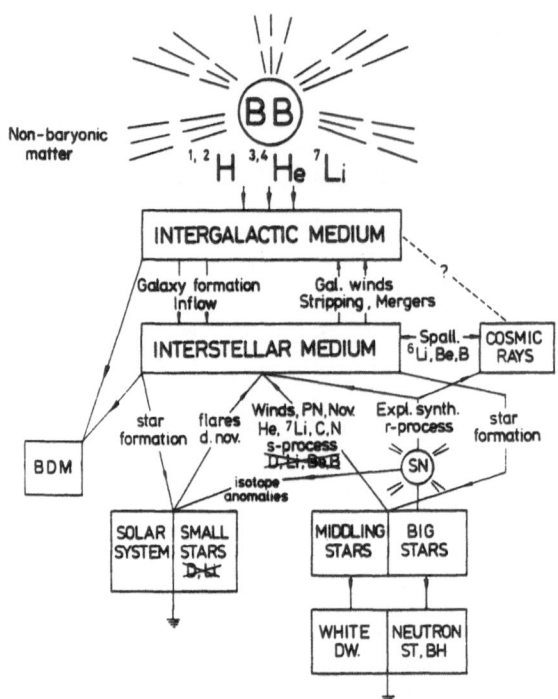

Fig. 1. A possible scheme for Galactic chemical evolution (adapted from [Pag81a])

can be confidently expected to exceed the D and ^{3}He in the early Solar System (deduced mainly from Solar wind samples) or D/H in the present-day interstellar medium (deduced mainly from *Copernicus* observations of Lyman lines) because the effect of recycling of diffuse material through stars ("astration") generally leads to destruction of deuterium and not creation. These abundances therefore give a conservative upper limit to the density. Yang et al. [Yan84] also deduced an upper limit of 10^{-4} to primordial $(D + {}^3He)/H$ from the very reasonable argument that when deuterium is destroyed it becomes ^{3}He of which at least 1/4 survives further astration, and this argument is supported by a wide range of numerical GCE models [Van88]. The resulting lower limit on baryon density, which is of importance for some of the arguments which follow, as well as requiring existence of substantial baryonic dark matter [Pag90], thus seems to be very well established; eventually it may receive an independent test from observations of suitable high red-shift absorption line systems of low metallicity on the line of sight to QSO's [Web91].

Helium 4 is observed in a variety of objects where it is mostly somewhat enhanced in abundance as a result of astration. The most precise primordial value is obtained, following the method proposed by Peimbert and Torres-Peimbert [PTP: Pei74, Pei76], by observing hydrogen and helium emission lines in extragalactic H II regions with differing and low abundances of measurable heavier elements (found mostly in gas-rich dwarf galaxies) and plotting a linear regression which is extrapolated to zero heavy-element content. PTP and Lequeux

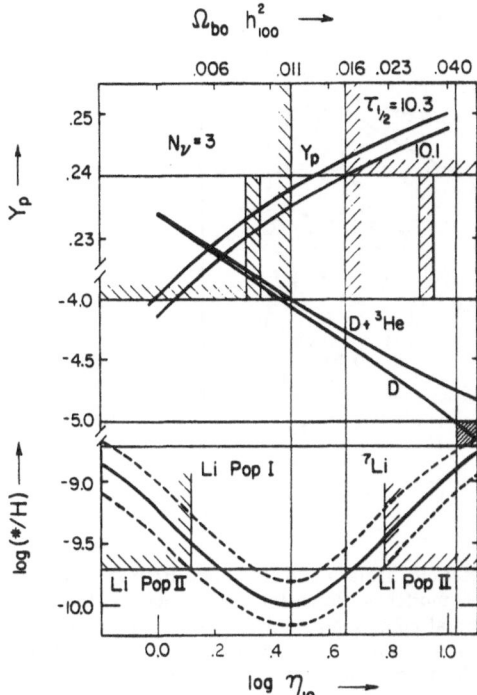

Fig. 2. Primordial abundances from BBNS theory after [Yan84, Del90, Oli90] as a function of baryonic density parameters and the neutron half-life. Horizontal lines show limits based on observations or reasonable extrapolations thereof and tall vertical lines show resulting upper and lower limits to baryon density from the standard (homogeneous) BBNS model while shorter double vertical lines show (roughly) the somewhat broader limits resulting from plausible mildly inhomogeneous models, adapted from [Kur90]

et al. [Leq79] obtained regressions against oxygen leading to a primordial mass fraction $Y_p = 0.23$ which, after many controversies [Boe85, Pag87], has been confirmed by more recent work using regressions against both oxygen and nitrogen [Pag87, Pag91, Pag89a, Sim90]. The regression against nitrogen looks slightly better, perhaps because of local pollution in some cases by winds from Wolf-Rayet stars or their red-giant progenitors. Since the statistical and systematic errors together seem to be ± 0.01 or less, the 95 per cent confidence upper limit to Y_p is 0.240, which gives an upper limit to the baryon density and also requires the existence of no more than 3 light neutrino families for consistency with SBBN theory and the upper limit to $(D + {}^3He)/H$. The limit on the number of neutrino families, N_ν, has been nicely confirmed by the famous accelerator experiments on the Z° [Ell90], but the two results leave slightly different loopholes: the cosmological argument goes further in excluding certain classes of hypothetical light exotic particles coupling to photons but not to the Z°, whereas the accelerator experiments go further in excluding heavy neutrinos up to 45 GeV.

The primordial Lithium 7 abundance is certainly at least the amount near the minimum of the curve in Fig. 2 discovered by Spite and Spite [Spi82] in sub-

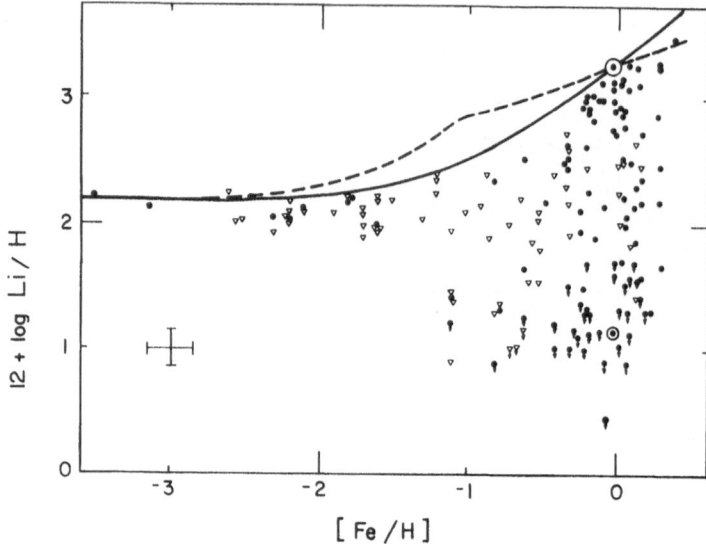

Fig. 3. Plot of stellar ^{7}Li abundances after Rebolo, Molaro and Beckman [Reb88]. The solid curve shows the prediction of a simple GCE model in which the abundance of ^{7}Li in the interstellar medium is the sum of a primordial component and an additional component proportional to iron. The broken curve is similar but with the additional component proportional to oxygen. The two "Sun" symbols refer to meteoritic (identified as proto-Solar) and photospheric (depleted) values respectively

dwarf stars, which have very low abundances of heavier elements, and probably not significantly more (cf. Fig 3), although the latter is less certain [Mat90a, Kra90]. Assuming the true primordial ^{7}Li abundance to be given by the line labelled "Li Pop II" in Fig. 2, one obtains a nice confirmation of the results deduced from D, ^{3}He and ^{4}He and consistency with SBBN theory over 9 orders of magnitude in abundances.

Is this all too good to be true? Several authors have developed inhomogeneous BBN models with isothermal density fluctuations caused by the quark-hadron phase transition (if this is indeed first order) and resulting spatial variations in the neutron-proton ratio caused by differential diffusion, and tried to fit the abundances with $\Omega_b = 1$. The latter now appears to be ruled out because it predicts too much ^{4}He and probably too much ^{7}Li, but more mildy inhomogeneous models are still acceptable [Kur90, Ree90] and give somewhat wider limits on baryonic density than does SBBN theory; a rough indication of these limits, adapted from Kurki-Suonio et al. [Kur90], is given by the shorter double vertical lines in Fig. 2. A remote possibility exists that there could be a universal "floor", at some low level, to the abundances of ^{9}Be and some heavier elements attributable to nucleosynthesis in an inhomogeneous Big Bang, but up to now there is no evidence for this [Rya90a, Pag91].

3. Abundance Trends Between and Across Galaxies

Evidence on the composition of external galaxies comes from emission lines in H II regions seen in gas-rich disks with current star formation (e.g. [Edm84]) or from colours and strong absorption-line features in the integrated light of stellar populations (e.g. [Bic88]). The "Simple" model of GCE predicts that the abundance of an element like oxygen (for which instantaneous recycling may be a good approximation) is proportional to the logarithm of the (inverse) gas fraction [Sea72] and there is some support for this from observations of gas-rich dwarf irregular and blue compact galaxies [Leq79, Axo88] (see Fig. 4) although the corresponding yield is rather low compared to solar abundances and there are uncertainties both in the mass of gas (which may be affected by unseen molecules) and in the total mass (which is affected by dark matter that presumably is not directly involved in chemical evolution).

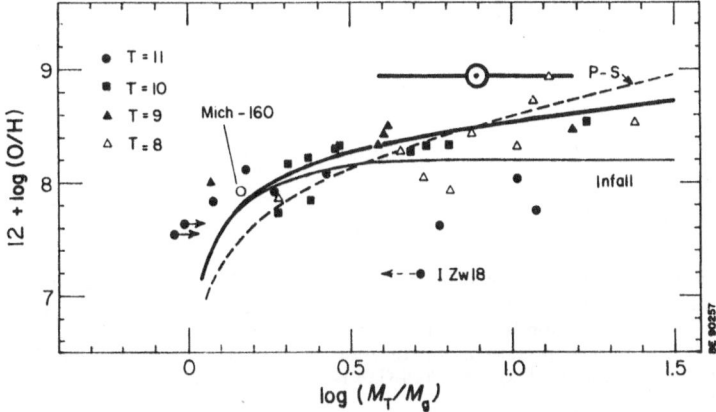

Fig. 4. Relation between oxygen abundance and ambient (H I) gas fraction in extragalactic H II regions and the Sun, adapted from [Axo88]. Symbols represent the morphological types of the host galaxies. Full-drawn curves represent a "Simple" model with oxygen yield of 0.2 times solar abundance and its modification by infall of unprocessed gas at a constant rate. The broken line represents a model in which the yield increases with metallicity [Pei82]

More striking is a relationship between abundances and total luminosity [Ski89] (see Fig. 5) and presumably mass including any dark matter. This relation, which continues qualitatively up to the biggest galaxies with blue absolute magnitude −24, is attributed to escape of gas from shallow potential wells, either as a terminal galactic wind [Lar74, Dek86, Yos87] or possibly as a continuous wind (cf. [Har76]); alternatively, hot supernova remnants may escape preferentially as a metal-enhanced wind [Vad87, Lyn91]. Owing to selection effects, and some ambiguities in the interpretation of colours and spectral features of stellar populations with a mixture of ages and compositions, it is not entirely clear that composition is just a function of total luminosity (or mass). Elliptical galaxies show a biparametric relation defining a "fundamental plane" in the three-dimensional space of luminosity, surface brightness and velocity dispersion [Fab87, Djo87] and their line strengths have been found to depend on velocity dispersion as well as luminosity [Ter81]. Several authors have suggested that surface brightness is actually more important than luminosity in determining the composition [Bot88, Dav88, Dav89, Edm89, Phi90] and recent results on a low surface brightness galaxy [Ber90] may support this (if $h_{100} <$ 0.7). Such an effect would arise naturally if there is a threshold gas surface density needed for the formation of massive stars [Dop85, Ski86, Hul87, Phi90] acting in conjunction with loss of gas from shallow potential wells.

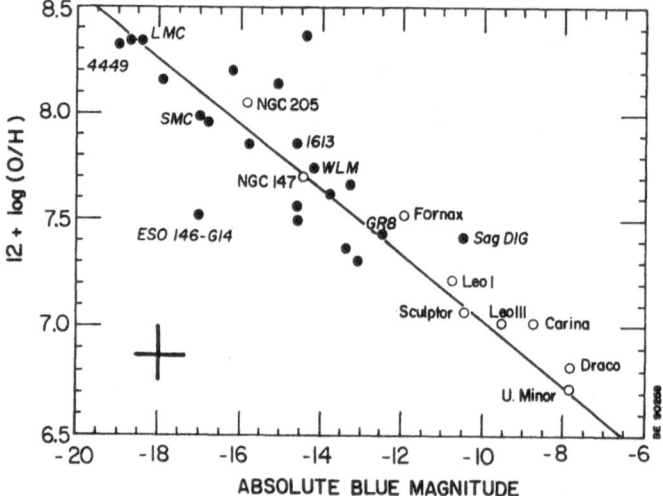

Fig. 5. Oxygen abundance as a function of luminosity in dwarf irregular, elliptical and spheroidal galaxies after Skillman, Kennicut and Hodge [Ski89]. The elliptical and spheroidal galaxies are plotted on the assumption that the mean [Fe/H] estimated for their stellar populations is equal to [O/H] in the interstellar medium or in the youngest stars. The data point for the low surface brightness galaxy ESO 146-G14 is taken from Bergvall et al. [Ber90] assuming $h_{100} = 0.5$

Radial abundance gradients in oxygen, nitrogen, sulphur and argon are found in spiral galaxies from observations of H II regions [Pag81b, McC82, Edm84, Pag85, Eva86, Gar87, Vil88, Tor89], with the strongest variations (about an order of magnitude decrease from central to outermost H II regions) tending to be seen in spirals of late morphological type (Scd). H II regions in our own Galaxy display a typical gradient of about 0.07 dex kpc^{-1} [Sha83b] which more or less agrees with metallicity gradients found from red giants [Jan79], open clusters [Pan81] and supergiants [Luc82], but there is a curious contradiction from abundance measurements in B-type stars belonging to young clusters which reveal virtually no gradients over a range of about 5 kpc centred on the Sun [Geh85, Fit90]. Gradients are smaller or non-existent in irregular galaxies and in parts of barred spirals. Differential trends between some elements probably exist, especially in N/O which shows a tendency to increase with O/H [Pag85, Rub88] with (probably partly real) scatter, and there is evidence that $^{13}C/^{12}C$ increases between here and the inner Milky Way, perhaps by a factor of 2 [Lan90]. Various possible causes of abundance gradients in spirals have been put forward including radial variations in gas fraction [Gar87], inflow of unprocessed material [Mat89, Pag89b] and radial flows due to mismatch of angular momentum of inflowing material [May81, Lac85, Pit89] or viscous processes [Som89, Cla89]. This whole subject has recently been reviewed by Matteucci [Mat91].

Radial abundance gradients are also deduced in at least some elliptical and S0 galaxies from radial variations in colour or strengths of major absorption features like Mg I – Mg H near $\lambda = 5200$ Å [Pag81b, Tho87, Efs85, Vad88]. Quantitative calibration of these trends in terms of (luminosity-weighted) mean abundances of particular elements is still rather uncertain, but there could be a close correlation with surface brightness in some ellipticals [Tho87], as there is with surface mass density in spirals [McC82, Edm84], or with local escape velocity [Vig81, Fra90].

4. Stellar Populations in Our Galaxy

It is customary to distinguish different stellar populations on the basis of common properties like age, spatial distribution, kinematics and chemical composition [Gil89], but there is some disagreement as to what kind of division is fundamental in the sense of revealing distinct evolutionary histories. Thus a given population may have a range of properties, overlapping those of other populations, owing to dynamical and chemical evolution, and artifical distinctions can result from selection effects, e.g. from the use of parallax and proper-motion catalogues; conversely, real discontinuities can be blurred by observational errors. I consider it useful to distinguish three stellar populations in our Galaxy, the disk (to which the Sun belongs), the Population II halo or spheroid and the bulge (within a few hundred parsecs of the Galactic centre)(see Fig. 6). In recent years much attention has been paid to the "thick disk" [Gil83], which is considered by many to be in some sense a separate population from the conventional young or old thin disk [Gil86a, San87, Car90, Nis91], but like Norris [Nor87] I am unconvinced that it is not more reasonable to regard all disk stars (including disk globular clusters) as a single population with a common evolutionary history [Pag89b, Pag89c].

Fig. 7 shows the distribution function of metallicities ([Fe/H] denotes the logarithm of iron abundance relative to the Sun) for globular clusters [Zin85, Arm88] and for field stars of the Galactic halo [Bee86, Bee87, Lai88, Sch88] with curves showing fits to some simple GCE models [Pag89b, Pag89c]. The curve in the left part of the diagram is based on the modified Simple Model of Hartwick [Har76] in which the effective yield is reduced by a constant factor due to continuous mass loss. The peak of the curve at [Fe/H] $= -1.6$ represents the effective yield in the halo, which is about a factor 10 below the true yield (defined as the mass of a newly synthesised element from a generation of stars in units of the corresponding mass of long-lived small stars and compact remnants) because of mass lost by the halo during its formation (presumably to the disk or maybe the bulge) and a further factor of a few below solar iron abundance because the latter includes a non-instantaneous contribution from Type Ia supernovae [Tin79, Mat86] which presumably exploded too late to affect abundances in the halo. Field stars, at least, are well fitted by this simple model so that there is no evidence for a basic "floor" to stellar metallicities, but the upper cutoff to the theoretical halo distribution at [Fe/H] = −1.1 neglects a few stars having higher metallicities but extreme kinematic properties that characterise them as belonging to the halo.

Fig. 8 shows the metallicity distribution in the galactic bulge – again a very broad distribution such as is expected from simple GCE models, but now with an iron yield noticeably higher than solar abundance, which justifies considering the bulge as a separate population from either the disk or the halo. The oxygen yield could be still higher (relative to solar) than the iron yield, if the bulge stars are old, and this population is of importance as a likely analogue of stellar populations in the nuclear regions of large galaxies in general [Fro88, Ric88, Jar90].

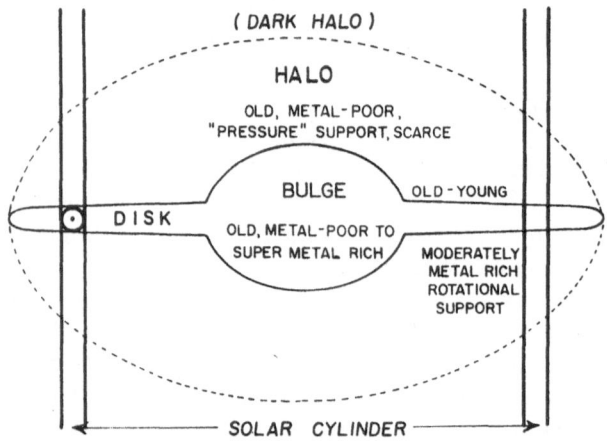

Fig. 6. A schematic cross-section through the Galaxy showing the different stellar populations

Fig. 9 shows the distribution function of oxygen abundances among G-dwarf stars in a cylinder perpendicular to the Galactic plane through the Sun after Pagel [Pag89c] and Sommer-Larsen [Som90]. The distribution is much narrower than the Simple-model type distributions in Figs. 7, 8 and this constitutes the well-known "G-dwarf problem" [Ber62, Sch63, Pag75, Lyn75, Lyn91, Pag89c]; this now needs to be discussed in terms of oxygen abundance (to which instantaneous recycling may be a good approximation because oxygen is believed to come predominantly from massive stars with short evolutionary lifetimes) rather than in the more traditional terms of [Fe/H] or metallicity. A natural solution to the "problem" is to suppose that the disk was formed by gradual

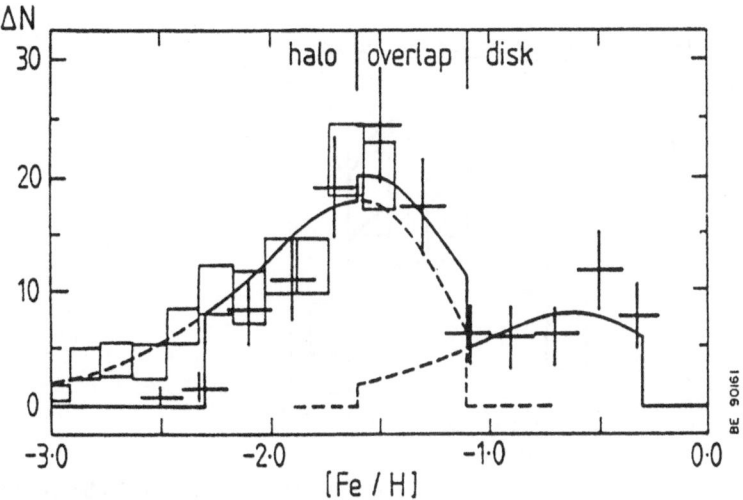

Fig. 7. Metallicity distribution function in globular clusters (crosses) and field stars of the halo population (boxes)

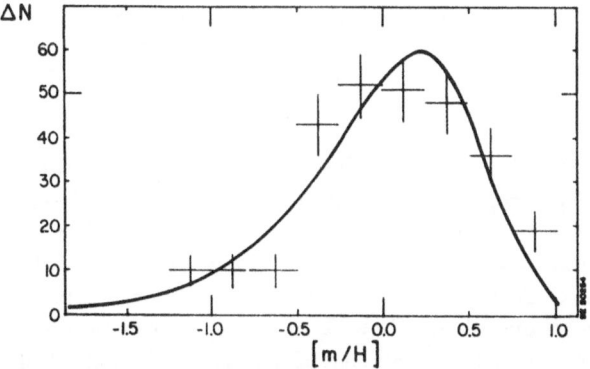

Fig. 8. Metallicity distribution function in the galactic bulge after Geisler and Friel [Gei90] as reported by Grenon [Gre90]. The curve gives the prediction of a "Simple" model with a metal yield 1.7 times solar abundance

accretion of unprocessed gas [Lyn75, Lar76, Cla85, Pag89c, Pag89b, Som90], but there is another type of solution assuming some sort of "initial spike" of heavy-element production or "prompt initial enrichment", e.g. from the halo and/or the bulge [Tru71, Ost75, Gil86b]. The difficulty with the latter type of solution is that such prior enrichment in an otherwise "Simple" (i.e. one-zone) model with constant yields of order solar abundance would need to make [O/H] $\geq$ −0.6 initially requiring a mass of stars in the bulge and halo of at least 1/4 of that in the disk, which is about an order of magnitude too large relative to actual star counts. However, with the large effective yield now seen to be present in the bulge, this type of prompt initial enrichment may just be a serious possibility after all [Köp90] although it is less easy to reconcile with existence of a "metal-weak thick disk" [Mor90].

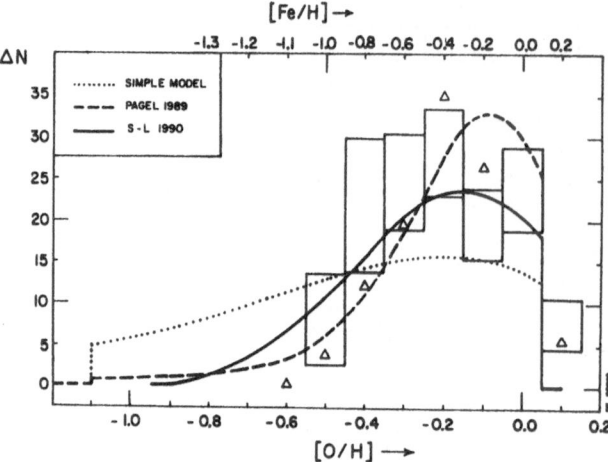

Fig. 9. Distribution function of oxygen abundance in G dwarfs of the Solar neighbourhood, corrected to the Solar cylinder after Pagel [Pag89c] (triangles) and after Sommer-Larsen [Som90] (boxes representing bin widths and error bars). Model curves are as follows: "Simple" model with plausible initial enrichment from the halo (dotted); inflow model of Pagel [Pag89b] (broken line curve); and dynamical model of Sommer-Larsen [Som90] (solid curve). Raw data have been corrected for an assumed Gaussian dispersion σ ([O/H]) = 0.075 resulting from a combination of observational errors and cosmic dispersion in the age-metallicity relation

A third view of the G-dwarf problem is that it is really a non-problem because the Solar cylinder is unrepresentative of all stars in the disk [Gre89, Gre90], but this seems to me to be rather extreme: the "silent majority" of stars not included in parallax and proper-motion catalogues seems to follow an exponential distribution of surface density with galactocentric radius and a fairly mild gradient in mean metallicity [Lew89] implying a correspondingly shallow variation of scale length with metallicity. If this is so, then the Solar cylinder should be quite typical of the Galactic disk as a whole.

The models mentioned here all assume that metallicity has increased, on average, with time, although by no means at the same rate in different stellar populations. Thus we have super metal-rich stars in the bulge that are probably older than quite metal-poor stars in the disk and the various populations can be spatially mixed. For the halo and bulge, enrichment in the course of time has to be taken on faith because metallicity-dependent systematic uncertainties in age, and real variations in age at a given metallicity [Van90], exceed 1 Gyr. In the case of the disk there is evidence for an age-metallicity relation [Nis85, Nis91], but with a large scatter and with remaining doubts as to the effect of selection biases. There is no evidence for enrichment of the local interstellar medium in oxygen or iron since formation of the Solar System 4.5 Gyr ago. On the other hand, the oldest disk stars are metal-deficient on average, and high red-shift absorption-line systems seen in front of quasars include a subset with high H I column density (damped Lyman-α systems) which may be proto-disks [Wol86, Bri89] and which at red-shifts greater than 2 show very low metallicities

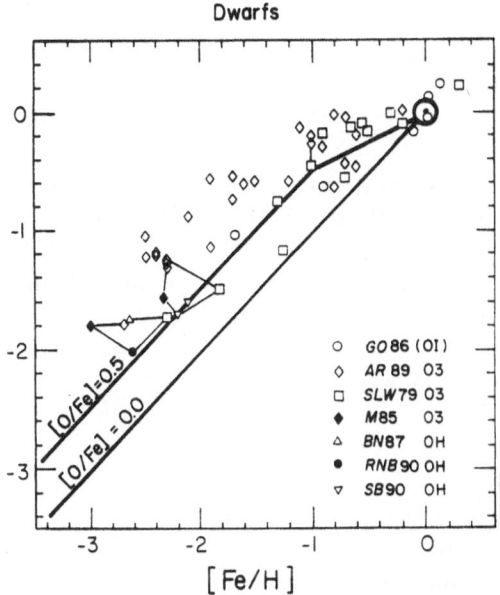

Fig. 10. Relation between oxygen and iron abundances in dwarf stars after Ryan, Norris and Bessell [Rya90b]. In this and subsequent figures, alternative determinations for the same star are joined by lines. See [Rya90b] for references to the original authors and methods

[Pet90, Rau90] and possibly low rates of star formation [Smi89] qualitatively consistent with inflow models of the sort shown in Fig. 9.

5. Differential Abundance Variations Among Elements

The previous discussion has concentrated mostly on oxygen in H II regions or metals (mainly Fe) in stars, for which the largest amount of information is available from relatively simple observations of emission lines in H II regions or colours and strong spectral features in the light from from individual stars or stellar populations. More refined observations give details of the behaviour of different elements, which is important to GCE theory because of their different nucleosynthetic sources, in high-mass, low-mass or interacting binary stars, which may enable a time scale for formation of the halo and disk to be derived, and because the yield for some (so-called "secondary") nuclear species may depend on the chemical composition of their progenitors as well as on the initial mass function and galactic evolutionary time scales. Information bearing on this issue has been comprehensively described by Wheeler, Sneden and Truran [Whe89]; in what follows I present a still more recent collection of data from Ryan, Norris and Bessell [Rya90b].

To make sense of the behaviour of different elements one first needs an abundance "clock" to measure the degree of enrichment of the interstellar medium or a stellar population and this should be a product of massive stars for which instantaneous recycling is at least a fair approximation. Oxygen is probably the best candidate for such a "clock" [Whe89], but unfortunately there are major uncertainties in its abundance in low-metallicity stars (see Fig. 10). For stars with [Fe/H] ≥ -1, [O/Fe] ≈ -0.5 [Fe/H]; at lower metallicities corresponding to halo stars, measurements of the forbidden [OI] line, mostly referring to giant stars, give a constant [O/Fe] ≈ 0.5 [Whe89, Bar88] whereas measurements of permitted lines in dwarf stars give larger values of [O/Fe] flattening off (if at all) near [O/Fe] $= 1.0$ [Abi89]. The values derived from permitted lines are extremely sensitive to stellar effective temperature and are probably about 0.2 dex too high owing to departures from local thermodynamic equilibrium [Kis90], but a serious unresolved discrepancy remains.

Fig. 11 shows the behaviour of elements with α-particle nuclei with respect to iron, together with a theoretical relation which assumes that these elements (and oxygen and r-process elements) are formed in instantaneous recycling but that a major portion of iron (2/3 of the abundance in the Solar System) is formed after a time delay comparable to the formation time of the halo [Pag89b]; this model is more or less equivalent to the more elaborate numerical models of Matteucci and colleagues [Mat86, Mat89] and it fits the data fairly well within their large scatter. How much of the scatter is real is not clear. Fig. 12 shows a plot of aluminium against magnesium which shows some tendency towards the "secondary" behaviour predicted in some explosive nucleosynthesis models. However, the scatter is large (mainly owing to difficulties in determination) and the ratio Al/Fe is constant within errors as is C/Fe; N/Fe has a large scatter part of which is probably real. Iron-group elements show insignificant variations relative to iron itself except in the case of Mn which is over-deficient by up to a factor of 2 or so in stars of the lowest metallicities.

Heavy elements above the iron group generally have contributions from s-process and r-process. Sites of the r-process are uncertain, but like oxygen and α-elements its products are believed to come out quite promptly as a result of massive-star evolution. Fig. 13, after Mathews and Cowan [Mat90b], shows the stellar data for europium against iron with some curves based on theoretical models. The best fit is given by Mathews and Cowan's model "Low-mass SN II" which implies a small time delay for the r-process compared to oxygen and α-elements. The situation for elements like Sr, Y, Zr and Ba, generally ascribed in disk stars like the Sun predominantly to the s-process, is complicated by the fact that the r-process contribution is not negligible and may even dominate in very metal-deficient stars because of time-delays and/or metallicity dependence of the s-process yield [Tru81, Gil88]. Pagel [Pag89b] assumed that the s-process contribution to barium was independent of metallicity but subject to a small time delay relative to the r-process, thereby achieving a good fit to the relation between Ba and Eu described by Lambert [Lam89] who had noted that the Ba/Eu ratio tends to the pure r-process ratio estimated by Cameron [Cam82] for the Solar System. That ratio, and the corresponding ratios for Sr and Y, form the basis for the horizontal lines marked "pure r" in Fig. 14 which give an indifferent fit to the data, and in the case of barium this line is continued

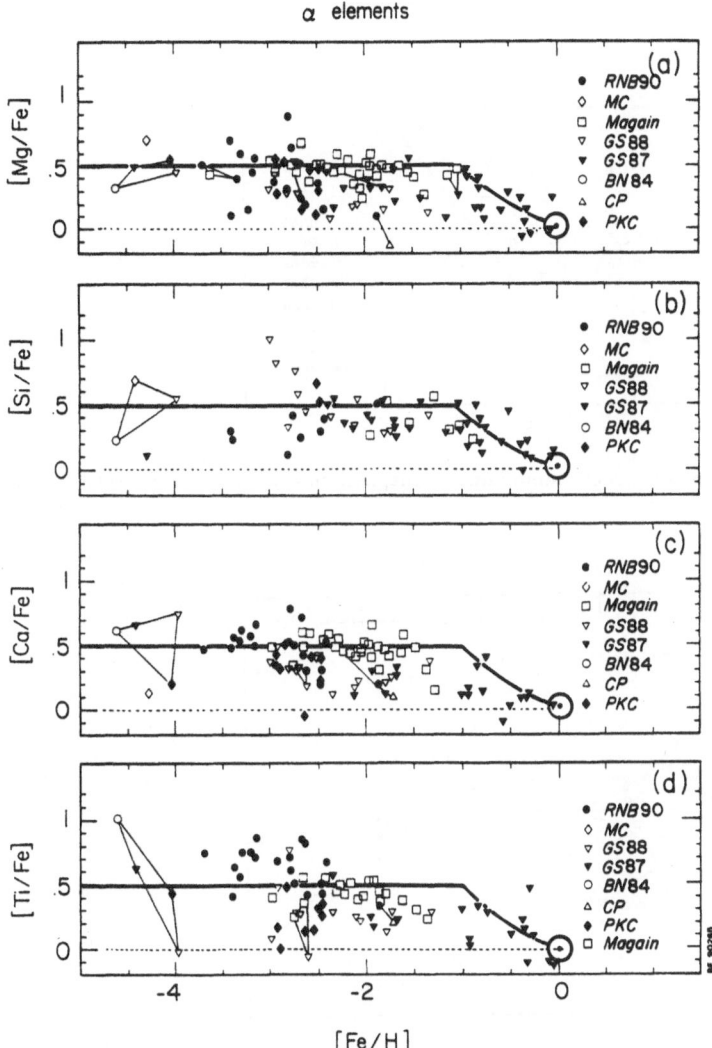

Fig. 11. Relation between stellar α-particle element and iron abundances after Ryan, Norris and Bessell [Rya90b]. The horizontal line and descending curve (identical for all four panels) represents the analytical GCE model of Pagel [Pag89b]

rightwards to give the predictions of my model, which evidently fail very badly for disk stars with [Fe/H] > −1. The assumption of a single time delay for the s-process (and possibly of its "primary" nature; cf. [Pra90]) is evidently oversimplified and in any case the r/s ratios in the Solar System are probably due for revision in the light of new data [Käp89]. Furthermore, Ryan, Norris and Bessell [Rya90b] draw attention to the existence of significant real scatter in the Sr/Fe ratio among different stars of low metallicity as shown in the top panel of Fig. 14, with some stars even falling below the "pure r" line which I

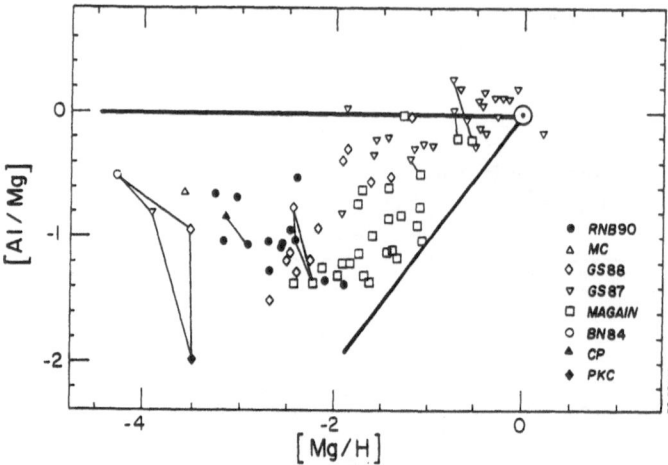

Fig. 12. Relation between stellar aluminium and magnesium abundances after Ryan, Norris and Bessell [Rya90b]

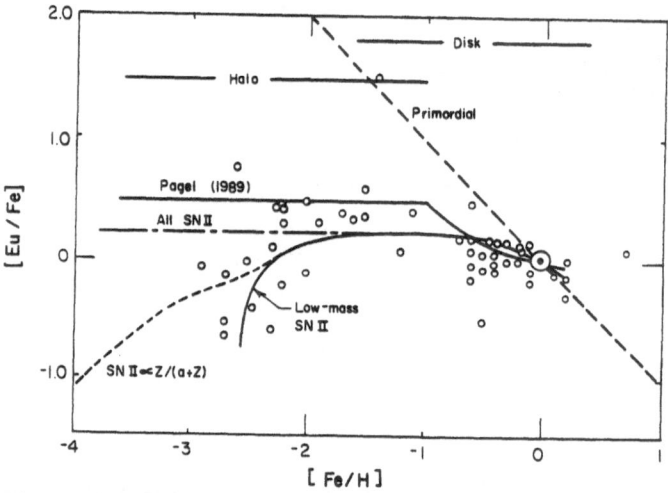

Fig. 13. Relation between stellar europium and iron abundances, with various model fits, adapted from Mathews and Cowan [Mat90b]. The curve marked Pagel (1989) is based on the simple assumption that Eu, O and α-particle elements all vary in lockstep

have assumed. The whole question of the evolution of these heavy elements in the Galaxy deserves further investigation.

Finally, one should mention that the abundance trends among local stars in our Galaxy, discussed above, having a particular (if not yet very well defined) relation between age, kinematics and metallicity, are not necessarily the same as those in other galaxies such as the Magellanic Clouds which may have an entirely different evolutionary history. Existing data for stars in the Magellanic Clouds are confined (for weighty technical reasons) to supergiants which may have undergone some internal modification of C, N, O and s-process elements,

Neutron capture elements

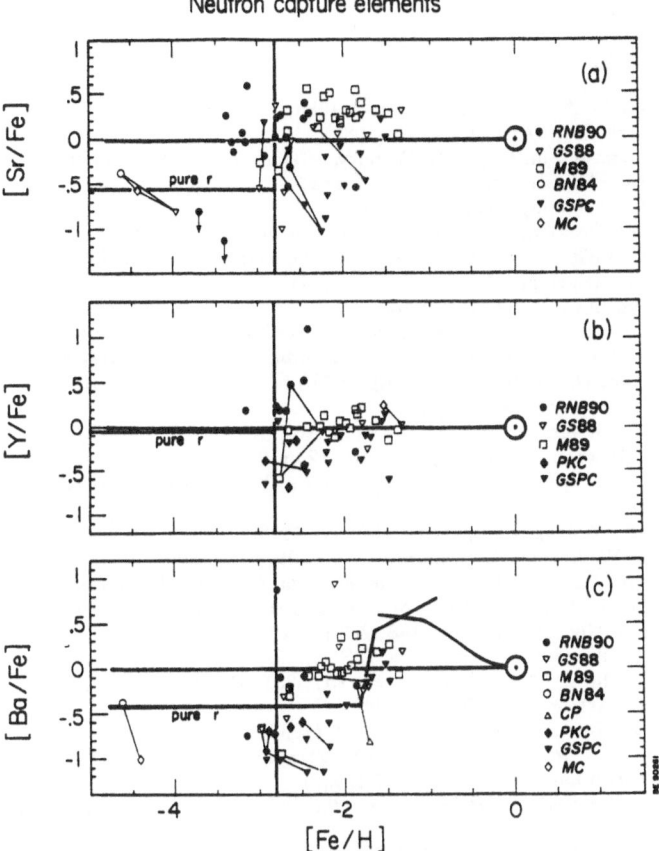

Fig. 14. Relation between Sr, Y and Ba abundances and iron abundance after Ryan, Norris and Bessell [Rya90b]. Lines and curves are explained in the text

as well as having atmospheres that present some difficulties in modelling, a problem that is overcome to some extent by comparing with supergiants in our own Galaxy [Rus89, Spi89a, Spi89b]. From the rather sparse data available up to now, both [Fe/H] and [O/H] in stars of the Large and Small Clouds are about −0.3 and −0.65 respectively, similar to [O/H] measured in H II regions, and there is no evidence that [O/Fe] is enhanced, which suggests that enrichment in the Clouds has been slow enough for Type Ia supernovae to "catch up". C/O is measured to be solar in stars but lower in H II regions [Duf87]; reasons for this discrepancy are not known. Heavy neutron-capture elements La, Ce, Nd, Eu and Sm are apparently slightly overabundant in both Clouds relative to iron. The 8m and larger class telescopes of the future should enable these trends to be confirmed (or otherwise) in stars of lower luminosity.

I thank Elisabeth Grothe of the Niels Bohr Institute for her meticulous work in preparing the many diagrams accompanying this paper and Anne Lumholdt for her expert assistance in producing the typescript.

References

Abi89   C. Abia, R. Rebolo: Astrophys. J. **347**, 186 (1989)
Arm88   T. Armandroff, R. Zinn: Astr. J. **96**, 92 (1988)
Axo88   D.J. Axon, L. Stavely Smith, R.A.E. Fosbury, J. Danziger, A. Boksenberg, R.D. Davies: Mon. Not. R. astr. Soc. **231**, 1077 (1988)
Bar88   B. Barbuy: Astr. Astrophys. **191**, 121 (1988)
Bee86   T.C. Beers, G.W. Preston, S.A. Shectman: Astr. J. **90**, 2089 (1986)
Bee87   T.C. Beers: In *Nearly Normal Galaxies, From the Planck Time to the Present*, ed. by S.M. Faber (Springer, New York 1987) p.41
Ber62   S. van den Bergh: Astr. J. **67**, 486 (1962)
Ber90   N. Bergvall, J. Rönnback, L. Johansson, E. van Groningen: In *Astrophysical Processes and Structures in the Universe (Nordic-Baltic Astronomy Meeting)*, ed. by C.I. Lagerkvist, D. Kiselman, M. Lindgren (Uppsala Observatory, 1990)
Bic88   E. Bica: Astr. Astrophys. **195**, 76 (1988)
Boe85   A. Boesgaard, G. Steigman: Ann. Rev. Astr. Astrophys., **23**, 319 (1985)
Bot88   G.D. Bothun, J.R. Mould: Astrophys. J. **324**, 123 (1988)
Bri89   F.H. Briggs, A.M. Wolfe, H.S. Liszt, M.M. Davis, K.L. Turner: Astrophys. J. **341**, 650 (1989)
Cam82   A.G.W. Cameron: Astrophys. Space Sci. **82**, 123 (1982)
Car90   B.W. Carney, D.W. Latham, J.B. Laird: Astr. J. **99**, 572 (1990)
Cla85   D.D. Clayton: In *Nucleosynthesis: Challenges and New Developments*, ed. by W.D. Arnett and J.W. Truran (Univ. of Chicago Press 1985) p.65
Cla89   C.J. Clarke: Mon. Not. R. astr. Soc. **238**, 283 (1989)
Dav88   J.I. Davies, S. Phillipps: Mon. Not. R. astr. Soc., **233**, 533 (1988)
Dav89   J.I. Davies, S. Phillipps: Astrophys. Sp. Sci. **157**, 291 (1989)
Dek86   A. Dekel, J. Silk: Astrophys.J **303**, 39 (1986)
Djo87   S. Djorgowski, M. Davis: Astrophys J. **313**, 59 (1987)
Del90   C.P. Deliyannis, P. Demarque, S.D. Kawaler: Astrophys. J. Suppl. **73**, 21 (1990)
Dop85   M.A. Dopita: Astrophys. J. Lett. **295**, L 5 (1985)
Duf87   R.J. Dufour: In *Exploring the Universe with the IUE Satellite* ed. by Y. Kondo (Reidel 1987) p.577
Edm84   M.G. Edmunds, B.E.J. Pagel: Mon. Not. R. astr. Soc. **211**, 507 (1984)
Edm89   M.G. Edmunds, S. Phillipps: Mon. Not. R. astr. Soc. **241**, 9P (1989)
Efs85   G. Efstathiou, J. Gorgas: Mon. Not. R. astr. Soc. **215**, 37P (1985)
Ell90   J. Ellis, P. Salati, P.A. Shaver (eds): *ESO-CERN Topical Workshop on LEP and the Universe* (CERN-TH 5709/90, Geneva 1990)
Eva86   I.N. Evans: Astrophys. J. **309**, 544 (1986)
Fab87   S.M. Faber, A. Dressler, R.L. Davies, D. Burstein, D. Lynden-Bell, R. Terlevich, G. Wegner: In *Nearly Normal Galaxies: From the Planck Time to the Present*, ed. by S.M. Faber (Springer, New York 1987) p.175
Fit90   A. Fitzsimmons, P.J.F. Brown, P.L. Dufton, D.J. Lennon: Astr. Astrophys. **232**, 437 (1990)
Fra90   M. Franx, G. Illingworth: Astrophys. J. Let. **349**, L41 (1990)
Fro88   J.A. Frogel: Ann. Rev. Astr. Astrophys. **26**, 51 (1988)
Gar87   D. Garnett, G. Shields: Astrophys. J. **317**, 82 (1987)
Geh85   T. Gehren, P.E. Nissen, R.P. Kudritzki, K. Butler: In *Production and Distribution of the CNO Elements*, ed. by I.J. Danziger, F. Matteucci, K. Kjär (ESO, Garching 1985) p.171
Gei90   D. Geisler, E.D. Friel: In *Bulges of Galaxies*, ed. by B.J. Jarvis, D. Terndrup (ESO Conf. and Workshop Proceedings no. 35, Garching 1990) p.77
Gil83   G. Gilmore, I.N. Reid: Mon. Not. R. astr. Soc. **202**, 1025 (1983)
Gil86a   G. Gilmore, R. Wyse: Astr. J. **90**, 2015 (1986)
Gil86b   G. Gilmore, R. Wyse: Nature, London, **322**, 806 (1986)
Gil88   K.K. Gilroy, C. Sneden, C. Pilachowski, J.J. Cowan: Astrophys. J. **327**, 298 (1988)
Gil89   G. Gilmore, R. Wyse, K. Kuijken: Ann. Rev. Astr. Astrophys. **27**, 555 (1989)
Gre89   M. Grenon: Astrophys. Sp. Sci. **156**, 29 (1989)

Gre90   M. Grenon: In *Bulges of Galaxies*, ed. by B.J. Jarvis, D.M. Terndrup (ESO Conf. and Workshop Proc. no.35, Garching 1990) p.143

Har76   F.D.A. Hartwick: Astrophys. J. **209**, 418 (1976)

Hul87   J.M. van der Hulst, E.D. Skillman, R.C. Kennicutt, G.D. Bothun: Astr. Astrophys. **177**, 63 (1987)

Jan79   K.A. Janes: Astrophys. J. Suppl. **39**, 135 (1979)

Jar90   Ed. by B.J. Jarvis, D.M. Terndrup: *Bulges of Galaxies* (ESO Conf. and Workshop Proc. no. **35**, Garching 1990)

Joh89   Ed. by H.R. Johnson, B. Zuckerman: *Evolution of Peculiar Red Giant Stars (IAU Coll. no. 106)*, (Cambridge Univ. Press, 1989)

Käp89   F. Käppeler, H. Beer, K. Wisshak: Rep. Prog. Phys. **52**, 945 (1989)

Kis90   D. Kiselman: In *Astrophysical Processes and Structures in the Universe, (Nordic-Baltic Astronomy Meeting)*, ed. by C.I. Lagerkvist, D. Kiselman, M. Lindgren (Uppsala Observatory 1990)

Kra90   L.M. Krauss, P. Romanelli: Astrophys. J. **358**, 47 (1990)

Köp90   J. Köppen, N. Arimoto: Astr. Astrophys. **240**, 22 (1990)

Kur90   H. Kurki-Suonio, R.A. Matzner, K.A. Olive, D.N. Schramm: Astrophys. J. **353**, 406 (1990)

Lac85   C.G. Lacey, S.M. Fall: Astrophys. J. **290**, 154 (1985)

Lai88   J.B. Laird, B.W. Carney, D.W. Latham: Astr. J. **95**, 1843 (1988)

Lam89   D.L. Lambert: In *Cosmic Abundances of Matter*, ed. by C.J. Waddington (Amer. Inst. Phys., New York 1989) p.168

Lan90   W.D. Langer, A. Penzias: Astrophys. J. **357**, 477 (1990)

Lar74   R.B. Larson: Mon. Not. R. astr. Soc. **166**, 585 (1974)

Lar76   R.B. Larson: Mon. Not. R. astr. Soc. **176**, 31 (1976)

Leq79   J. Lequeux, M. Peimbert, J.F. Rayo, A. Serrano, S. Torres-Peimbert: Astr. Astrophys. **80**, 155 (1979)

Lew89   J.R. Lewis, K.C. Freeman: Astr. J. **97**, 139 (1989)

Luc82   R.E. Luck: Astrophys. J. **256**, 177 (1982)

Lyn75   D. Lynden-Bell: Vistas in Astr. **19**, 299 (1975)

Lyn91   D. Lynden-Bell: In *31st Herstmonceux Conference: Elements and the Cosmos*, ed. by M.G. Edmunds, B.E.J. Pagel, R.J. Terlevich (Cambridge Univ. Press., 1991)

Mat86   F. Matteucci, L. Greggio: Astr. Astrophys. **154**, 279 (1986)

Mat89   F. Matteucci, P. François: Mon. Not. R. astr. Soc. **239**, 885 (1989)

Mat90a   G.J. Mathews, C. Alcock, G.M. Fuller: Astrophys J. **349**, 449 (1990)

Mat90b   G.J. Mathews, J.J. Cowan: preprint (1990)

Mat91   F. Matteucci: In *Morphological and Physical Classification of Galaxies*, ed. by G. Longo (Kluwer 1991)

May81   M. Mayor, L. Vigroux: Astr. Astrophys. **98**, 1 (1981)

McC82   M. McCall: Thesis, University of Texas at Austin (1982)

Mor90   H.L. Morrison, C. Flynn, K.C. Freeman: Astr. J. **100**, 1191 (1990)

Nil85   P.E. Nissen, B. Edvardsson, B. Gustafsson: In *Production and Distribution of the CNO Elements*, ed. by I.J. Danziger, F. Matteucci, K. Kjär (ESO, Garching 1985) p.131

Nis91   P.E. Nissen: In *Elements and the Cosmos, (31st Herstmonceux Conference)*, ed. by M.G. Edmunds, B.E.J. Pagel, R.J. Terlevich (Cambridge Univ. Press, 1991)

Nor87   J. Norris: Astrophys. J. Let. **314**, L39 (1987)

Oli90   K.A. Olive, D.N. Schramm, G. Steigman, T.P. Walker: Phys. Rev. Lett. B **236**, 454 (1990)

Ost75   J.B. Ostriker, T.X. Thuan: Astrophys. J. **202**, 353 (1975)

Pag75   B.E.J. Pagel, B.E. Patchett: Mon. Not. R. astr. Soc. **172**, 13 (1975)

Pag81a   B.E.J. Pagel: In *The Structure and Evolution of Normal Galaxies*, ed. by S.M. Fall, D. Lynden-Bell (Cambridge University Press, 1981) p.211

Pag81b   B.E.J. Pagel, M.G. Edmunds: Ann. Rev. Astr. Astrophys. **19**, 77 (1981)

Pag82   B.E.J. Pagel: In *The Big Bang and Element Creation*, ed. by D. Lynden-Bell, Phil. Trans. R. Soc. London A, **307**, 19 (1982)

Pag85   B.E.J. Pagel: In *Production and Distribution of the CNO Elements*, ed. by J. Danziger, F. Matteucci, K. Kjär (ESO, Garching 1985) p.155

Pag87   B.E.J. Pagel: In *A Unified View of the Macro and the Micro-Cosmos*, ed. by
           A. de Rujula, D.V. Nanopoulos, P.A. Shaver (World Scientific, Singapore 1987)
           p.399
Pag89a   B.E.J. Pagel, E.A. Simonson: Rev. Mex. Astr. Astrofis. **18**, 153 (1989)
Pag89b   B.E.J. Pagel: Rev. Mex. Astr. Astrofis. **18**, 161 (1989)
Pag89c   B.E.J. Pagel: In *Evolutionary Phenomena in Galaxies*, ed. by J.E. Beckman,
           B.E.J. Pagel (Cambridge University Press 1989) p.201
Pag90   B.E.J. Pagel: In *Baryonic Dark Matter*, ed. by D. Lynden-Bell, G. Gilmore (Rei-
           del 1990) p.237
Pag91   B.E.J. Pagel: Phys. Scripta, in press (1991)
Pan81   N. Panagia, M. Tosi: Astr. Astrophys. **96**, 306 (1981)
Pei74   M. Peimbert, S. Torres-Peimbert: Astrophys J. **193**, 327 (1974)
Pei76   M. Peimbert, S. Torres-Peimbert: Astrophys. J. **203**, 581 (1976)
Pei82   M. Peimbert, A. Serrano: Mon. Not. R. astr. Soc. **198**, 563 (1982)
Pet90   M. Pettini A. Boksenberg, R.W. Hunstead: Astrophys. J. **348**, 48 (1990)
Phi90   S. Phillipps, M.G. Edmunds, J.I. Davies: Mon. Not. R. Ast. Soc. **244**, 168 (1990)
Pit89   E. Pitts, R.J. Tayler: Mon. Not. R. astr. Soc. **240**, 373 (1989)
Pra90   N. Prantzos, K. Hashimoto, K. Nomoto: Astr. Astrophys. **234**, 211 (1990)
Rau90   M. Rauch, R.F. Carswell, J.G. Robertson, P.A. Shaver, J.K. Webb: Mon. Not.
           R. astr. Soc. **242**, 698 (1990)
Reb88   Rebolo R., P. Molaro, J.E. Beckman: Astr. Astrophys. **192**, 192 (1988)
Ree90   H. Reeves: Physics Reports, in press (1990)
Ric88   R.M. Rich: Astr. J. **95**, 828 (1988)
Rub88   R.H. Rubin, J.P. Simpson, F. Erickson, M.R. Haas: Astrophys. J. **327**, 377 (1988)
Rus89   S.C. Russell, M.S. Bessell: Astrophys. J. Suppl. **70**, 865 (1989)
Rya90a   S.G. Ryan, M.S. Bessell, R.S. Sutherland, J. Norris: Astrophys. J. Lett. **348**, L57
           (1990)
Rya90b   S. Ryan, J. Norris, M.S. Bessell: submitted to Astrophys. J. (1990)
San87   A.R. Sandage, G. Fouts: Astr. J. **93**, 74 (1987)
Sch63   M. Schmidt: Astrophys. J. **137**, 758 (1963)
Sch88   W.J. Schuster, P.E. Nissen: Astr. Astrophys. Suppl. **73**, 225 (1988)
Sea72   L. Searle, W.L.W. Sargent: Astrophys. J. **173**, 25 (1972)
Sha83a   Ed. by P.A. Shaver, D. Kunth, K. Kjär: *Primordial Helium* (ESO, Garching 1983)
Sha83b   P.A. Shaver, R.X. McGee, A.C. Danks, S.R. Pottasch: Mon. Not. R. astr. Soc.
           204, 53 (1983)
Sim90   E.A. Simonson: Thesis, Sussex University (1990)
Ski86   E.D. Skillman: In *Star Formation in Galaxies*, ed. by C.J. Lonsdale-Persson
           (NASA, Washington 1986) p. 263
Ski89   E.D. Skillman, R.C. Kennicutt, P.W. Hodge: Astrophys. J. **347**, 875 (1989)
Smi89   H.E.Smith, R.D. Cohen, J.E. Burns, D.J. Moore, B.A. Uchida: Astrophys. J. **347**,
           87 (1989)
Som89   J. Sommer-Larsen, Y. Yoshii: Mon. Not. R. astr. Soc. **238**, 133 (1989)
Som90   J. Sommer-Larsen: submitted to Mon. Not. R. astr. Soc. (1990)
Spi82   M. Spite, F. Spite: Astr. Astrophys. **115**, 357 (1982)
Spi89a   F. Spite, M. Spite, P. François: Astr. Astrophys. **210**, 25 (1989)
Spi89b   M. Spite, B. Barbuy, F. Spite: Astr. Astrophys. **222**, 35 (1989)
Ter81   R.J. Terlevich, R. Davies, S. Faber, D. Burstein: Mon. Not. R. astr. Soc. **196**,
           381 (1981)
Tho87   B. Thomsen, W.A. Baum: Astrophys. J. **315**, 460 (1987)
Tin79   B.M. Tinsley: Astrophys. J. **229**, 1046 (1979)
Tor89   S. Torres-Peimbert, M. Peimbert, J. Fierro: Astrophys. J. **345**, 186 (1989)
Tru71   J.W. Truran, A.G.W. Cameron: Astrophys. Space Sci. **14**, 179 (1971)
Tru81   J.W. Truran: Astr. Astrophys. **97**, 391 (1981)
Vad87   J.P. Vader: Astrophys J. **317**, 128 (1987)
Vad88   J.P. Vader, L. Vigroux, M. Lachièze-Rey, J. Souviron: Astr. Astrophys. **203**, 217
           (1988)
Van88   E. Vangioni-Flam, J. Audouze: Astr Astrophys. **193**, 81 (1988)
Van90   D.A. VandenBerg, M. Bolte, P.B. Stetson: Astr. J. **100**, 445 (1990)
Vig81   L. Vigroux, J.P. Chièze, B. Lazareff: Astr. Astrophys. **98**, 119 (1981)

Vil88   J.M. Vilchez, B.E.J. Pagel, A.I. Diaz, E. Terlevich, M.G. Edmunds: Mon. Not.
           R. astr. Soc. **235**, 633 (1988)
Web91   J.K. Webb, R.F. Carswell, M.J. Irwin, M.V. Penston: Mon. Not. R. astr. Soc., in
           press. (1991)
Whe89   J.C.Wheeler, C. Sneden, J.W. Truran: Ann. Rev. Astr. Astrophys. **27**, 279 (1989)
Wol86   A.M. Wolfe: Phil. Trans. R. Soc. A. **320**, 503 (1986)
Yan84   J. Yang, M.S. Turner, G. Steigman, D.N. Schramm, K.A. Olive: Astrophys. J.
           281, 493 (1984)
Yos87   Y. Yoshii, N. Arimoto: Astr. Astrophys. **188**, 13 (1987)
Zin85   R. Zinn: Astrophys. J. **293**, 424 (1985)

Nuclei in Cosmic Rays

Gerd Schatz

Kernforschungszentrum Karlsruhe, Institut für Kernphysik Postfach 3640,
D-7500 Karlsruhe, Germany

1. Introduction

Cosmic rays were discovered almost 80 years ago by Victor Hess [Hes12] who
was at that time working at the University of Vienna. He searched for the
cause of the residual ionization of air which persisted even behind the most
massive shielding. For this purpose he performed several balloon flights with
an instrument which we should now call an ionization chamber. Most of these
flights were launched from the site of the "K. u. K. Österreichischen Aero
Club" in the Prater at the outskirts of Vienna although the most successful
flight started from Aussig in northern Bohemia, went north and ended some 50
km east of Berlin [Hes12]. The results of the measurements during this flight
showed unambiguously that the ionization of air passed through a flat minimum
between 1000 and 2000 m above ground and then increased steeply to values
much above ground level. This clearly indicated an extraterrestrial source of
radiation.

Research during subsequent decades showed cosmic radiation to be a rather
complex phenomenon. It spans a tremendous range of energies from almost
thermal to the highest values known to be carried by an atomic particle somewhere in the universe, some 10^{20} eV. The latter value corresponds to the kinetic
energy of a tennis ball at 50 km/h. In the lower energy range most cosmic ray
particles originate from the sun, but the sun only seldomly accelerates particles
to above 10 GeV. Most particles with energy larger than 1 GeV/amu originate
from outside the solar system. They represent in fact the only such material
accessible to direct measurements.

The present review will be confined to this material from outside the solar
system, i.e. to energies above 1 GeV/amu. Below this range the fraction of
solar particles increases quickly, and in addition the extra solar component is
intensely modulated by interplanetary magnetic fields. These fields are carried
away from the sun by the solar wind and are highly variable in time and space.

The cosmic ray particles in the energy range of interest are known to be
positively charged from their deflection by the magnetic field of the earth. Also,

a considerable number of hadrons can be observed at ground level when a high energy particle bombards the atmosphere, in fact quite a number of strongly interacting elementary particles have been detected in cosmic rays. Hence these projectiles can be safely assumed to be atomic nuclei, and then the question of chemical composition arises. This point is of great importance for the problem of the origin of cosmic rays which is in no way completely clear in the energy range of interest. Since the particles are charged they are deflected by the magnetic fields which are known to exist in the Galaxy from astronomical observations. These fields are roughly parallel to the galactic plane, though with a lot of irregularities. Their strength is of order $5\,\mu G$. This means that the radius of curvature of a 10^{16} eV proton is approximately 1 pc, i. e. the distance between the sun and its next neighbouring fixed star. The cosmic ray particles are on the one hand bound to the Galaxy by these magnetic fields, on the other hand they are scattered by field irregularities and after a while attain an isotropic velocity distribution. An important consequence of these facts is that the direction of incidence of cosmic rays on the earth does not tell us anything about their point of origin. So we have to rely on indirect evidence only for the question of origin, and the most important information in this respect are energy distribution and chemical nature of the primary particles.

In the next section, two methods of measuring the nuclear charge of primary cosmic ray particles directly above the atmosphere are described. Sect. 3 then describes and discusses the results of such measurements. Attempts to extend our knowledge to higher energies where these techniques cannot be applied are dealt with in Sect. 4.

2. Techniques for Measuring Composition from Satellites and Balloons

The nuclear charge of a primary cosmic ray particle is best measured above the atmosphere before there was any chance of interaction with air atoms. Such interactions would be expected to change the nature of the particle, by nuclear break-up e.g. (Nuclear break-up due to collisions in the interstellar medium do occur and will be discussed in Sect. 3.) This requires experiments on either satellites or high-altitude balloons. The latter can reach altitudes where the residual overburden is less than 1 % of the total atmosphere, i.e. less than $10\,g/cm^2$. This is less than the average mean free path of a high energy iron nucleus in air. So most nuclei will reach the detector without prior interaction although corrections for such collisions are required at least for heavier incident nuclei.

Basically there are two experimental approaches to measuring the nuclear charge of an incident cosmic ray particle. One method uses electronic detectors similar to those employed in high energy or relativistic heavy ion experiments in the laboratory. The raw data are transmitted to a ground station and processed further in the laboratory. This method does not require the detector to be recovered which is unfeasible for most satellite experiments. The second method is based on photographic emulsions which have a long tradition in the study of cosmic rays, and special plastic foils which show the track of a charged particle after etching. Stacks of sheets of material sensitive to the passage of high energy particles are exposed to the primary flux. The sheets are usually separated by layers of other materials with which the particles interact. These interactions can be identified and may give additional information about the nature of the particle. The sensitive layers have to be recovered and processed chemically, either by photographic development or etching, and then the tracks become visible. Rather than discuss these two methods in general terms we will describe a recent example of either technique in the subsequent subsections.

2.1 Electronic Detectors: The "Chicago Egg"

This instrument was flown on the Space Shuttle Challenger in 1985. It is described in detail by L'Heureux et al. [LHe90]. Fig. 1 gives a cross section of the detector. The two spherical endcaps are Cherenkov detectors. Their signal is proportional to the square of the nuclear charge Z and increases with energy from a threshold near 40 GeV/amu and saturates at about 150 GeV/amu. The output of the scintillation counters increases in proportion to Z^2, but is almost independent of energy in the range of interest.

The central part is formed by a stack of multiwire proportional counters between sheets of plastic fibers. It serves a dual purpose. It determines the direction of the particle with respect to the detector. This is necessary in order to correct for the path length in the scintillation and Cherenkov detectors. In addition it measures the particle energy by a combination of ionization loss and transition radiation. The latter consists of soft x-rays which are produced when highly relativistic particles enter or leave a dielectric medium which is in this case formed by the plastic fibers. The resulting Z resolution of the instrument is shown in the lower part of Fig. 1.

The detector was flown on the Space Shuttle Challenger for about a week and took data for only several days. The results have been published ([Gru88], [Mue90]). They extend to about 1500 GeV/amu and are limited by statistics only in the high energy range. A longer flight of the instrument, of order one year, which is technically feasible, could extend this limit by about an order magnitude. A disadvantage of this experiment is that the detector is insensitiv to elements below carbon.

2.2 An Emulsion Experiment: JACEE

JACEE stands for Japanese American Cooperative Emulsion Experiment. It uses stacks of nuclear emulsions and etchable plastic foils carried by high-altitude balloons to a residual atmosphere of between 4 and 10 g/cm^2 . The design of the detector is described in detail by Burnett et al. [Bur86]. Fig. 2

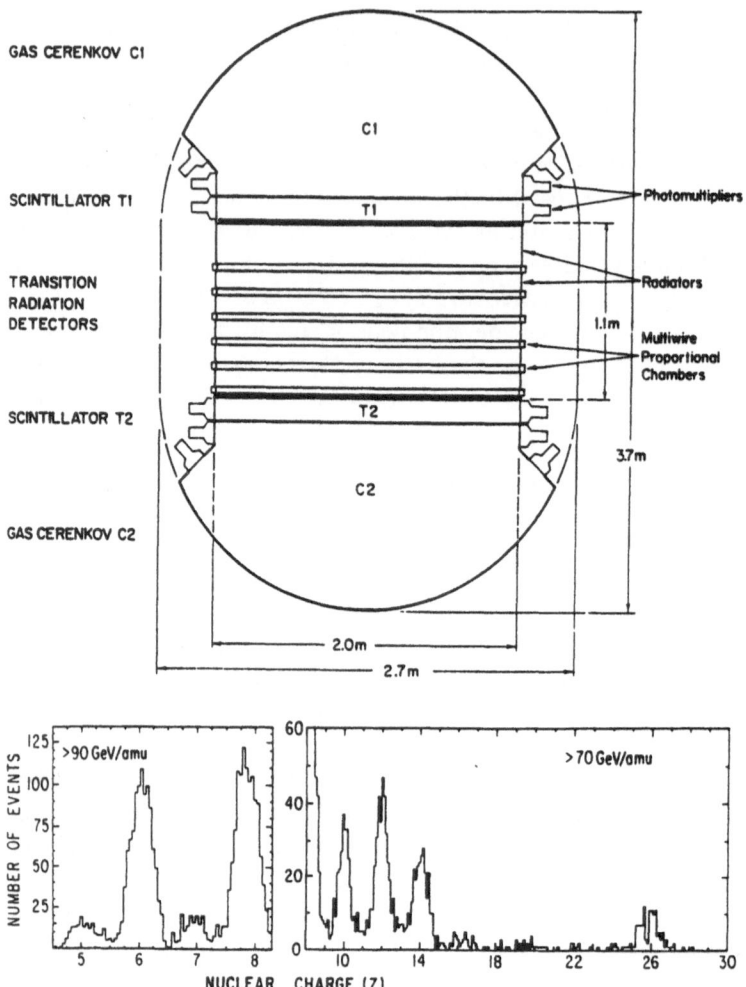

GAS CERENKOV C1

SCINTILLATOR T1

TRANSITION
RADIATION
DETECTORS

SCINTILLATOR T2

GAS CERENKOV C2

Fig. 1. Upper part: Cross section of the 'Chicago Egg' satellite detector. The different subsystems are indicated. Lower part: Demonstration of the achieved charge resolution (from [Mue90])

gives a cross section of the instrument. The stack consists of four main sections: charge detector, target, spacer, and calorimeter. The charge detector contains several nuclear emulsions of different sensitivities in addition to a small number of plastic foils. The nuclear emulsions essentially measure the ionization density. For low densities this is achieved by determining the average separation between the developed silver grains. At high ionization density neighbouring grains merge even in low sensitivity emulsions. For higher Z therefore the number of side tracks due to recoil electrons ("delta rays") are counted. The first section of the instrument thus allows to determine the nuclear charge of the projectile reliably. The charge resolution achieved is displayed in the lower part of Fig. 2.

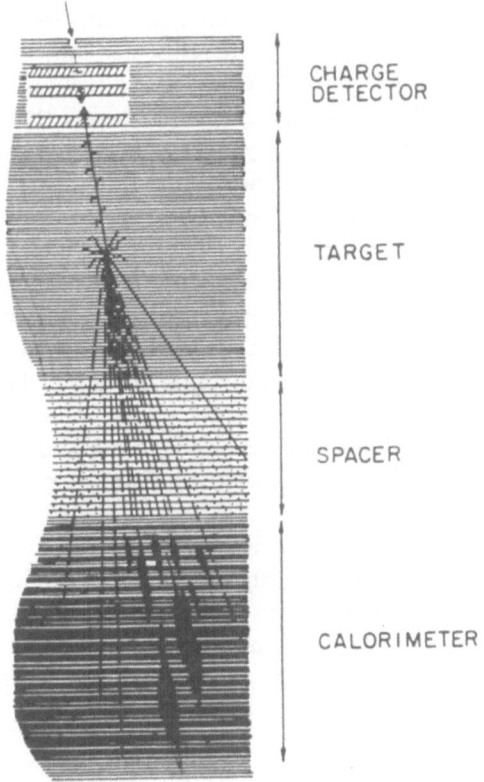

CHARGE
DETECTOR

TARGET

SPACER

CALORIMETER

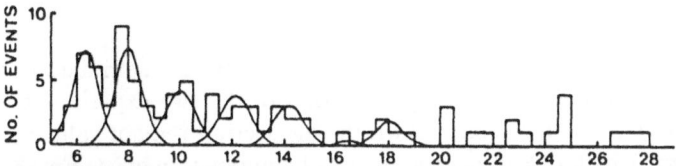

Fig. 2. Upper part: Cross section of the JACEE emulsion stack. Each of the four sections consists of several sheets of nuclear emulsions and special plastic foils interleaved with sheets of other material. The purpose of the four sections is described in the text. Lower part: Demonstration of the achieved charge resolution (from [Bur86] and [Bur90])

The other three sections serve mainly to determine the energy of the particle. The basic procedure for this is the following: In the target section, a large number of nuclear emulsions and plastic foils is interleaved with plastic plates of c. 1 mm thickness. The total thickness of this section is chosen such that every fourth proton and a larger fraction of heavier nuclei undergo a nuclear interaction. Gamma rays produced in this collision (mainly via the decay of neutral pions) are measured in the calorimeter section. The latter consists of each c. 25 nuclear emulsions and x-ray films interleaved with lead plates of 1 and 2.5 mm thickness. The high energy gamma rays are converted in the lead plates and produce an electromagnetic cascade whose intensity and development can be determined from the darkening of the x-ray films and emulsions. (The total thickness of the calorimeter is c. 7 radiation lengths, i.e. the conversion probability is practically 100 %.) The purpose of the space section is to obtain a better spatial separation of the individual gamma rays and their associated cascades though it is also provided with 20 layers of emulsions. This kind of energy measurement has a considerable drawback. What is primarily measured with satisfactory precision is the energy converted into gamma rays during the nuclear interaction of the projectile. It is well known that the inelasticity, i.e. the fraction of the projectiles' energy dissipated by the production of secondary particles fluctuates between almost zero and near 100 %. In addition the average energy fraction converted into gammas is not well known in the high energy range. So the energy determined for an individual event can be uncertain by more than an order of magnitude. This does not prevent the determination of the energy spectra for individual elements although one should bear in mind that the high energy end of the spectra where statistics is poor may be inaccurate.

The lateral dimensions of one JACEE stack are usually 40×50 cm^2 and up to four of these stacks have been exposed simultaneously in one flight. So far about a dozen flights have been performed with durations between 15 and 150 h [Wil90]. A summary of their results has recently been published by Burnett et al. [Bur90].

3. Results of Satellite and Balloon Experiments

The results of various experiments of the type described above are summarized in Fig. 3. They refer to a lower energy range than that of the two experiments discussed in Sect. 2 but the conclusions drawn in this section remain valid to the highest energies for which direct measurements of composition exist. Fig. 3 only includes elements up to and including the iron group. Heavier elements wil be discussed briefly later on.

Fig. 3 compares the results with the average composition of the solar system which is well known from the analysis of the solar spectrum and the chemical analysis of meteorites (cf. [And89]). The conspicuous feature of this comparison is a good agreement between the two data sets for abundant elements and severe discrepancies, up to orders of magnitude, for rare elements. This can be understood in detail by spallation reactions which the particles undergo on their path from source to earth. They collide with nuclei of the interstellar medium, mostly protons and helium nuclei. As is known from laboratory experiments (of protons bombarding nuclei) a number of nucleons are lost in such collisions. After the interaction the residual nucleus proceeds at the original energy per nucleon. So the cosmic ray nuclei are successively converted into lighter ones, and the composition measured near the earth does not represent the one at the

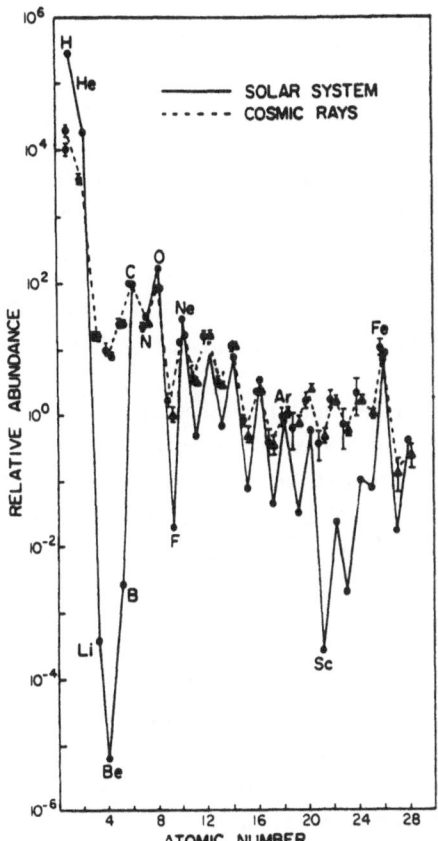

Fig. 3. Results of direct measuremernts of the elemental composition of primary cosmic rays compared to the average composition of the solar system. The conspicuous differences for elements of low solar abundance are attributed to spallation reactions between source and earth. (From [Aus81])

source. This explains at least qualitatively the overabundance of rare elements in cosmic rays: Li, Be, and B are the spallation products of C, N, and O nuclei which are much more abundant, and the same holds for F (which is produced from Ne, Mg, and S) and Sc to Mn (produced from Fe).

Since only the composition at the source is of relevance for the question of origin one has to correct the measured results for the effects of reactions during propagation through the interstellar medium. For a given path length (measured in g/cm^2) the change of composition can be calculated if the relevant spallation cross sections are known. Experiments with relativistic heavy ions have brought considerable progress in this field during the last few years. (For a review of results cf. Silberberg and Tsao [Sil90a].)

Numerous calculations have been performed along these lines with models of increasing sophistication. The problem of propagation turns out to be rather complex if studied in detail. It has recently been reviewed by Berezinsky [Ber90]. For our purpose a very simple model, referred to as the "leaky box" model, gives a very satisfactory description of the observational results. It is based on the following assumptions:

- Cosmic ray particles undergo spallation reactions as described above on their way from source to earth.
- The path lengths between source and earth are not uniform, but show a broad distribution. This is of course to be expected because propagation in the interstellar medium is a diffusionlike process and there will probably be many sources distributed in the Galaxy. It turns out that an exponential path length distribution describes the observations satisfactory although more complicated distributions have been proposed and are probably required to account for all details [Gar87].
- Particles "leak out" of the Galaxy at a rate which increases with magnetic rigidity, i.e. the mean life or residence time of cosmic ray particles depends on E/Z. This assumption appears reasonable because the particles are bound to the Galaxy by magnetic fields and they should all start to escape at the same radius of curvature. This results in a decrease of the mean path length (which is the only free parameter of an exponential distribution) with increasing rigidity.

For a given source composition the mean path length and its rigidity dependence are the only free parameters.

When this model is applied to the experimental data the gross differences between cosmic ray source composition and that of the solar system are reduced to less than a factor of ten for all elements and vanish for quite a number of them, i.e. the composition at the source is at least very similar to the solar one (termed "local galactic" in cosmic ray physics). The mean free path deduced this way is c. 4 g/cm^2 near 10 GeV/amu and decreases with magnetic rigidity R according to $R^{-\gamma}$, where $\gamma \approx 0.6$. This corresponds to a distance of 1 Mpc if the particles spend all their time in the galactic disc (with density 1 atom/cm^3 or $2 \cdot 10^{-24}$ g/cm^3). Since the particles travel at the speed of light this would correspond to a life time of c. $2 \cdot 10^6$ years. It should be mentioned that there is an independent method of determining the mean life of cosmic ray particles. This makes use of the measured abundance of beta unstable nuclei, notably ^{10}Be, in the primary radiation. These are also produced by spallation reactions and decay during their further travel, their abundance therefore depends on the time the particles spend in the interstellar medium. The mean life determined this way is larger by about a factor of 5. One therefore has to conclude that cosmic ray particles spend most of their life in regions of space where the density is lower than in the galactic disc, probably in the galactic halo.

The differences between cosmic ray source and solar system abundances which remain after correction for spallation can be summarized as follows:

- Elements with a high ionization potential are underabundant in cosmic rays. This appears to be a feature of the very beginning of the acceleration process (called "injection") and is not well understood. It is demonstrated in Fig. 4 which displays the ratio of cosmic ray source to local galactic abundances. A similar effect occurs among solar energetic particles which are accelerated from the surface of the sun to energies of MeV to GeV. These particles are probably accelerated by the same basic mechanism. The depletion of elements which are difficult to ionize therefore seems to be a very general phenomenon.

- The two lightest elements hydrogen and helium, and possibly also nitrogen, are further depleted by about an order of magnitude. This has been proposed to indicate that a considerable fraction of the cosmic ray material originate from supernovae of type I which are known to contain only very small amounts of the lightest elements [Yan90].

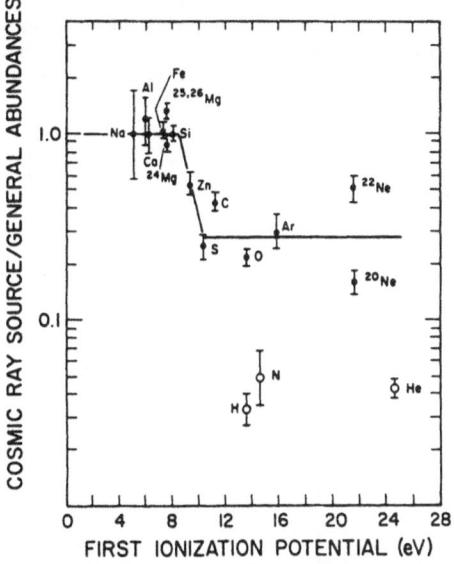

Fig. 4. Ratio of the inferred elemental abundances at the cosmic ray source to those in the solar system plotted versus the first ionization potential. The drop of the abundance ratio near 9 eV ionization energy is obvious (from [Sil90b])

The discussion so far only referred to the elements up to and including the iron group. Our knowledge about heavier elements is considerably less comprehensive, mainly due to the low abundance of these elements everywhere in the universe which results in poor statistics of the measurements. It results mainly from the satellites ARIEL 6 [Fow87] and HEAO 3 [Bin89]. Fig. 5 shows the HEAO 3 results. The measurements have been corrected for spallation in the interstellar medium and for the influence of the first ionization potential (by applying an empirical correction derived from data as shown in Fig. 4). As can be seen from Fig. 5 there is good agreement with the solar system composition for the lighter elements up to $Z = 60$, whereas the inferred source composition for the heaviest elements may deviate from solar.

So the main conclusion to be drawn from cosmic ray composition measurements is that, except for the depletion of the two most abundant elements and possibly the heaviest, the source material is very similar to that inside the solar system which again is probably typical for the composition of the galactic disc.

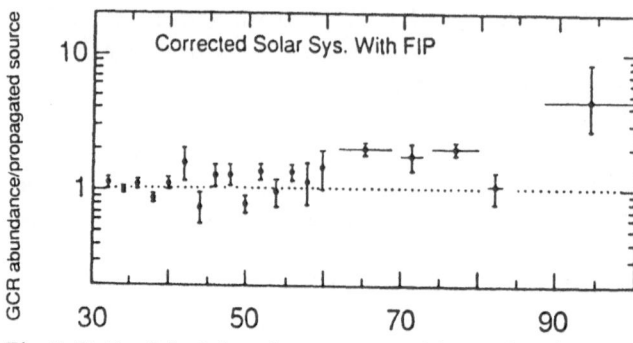

Fig. 5. Ratio of the inferred source composition to the solar system composition for elements above the iron group. The data have been corrected for spallation reactions in the interstellar medium and for the empirical depletion with increasing first ionisation potential. The term "corrected solar system" refers to a correction of some solar elemental abundances for which more recent results exist. (From [Bin89])

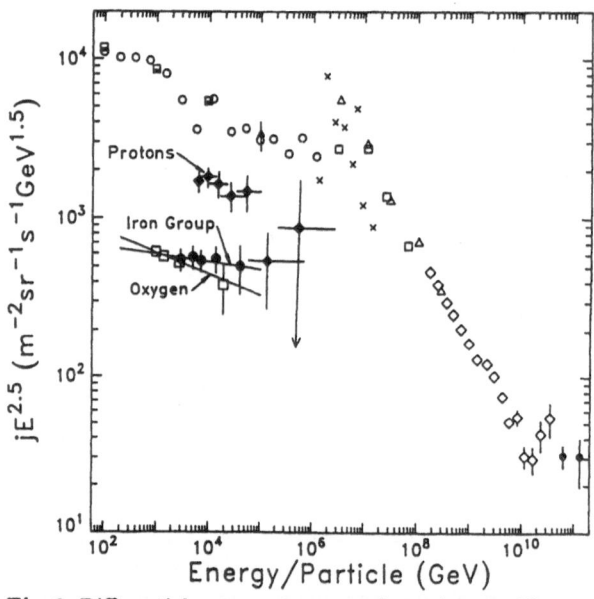

Fig. 6. Differential energy spectrum of cosmic rays. The uppermost data represent the total spectrum, the other points individual elements. The latter are the highest energy results known at present from direct measurements. All data have been multiplied by a power of the energy in order to compress the vertical scale. The total spectrum data have been compiled by Linsley [Lin83], the proton results are from JACEE [Bur90], those for oxygen and iron from [Gru88]

4. How to Measure Cosmic Ray Composition above 10^{15} eV

Direct composition measurements now extend to some 10^{14} eV. The highest energy data points are shown in Fig. 6 together with measurements of the total spectrum. All data have been multiplied by $E^{2.5}$ in order to compress the ordinate scale. The total differential spectrum follows a power law with exponent -3 up to some point between 10^{15} and 10^{16} eV where the slope steepens by about 0.5. This means that the integral intensity decreases by at least a factor of 100 if the energy increases by a factor of 10. Due to the weight and size limitations on satellites and balloons it is not to be expected that this limit will be extended to considerably higher energies in the forseeable future, with 10^{15} eV probably being the utmost limit one may hope to reach. On the other hand the question of composition is of special interest in the higher energy range beyond the general arguments given in the introduction:

- The change of the slope of the total cosmic ray spectrum, the so called "knee", is not understood although it has been known for a long time. It is speculated that it may be due to a change of source or acceleration mechanism. This might show up in a change of composition.
- There seems to be agreement today that the bulk of cosmic rays is accelerated by shock waves caused by supernovae in the interstellar medium (for reviews of this mechanism cf. [Voe87] and [Ell90]). The maximum energies which can be obtained by this acceleration mechanism are limited, but the precise limit is not well known. While it was originally believed to lie at some 10^{14} eV [Lag83] there are more recent considerations that try to extend it to near 10^{16} eV [Voe88].

In the energy range of interest here only experiments on ground level are possible for the reasons given above. The obvious and severe disadvantage of this is that cosmic ray particles will interact with air atoms during their passage through the atmosphere. The probability of a high energy proton to reach sea level without nuclear interaction is less than 10^{-5} and several orders of magnitude smaller for complex nuclei. So only reaction products can be observed and it may appear very doubtful whether it is possible to distinguish between complex nuclei and single nucleons by measurements of the secondary particles. Before discussing this point in more detail let us have a closer look at what happens when a high energy cosmic ray particle enters the atmosphere.

4.1 Outline of Extensive Air Showers

The interaction mean free path of a high energy proton in air is c. 80 g/cm^2. So a proton will on average interact more than 10 times before it reaches sea level. In each of these collisions secondary particles will be produced, mainly pions but also heavier mesons, hyperons, antinucleons etc. The neutral pions decay immediately into gamma rays which have a mean free path of only c. 35 g/cm^2 and therefore soon interact to produce an electron-positron pair. Both these particles produce bremsstrahlung whose energy (at least in the upper part of the atmosphere) is sufficiently high to create further pairs etc. In this way each gamma ray initiates an avalanche of electromagnetic particles whose number quickly increases until their energy has dropped so far that ionization loss becomes important, the charged particles are slowed down and the positrons eventually annihilate. The charged pions and other hadrons either decay into high energy muons or interact producing more secondary hadrons. The muons interact only weakly within the atmosphere and have a rather large half life, due to relativistic time dilatation, and therefore reach ground level with a high probability. The whole phenomenon is called an extensive air shower (EAS) and was discovered by Auger during the thirties [Aug38].

On observational level an extensive air shower consists of three components: electromagnetic particles which interact frequently and exhibit an equilibrium between electrons, positrons and gammas; muons; and hadrons. Typically on ground level 90 % of all particles are electromagnetic, 10 % muons and 1 % hadrons. On average approximately one secondary particle is observed at sea level for every 10 GeV of primary energy. The three components exhibit very different lateral distributions: The hadrons are confined to the vicinity of the axis defined by the momentum of the primary particle to within 5 to 10 m. A typical radius of the electromagnetic component (at sea level) is 80 m; this is a consequence of multiple Coulomb scattering of electrons and positrons in the atmosphere. The muons reach out to radii of several hundred meters because they mostly originate from high up in the atmosphere and therefore even a small transverse momentum which they (or their parent hadrons) obtain at birth results in sizable displacements on the ground. The lateral development of an extensive air shower if plotted versus the depth in the atmosphere in g/cm^2 , is charaterized by a rapid increase, a flat maximum and an exponential fall-off. The maximum lies near sea level for the very highest energies observed and at about 6000 m for 10^{15} eV primaries. In view of the large lateral extent of an extensive air shower only a small fraction of particles can be measured. The distributions are sampled by arrays of detectors which are spread out over the size of a shower with sampling fractions usually well below 1 % for most previous experiments.

4.2 Extensive Air Shower Properties Sensitive to Primary Particle Mass

In view of the complexity of an extensive air shower it may seem hopeless to try to determine the mass of the primary particle from a measurement of shower properties at ground level. Yet there is a number of observables which can be expected to show the required (and desired) dependence. This can be understood qualitatively on the basis of high energy nucleon nucleon interactions. In this field considerable progress has been made during the last few years by experiments using the proton-proton and proton-antiproton colliders at CERN and the Fermilab. The results of these experiments have been recently reviewed by Geich-Gimbel [Gei89].

First one should realize that the total binding energy of even a rather heavy nucleus such as Fe which amounts to c. 0.5 GeV is small as compared to the total energy available in the centre of mass system. Therefore it seems justified to consider an air shower induced by a high energy Fe nucleus as the superposition of 56 independent showers each initiated by a nucleon of 1/56 of the primary energy ("superposition model"). There is of course another difference between nucleon-nucleus and nucleus-nucleus collisions namely the higher spatial density of particles and there are theoretical predictions that the large number of individual interactions taking place in the small volume of a nucleus may result in the formation of a completely new state of matter called "quark gluon plasma". We will disregard this possibility in our discussion because there is so far no experimental evidence for this new phenomenon and, more important, it is only expected to occur for central collisions which represent only a small fraction of all nucleus-nucleus interactions. Peripheral collisions which are much more frequent are not expected to be much influenced by this intriguing possibility. If the superposition model is taken for granted there are two main points of difference between proton and nucleus induced showers. The first of these is the fact that a shower initiated by, e.g., an iron nucleus is the superposition of 56 individual showers of the same kind. This should reduce the fluctuations of all observables. The other one is due to the difference in the energy per nucleon, i.e. to the energy dependence of strong interaction. The relevant features of the latter can be summarized as follows [Gei89]:

- The inelastic interaction cross section rises in proportion to the logarithm of the energy squared.
- The average number of secondary particles produced during an inelastic collision increases in proportion to the logarithm of the energy, i.e. clearly slower than in proportion to the energy. This implies that one nucleon of energy E produces fewer particles than A nucleons of energy E/A.
- The inelasticity, i.e. the mean fraction of the energy dissipated by secondary particle production is almost independent of energy. Consequently the mean energy of the secondaries is higher for a proton induced reaction than for one induced by a nucleus.
- The mean transverse momentum imparted to the secondaries at their production rises only very slowly with increasing energy. Since the longitudinal momentum is larger in an interaction initiated by a proton their angle with respect to the direction of the primary is smaller in proton induced reactions.

These arguments hold in a large energy range and hence also for reactions by secondary hadrons. Therefore one can expect the following differences between showers initiated by protons and nuclei:

- The number of muons is expected to be smaller in proton induced showers. This results from the smaller number of secondary particles which are mostly pions by whose decay most of the muons are produced.
- The energy of hadrons at observational level should be higher for proton induced events because the argument given above holds for all successive interactions.
- The width of all lateral distributions will be smaller in proton showers due to the smaller production angle of secondaries.

So an experiment capable of measuring the lateral distributions of the electromagnetic, muonic and hadronic components on observation level and the energy of hadrons can be expected to yield information on the mass of the primary nucleus. It should be pointed out, though, that all quantities observable in an extensive air shower are subject to substantial fluctuations. This is not surprising for a situation where measurements are only possible after ~ 10 interactions and in addition the output of each of these and their sites in the atmosphere fluctuate considerably.

It should be mentioned that a considerable number of experiments along these or similar lines have been performed in the past. Most of them are described in the proceedings of the International Cosmic Ray Conferences the last two of which were held in Moscow 1987 and Adelaide 1990. The results are not at all consistent. There is not even agreement on such qualitative features as to whether the fraction of heavy nuclei increases with energy or not. I will discuss possible reasons for this situation at the end of the next subsection.

4.3 A Proposed Experiment: KASCADE

As in Sect. 2 we want to describe here a specific experiment aimed at determining the primary cosmic ray composition near 10^{15} eV rather than giving a list of the large number of attempts which have been made in this field in the past, but have not yielded unambiguous results. The KASCADE (KArlsruhe Shower Core and Array DEtector) experiment is being set up in the Karlsruhe Nuclear Research Centre and expected to start taking data in early 1994. A plan of the detector arrangement is shown in Fig. 7. There are 316 detector stations on a total area of 200×250 m^2. Each of these station contains detectors for electrons and muons each with a sensitive area of c. 3 m^2. Sixteen detector stations are combined to form a cluster with a common electronic supply container (black rectangles in Fig. 7). In addition there is a central detector which is capable of measuring the energy of hadrons above c. 10 GeV, their position to c. 0.5 to 1 m and the number of muons within its area of 16×20 m^2. The experiment is therefore capable to measure electrons, muons and hadrons, their lateral distributions and the energies of hadrons, as required for a determination of the primary mass according to the arguments given above. A unique feature of KASCADE is the high degree of sampling, 2.5 % for muons and 2 % for electrons. The experiment has been described in more detail in a report by Doll et al. [Dol90].

Fig. 8 shows the expected mass resolution of the KASCADE experiment. It is based on extensive Monte Carlo simulations of the development of extensive air showers in the atmosphere. The relative intensities have been obtained by extrapolation of the Chicago [Mue90] and JACEE [Bur90] data. The abscissa is a combination of mainly three observables, the muon to electron ratio, a parameter characterizing the slope of the radial muon distribution, and the total energy of hadrons. As can be seen from the figure a clear separation of heavier nuclei from the lightest elements can be expected.

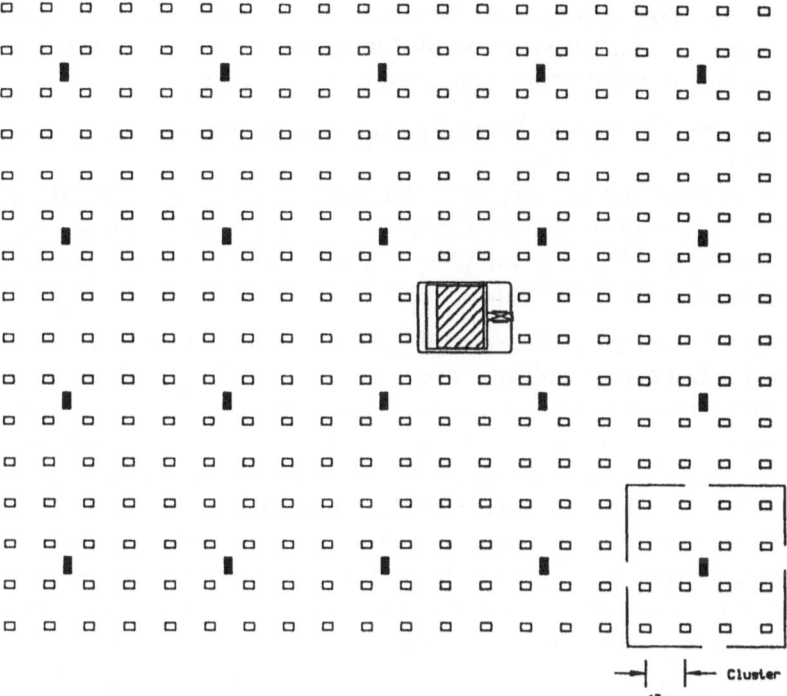

Fig. 7. Lay-out of the KASCADE air shower experiment. It consists of a central detector and 316 detector stations. The latter are indicated by open rectangles and capable of measuring electrons and muons, whereas the central detector is a hadron calorimeter and in addition registers muons (cf. [Dol90]). Sixteen detector stations are combined to a cluster and supplied by one electronic station shown as black rectangle

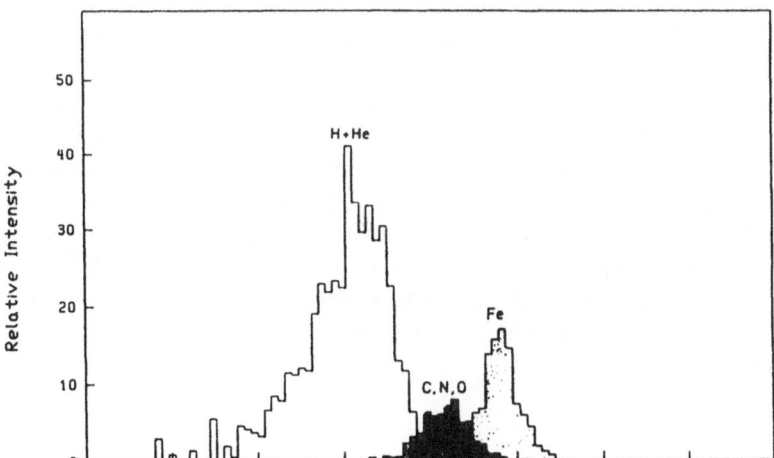

Fig. 8. Expected mass resolution of the KASCADE experiment near 1 PeV. The relative intensities of the three groups of elements have been extrapolated from satellite and balloon data

A comment is now appropriate concerning the comparison with previous similar experiments which have not yielded unambiguous results as was mentioned above. In our opinion this is mainly due to two facts:

- Most previous experiments have only measured one parameter of each individual shower.
- The rate of sampling was very low. This holds especially for muon detection. As a consequence the fluctuations of the observed quantities were gouverned by the degree of sampling and hence larger than the intrinsic fluctuation of shower development.

Therefore the sensitivity of measured shower properties to primary mass is diminished considerably and only averages of observed quantities can be compared with the corresponding results of simulation calculations assuming different primary compositions. In a number of cases a change of composition could be compensated by modifications of the assumed model of high energy interaction within the range of observational uncertainties. These uncertainties can be expected to decrease if several quantities each of which is sensitive to primary mass, are registered for each shower and if the statistical fluctuations are reduced to the unavoidable limit of the intrinsic shower fluctuations. This is borne out by the results of detailed simulations as displayed in Fig. 8.

5. Summary

Cosmic rays consist mainly of atomic nuclei. Above some 100 MeV/nucleon they originate almost exclusively from outside the solar system. Up to energies of several 100 TeV their composition is known from experiments on satellites and balloons. These measurements do not represent the composition at the source because cosmic ray particles undergo spallation reactions on their way through the interstellar medium from source to earth. If this is accounted for most elemental abundances are very similar to the mean composition of material in the Galactic disc except for a systematic depletion of elements with an ionization potential above c.9 eV. Discrepancies between cosmic ray source and average Galactic compositions exist for the two lightest elements hydrogen and helium which are clearly underabundant in cosmic rays, and probably for heavy elements with atomic number $Z > 60$. Neither of these discrepancies is well understood at present.

References

And89   E. Anders, N. Grevesse: Geochim. Cosmochim. Acta **53** 197 (1989)

Aug38   P. Auger, R. Maze, T. Grivet-Meyer: Comptes Rendus **206** 119 (1938)

Aus81   S. M. Austin: Progr. Part. Nucl. Phys. **7** 1 (1981)

Ber90   V. S. Berezinsky: Proceedings 21st Internat. Cosmic Ray Conference, Adelaide 1990, **11** 115

Bin89   W. R. Binns et al.: Astrophys. J. **346** 997 (1989)

Bur86   T. H. Burnett et al.: Nucl. Instr. Meth. **A251** 583 (1986)

Bur90   T. H. Burnett et al.: Astrophys. J. **349** L25 (1990)

Dol90   P. Doll et al.: KfK Report 4686 (1990)

Ell90   D. C. Ellison: Proceedings 21st Internat. Cosmic Ray Conference, Adelaide 1990 **11** 133

Fow87   P. H. Fowler et al.: Astrophys. J. **314** 739 (1987)

Gar87   M. Garcia-Munoz et al.: Astrophys. J. Suppl. **64** 269 (1987)

Gei89   C. Geich-Gimbel: Internat. J. Mod. Phys. **A4** (1989)

Gru88   J. M. Grunsfeld, J. L'Heureux, P. Meyer, D. Müller, S. P. Swordy: Astrophys. J. **327** L31 (1988)

Hes12   V. F. Hess: Physikal. Zeitschrift **13** 1084 (1912)

Lag83   P. O. Lagage, C. J. Cesarsky: Astron. Astrophys. **125** 249 (1983)

Lin83   J. Linsley: Proceedings 18th Internat. Cosmic Ray Conference, Bangalore 1983, **12** 135

LHe90   J. L'Heureux, J. M. Grunsfeld, P. Meyer, D. Müller, S. P. Swordy: Nucl. Instr. Meth. **A295** 246 (1990)

Mue90   D. Müller, S. P. Swordy, P. Meyer, J. L'Heureux, J. M. Grunsfeld: Astrophys. J, in press

Sil90a   R. Silberberg, C. H. Tsao: Physics Reports **191** 351 (1990)

Sil90b   R. Silberberg, C. H. Tsao: Astrophys. J. **352** L49 (1990)

Voe87   H. J. Völk: *Proceedings* 20th *Internat. Cosmic Ray Conference,* (Moscow 198) **7** 157

Voe88   H. J. Völk, P. L. Biermann: Astrophys. J. **333** L65 (1988)

Wil90   R. J. Wilkes: *Proceedings* 21st *Internat. Cosmic Ray Conference,* (Adelaide 1990) **12** 166

Yan90   S. Yanagita, K. Nomoto, S. Hayakawa: *Proceedings* 21st *Internat. Cosmic Ray Conference,* (Adelaide 1990) **4** 44

Primordial Nucleosynthesis:
Beyond the Standard Model

Robert A. Malaney

Institute of Geophysics and Planetary Physics, University of California, Lawrence Livermore National Laboratory, CA 94550, U.S.A.

1. Introduction

Non-standard primordial nucleosynthesis merits continued study for several reasons. First and foremost are the important implications determined from primordial nucleosynthesis regarding the composition of the matter in the universe. Second, the production and the subsequent observation of the primordial isotopes is our most direct experimental link with the early ($t \lesssim 1$ sec) universe. Third, studies of primordial nucleosynthesis allow for important, and otherwise unattainable, constraints on many aspects of particle physics. Finally, there is tentative evidence which suggests that the Standard Big Bang (SBB) model is incorrect in that it cannot reproduce the inferred primordial abundances for a single value of the baryon-to-photon ratio.

Reviewed here are some aspects of non-standard primordial nucleosynthesis which mostly overlap with the author's own personal interest. For a more systematic (and balanced!) review of non-standard models the reader is referred to [Mal91].

We begin with a short discussion of the SBB nucleosynthesis theory, highlighting some recent related developments. Next we discuss some recent observations of helium and lithium abundances and why these observations are critical to the viability of the SBB model. We then discuss how the QCD phase transition, neutrinos, and cosmic strings can influence primordial nucleosynthesis. We conclude with a short discussion of the multitude of other non-standard nucleosynthesis models found in the literature, and make some comments on possible progress in the future.

2. Standard Primordial Nucleosynthesis

For a review of the SBB nucleosynthesis model the reader is referred to one of the many good review papers on the subject [eg. Boe85]. Here we will just discuss some recent developments which have impacted the field.

One of the most exciting of these developments is the results of the LEP and SLC e^+e^- colliders. Since the shape of the 91 GeV Z^o resonance is directly related to the number of light $(m_\nu < M_Z/2)$ neutrinos, an independent limit on their number can be obtained. From the results so far reported [Ale90, Opa90] one finds an upper limit to the number of neutrino species of $N_\nu \leq 3.2$ at the 2σ level.

A new technique has been developed to determine the neutron half-life based on the storage of ultra-cold neutrons [Mam89]. By combining this latest result with other recent results [Agu88] the neutron half-life is inferred to be 10.28 ± 0.05 mins. This is to be compared with the value of 10.6 ± 0.2 minutes employed in previous studies only a few years ago. The change in the neutron half-life slightly reduces the ^{4}He abundance.

New experimental studies and analyses of key reactions in the SBB nuclear reaction network have resulted in modified production rates. The most significant of these modified rates are those concerned with the production of ^{7}Li and to a lesser extent the D+^{3}He abundance. Reference to, and a compilation of the updated rates can be found in [Cau88, Mal91]. Also, there has been continued effort in the accurate determination of the primordial abundances of D, ^{3}He, ^{4}He and ^{7}Li as inferred from observation (see Sect. 3).

The combined effects of these new developments can be seen from the calculations of [Oli90] shown Fig. 1. Here the D, ^{3}He, ^{4}He and ^{7}Li abundances are shown as a function of η_{10}, and the baryon-to-photon ratio in units of 10^{-10}. The effect of the neutron half-life on the ^{4}He mass fraction and the effect of varying the number of neutrino species on ^{4}He and ^{7}Li are also shown. The allowed region of η_{10} is bounded from below by the constraint on the observed D+^{3}He abundance. We must use the sum of these two isotopes as D is processed to ^{3}He rapidly in stellar interiors via $D(p,\gamma)^3$He. The actual upper limit inferred for the D+^{3}He abundance is somewhat sensitive to the stellar and galactic chemical evolution models. Observations indicate the number abundance limit $(D +^3 He)/H < 1.1 \times 10^{-4}$, which leads to the constraint of $\eta_{10} > 2.6$. The upper limit on η_{10} is provided by the ^{7}Li abundance. Adopting the constraint $\log(\text{Li}/\text{H}) \leq 2.36$ one determines that $\eta_{10} < 4.8$ (for $N_\nu = 3$).

Using the relation

$$\Omega_b h^2 = 3.73 \times 10^{-3} \eta_{10} T_o^3 \quad , \tag{1}$$

where h is the Hubble parameter in units of 100 km s^{-1} Mpc^{-1}, and T_o is the microwave background temperature in units of 2.75 K, and adopting $0.4 < h < 0.65$ the above limits on η_{10} transform into a range in the dimensionless baryonic density parameter of $0.02 < \Omega_b < 0.12$ (this range can be extended slightly if reaction rate uncertainties are accounted for). It is from these limits on the

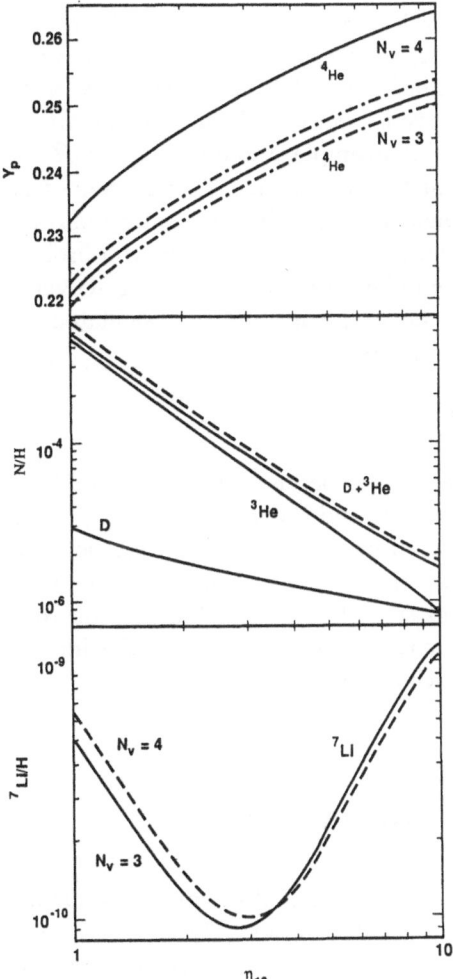

Fig. 1. Isotopic abundances as a function of η_{10} as predicted by the SBB model. For ^{4}He the solid curves represent the yield for $N_\nu = 3, 4$ and $\tau_n = 889.8$s. The dot-dashed curves show the $N_\nu = 3$ yield for 2σ variations in τ_n. For D, D+^{3}He,^{3}He and ^{7}Li the solid curves are for $N_\nu = 3$ and the dashed for $N_\nu = 4$ (adopted from [Oli90])

baryonic density that the most important conclusions from SBB nucleosynthesis theory are drawn.

The contribution of visual matter to the energy density of the universe is constrained to be $\Omega_v \lesssim 0.01$. From the lower bound to the baryonic density this implies that a significant fraction of baryons are in some from of dark matter. However, current estimates on the density parameter from stellar and galactic dynamical studies $0.1 \lesssim \Omega \lesssim 0.3$ [Tri87] are not inconsistent with the upper limit on Ω_b derived from SBB nucleosynthesis. As of this writing all claims of a determination of $\Omega \sim 1$ have been controversial and there is no compelling direct observational evidence for the existence of nonbaryonic dark matter.

The theoretical prejudice for an $\Omega = 1$ critically closed universe, which is naturally accounted for by an inflationary epoch in the early universe, would clearly be in violation of the bounds imposed by SBB nucleosynthesis. Apart from the presence of some form of baryonic matter which plays no role in nucleosynthesis, primordial black holes for example, the only way an $\Omega = 1$ universe can be reconciled with SBB nucleosynthesis is to invoke some form of non-baryonic matter. Massive neutrinos or some other elementary particles have been widely discussed in this context. A review of the dark matter problem is given in [Tri87] and a recent review of inflation models is given in [Olv90].

The number of neutrino species can be constrained by the calculated and inferred primordial ^{4}He mass fraction, Y_p, through

$$N_\nu = 3.0 - 0.8 \ln \eta_{10} + 19(\frac{Y_p - 0.228}{0.228}) - 15(\frac{\tau_n - 889.8}{889.8}) , \qquad (2)$$

where τ_n is the neutron mean life measured in seconds. Adopting the lower bound on $\eta_{10} \geq 2.6$, the lower 2σ bound on τ_n and the upper bound $Y_p \leq 0.24$ from observation we find that the number of neutrino species is constrained to be $N_\nu < 3.4$. Since the cosmological and the Z^o experimental determination are in such good agreement, this would seem to rule out the presence of a significant contribution to the energy density of the universe in the form of unknown relativistic particles at the epoch of nucleosynthesis.

3. Observational Tests of the Standard Model

A great deal of effort has gone into the determination of the primordial abundances of the light isotopes as inferred from observation. The most recent of these efforts are reviewed in [Pag90]. Here we will discuss just two aspects of these observational studies. The first of these, the determination of the primordial ^{4}He abundance, tentatively suggests that some storm clouds may be gathering for the SBB nucleosynthesis model.

3.1 Helium

The primordial ^{4}He abundance is often the most sensitive constraint on non-standard big bang models since its yield depends mostly on the universal expansion rate which in turn influences n/p ratio at the time of weak-reaction freeze-out.

In order to determine the primordial ^{4}He abundance one must correct for stellar processing. Such a correction is usually attempted by correlating the ^{4}He abundance with metallicity and extrapolating to zero metallicity. Assuming a linear correlation with metallicity, Z, we have

$$Y = Y_p + \frac{\Delta Y}{\Delta Z} Z , \qquad (3)$$

where Y is the observed ^{4}He mass fraction and Y_p is the actual primordial ^{4}He mass fraction.

The earlier observations were taken of highly-ionized H II regions in low-metallicity galaxies with oxygen as the metallicity tracer [Pei74]. However, oxygen may actually be a poor diagnostic of ^{4}He production (see [Ste89] and refs. therein) since ^{4}He is produced in stars of mass $M \gtrsim 2M_\odot$ whereas oxygen is produced only in stars of mass $M \gtrsim 12M_\odot$. Since it is expected that nitrogen and carbon are also produced in $M \gtrsim 2M_\odot$ stars, it is believed that these latter elements will represent an improved diagnostic of ^{4}He production.

The current maximum likelyhood regressions from the observed [Pag89] correlation of metallicity with helium abundance are,

$$Y = 0.224(\pm0.005) + 178(\pm45)O/H \qquad (4)$$

or

$$Y = 0.229(\pm0.004) + 3120(\pm780)N/H \ . \qquad (5)$$

Chemical evolution corrections are important and deserve careful attention, especially since the observed data does not go below 0.01 of solar metallicity. It has been noted [Ful90] that if only low-metallicity objects ($Z < 1/4$ solar) are used in the linear regression to zero metallicity, a lower Y_p is obtained. The lack of very-low metallicity observations also leaves open the possibility that some process which produces ^{4}He, but no significant metals, early in the galactic history (perhaps massive stars) could lead to an overestimate of Y_p.

If it would be possible to truly determine that the upper limit to the primordial helium abundance were less than 23.5% then the SBB model would be inconsistent with the abundance limits of the other primordial isotopes. We note that the above primordial values already begin to hint at such a problem. This point is shown more graphically in Fig. 2 where it can be seen that the upper limit to Y_p has tended to decrease over the past decade or so. However, it is also clear that systematic and other uncertainties in the determination of Y_p do not yet allow for any robust invalidation of the standard model. Within the allowed uncertainties the SBB theory remains valid. Hopefully, continued effort in the observational determination of Y_p will resolve this very important issue.

3.2 Lithium

Of the primordial isotopes, determination of the primordial abundance of ^{7}Li remains the most controversial. The problem has been that observations of the lithium abundance in the atmospheres of Pop I stars [Cay84, Hob88] and of the interstellar medium in the galactic disk [Fer84] result in a number fraction Li/H$\sim 10^{-9}$, whereas observations of Pop II stars [Spi82] result in Li/H$\sim 10^{-10}$ (all of the lithium observed is normally assumed to be in the form of ^{7}Li). The reason for the order of magnitude difference between the different sets of observations, and which measurement more closely represents the actual primordial abundance of lithium, has been the subject of much debate. The situation is

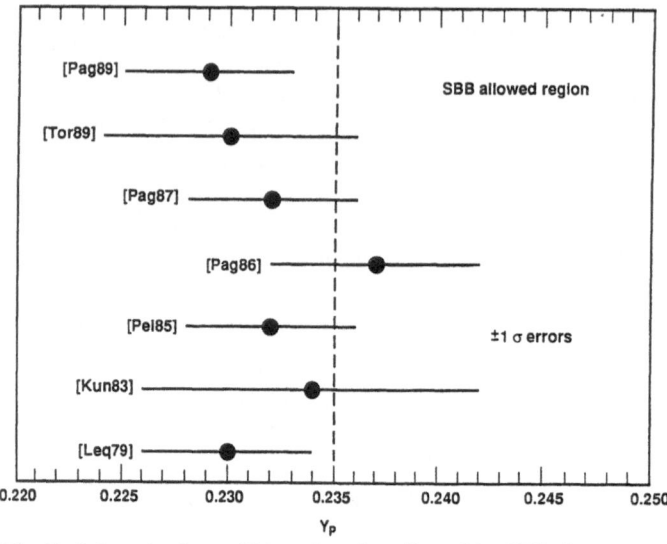

Fig. 2. Inferred values of Y_p and region allowed by SBB theory

further confused by galactic chemical and stellar evolution effects. In addition, interpretation of interstellar lithium observations requires complicated corrections for ionization, cosmic ray spallation, and grain depletion processes.

In spite of these difficulties it has been argued that the lower of the abundance determinations, namely the Pop II lithium abundance, represents the true primordial lithium value. The motivation for this view is that the Pop II stars are the oldest stars in the galaxy, and should therefore be representative of the abundances in the primordial material out of which these stars formed. Strong evidence for this view comes from the lithium abundance plateau determined in a wide spectral class of Pop II stars [Spi82], and from recent theoretical studies of lithium evolution in old halo stars [Del89].

An alternative view is that ^7Li is gradually removed from stellar atmospheres by some mechanism. In this scenario, the older stars would have destroyed more of their original lithium, thereby explaining the lower lithium abundance seen in the Pop II stars. The Pop I lithium abundance might then be more representative of the primordial lithium abundance. Any such depletion mechanism, however, must lead to a uniform depletion in all plateau halo stars. This would seem unlikely since the plateau halo stars may represent a range of stellar parameters and it is difficult to imagine any depletion mechanism which should be independent of such parameters [Del89].

Due to the confusion involved in interpreting stellar lithium data it would be more advantageous to observe the lithium abundance in the halo or intergalactic medium. In this regard, observation of interstellar lithium toward SN 1987A seemed, in principle, a promising new approach to the determination of the primordial lithium abundance. The fact that the supernova occurred in the LMC has several advantages. Due to the high galactic latitude of the LMC, the lithium abundance in the galactic halo, and not the galactic disk, can be investigated. Since the galactic halo is less influenced by stellar activity, the lithium abundance there should more closely represent the primordial value. Similarly, the low star formation rate in the LMC, as well as its low dust content, may allow for reduced stellar activity, and reduced grain depletion corrections, respectively.

Four searches for interstellar lithium toward SN 1987A were undertaken. The spectrum from one of these investigations [Mal90] is shown in Fig. 3, and the results from all of the studies are given in Table 1. The upper limits to equivalent widths and the corresponding upper limits to the primordial Li/H ratios are given. The early detection of interstellar lithium found in ref. [Vid87] was not verified by the later more detailed investigations. The much larger upper limit to the primordial Li/H ratio given by [Mal90] arises from a more conservative estimate of the present uncertainties involved in determining a lithium abundance from an interstellar Li I line toward SN1987A. These uncertainties primarily concern ionization corrections and depletion processes onto stellar grains.

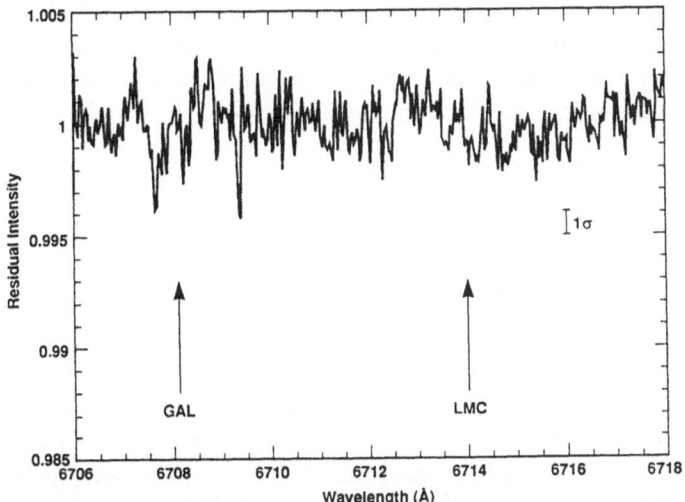

Fig. 3. The summed Reticon spectrum of the supernova 1987A in the spectral region of 6708 Å Li I, where the relative intensity, is plotted against the observed wavelength. The regions labeled GAL and LMC indicate the most likely velocity ranges for Li I absorption in the Galaxy and the LMC, respectively [Mal90]

Table 1. Upper Li/H limits toward SN1987A

Reference	Eq.Width(mÅ)	Interpretation
[Vid87]	0.2; 0.3; 0.5	detections?
[Baa88]	<0.15	Li/H< 4×10^{-10}
[Sah88]	<0.13	Li/H< 1.6×10^{-10}
[Mal90]	<0.25	Li/H< 4.4×10^{-9}

Since the more conservative estimate of Li/H$\lesssim 4 \times 10^{-9}$ encompasses the Li/H ratio observed in both Pop I and Pop II stars, this is somewhat disappointing. The observational data of interstellar lithium cannot presently resolve the issue of the true primordial lithium abundance unless one is willing to believe that the ionization and grain depletion processes are currently much better understood than as discussed in [Mal90].

Recently [Baa91] have attempted to better constrain the allowed depletion of interstellar lithium. They show how much tighter constraints result *if* several assumptions are made regarding such depletion processes. They have also combined all the data of Table 1 into to one spectra in an attempt to detect the interstellar lithium line. Although a detection remained elusive, a reduction in the [Mal90] upper limit by about a factor of two could be achieved.

The importance in determining the Li/H ratio toward SN1987A can be seen in the chemical evolution models of [Mat90]. One of the non-standard nucleosynthesis models we shall discuss (Sect. 4) predicts high primordial lithium. The chemical evolution models of [Mat90] can readily account for both a high and a low primordial lithium abundance. The key factor in discriminating between these models is the Li/H abundance in the Large Magellanic Cloud. Unfortunately, an upper limit of Li/H$\lesssim 2 \times 10^{-9}$ does not allow for any discrimination. If the Li/H ratio could be proven to be significantly below this value, however, the high-primordial lithium chemical-evolution models would be ruled out.

4. The QCD Phase Transition

There has been great interest in the past few years in the nucleosynthesis resulting from inhomogeneous models. These models are mostly based on the possible density inhomogeneities arising in the early universe ($\sim 10^{-5}$s) as a consequence of a first order QCD phase transition [Wit84] and the subsequent diffusion of neutrons from the high-density regions [App85]. The principal result of these ideas is that the limits on the baryonic density of the universe can be increased relative to the limits imposed by SBB nucleosynthesis theory. Another interesting aspect of these models is the production of heavy elements in the early universe. As this latter subject is reviewed in this volume by F. Thielemann [Thi91] it will not be addressed here. A detailed discussion of the QCD phase transition occurring in the early universe can be found in [Mal91]. Here we outline a simplified model.

A key requirement of the phase transition, in order for it to influence primordial nucleosynthesis, is that it be first order. The reason for this is that there must be a coexistence of the quark-gluon and hadron plasmas if baryon inhomogeneities are to arise. Coexistence between the different phases, however, requires a constant temperature, T_c, at which both phases are in pressure and chemical equilibrium. It is the latent heat release from the first order phase transition which compensates, for some finite time, the universal expansion thereby allowing the temperature to remain constant at T_c.

The thermodynamic potential, P, for each of the phases can be written

$$P = -T \ln(Z) \quad , \tag{6}$$

where Z is the grand partition function. The baryon number density n_b for each phase can be calculated from P through

$$n_b = -\frac{1}{V}\left(\frac{dP}{\partial \mu}\right)_{V,T} \quad , \tag{7}$$

where V is the volume and μ the chemical potential. After calculating the thermodynamic potential for each phase [Ful88] the ratio of the baryon number density of the quark-gluon plasma to that in the hadron plasma is given by

$$R \equiv \frac{n_b^q}{n_b^h} \approx \frac{2}{9}\left(\frac{\pi^3}{8}\right)^{1/2} \frac{e^{m/T_c}}{(m/T_c)^{3/2}} \quad , \tag{8}$$

where m is the baryon mass. The main reason for the resulting baryon inhomogeneities can be determined from eq. (8). Since baryon number is carried by light quarks in one plasma but by heavy hadrons in the other, the exponential mass term suppresses baryon number in the hadron phase.

Although T_c is expected to be in the vicinity of 100 MeV, uncertainties in our current understanding of the quark-gluon plasma do not allow for an accurate calculation. This lack of certainty has important implications for the determination of R, the ratio of baryon number densities. For example, if $T \sim$ 50 MeV then $R \sim 50$. Alternatively, if $T \sim 10$ MeV then $R \sim 10$. Such a difference in R has important consequences for nucleosynthesis.

The simple model outlined here shows how baryon inhomogeneities in the early universe could arise. However, even larger inhomogeneities than those indicated by eq. (8) could take place if baryon-chemical equilibrium between the two phases is not achieved. Since initially the hadron phase has zero net baryon number, then baryon-number transport across the phase boundary must take place. If this transport is not efficient then very large density contrasts such as $R \sim 10^4$ become feasible. Finally, we note that even if the phase transition went to completion before the universal expansion were to again cause cooling, the ever decreasing volume of the quark-gluon bubble would have an ever increasing baryon-number density. Upon completion of the transition this effect means a remnant inhomogeneity with a particular density profile would still be produced.

A schematic representation of these events is shown in Fig. 4, and the relevant times and temperatures are also indicated. As stated earlier, the key point as far as interesting nucleosynthesis is concerned, is the effect of neutron-proton segregation [App85]. Such a segregation occurs as a consequence of the proton's electric charge. This allows the protons to undergo large Coulomb scattering

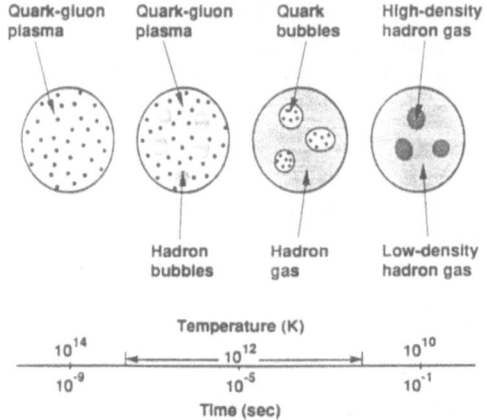

Fig. 4. Schematic representation of the QCD phase transition

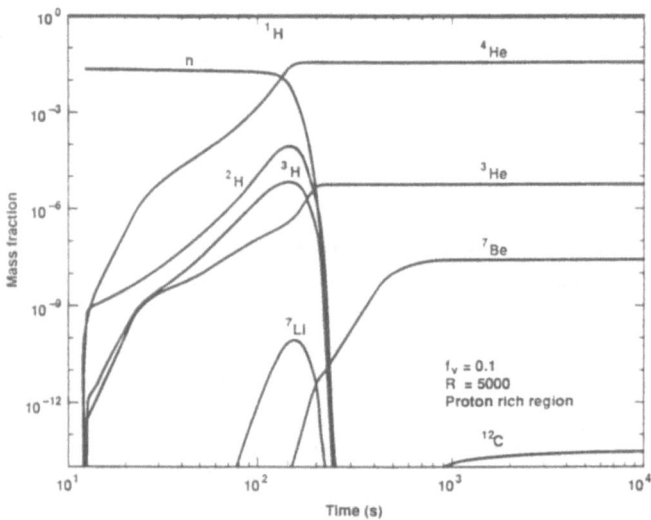

Fig. 5. Evolution of isotopic abundances (mass fractions) in proton-rich region [Kaw91]

effects, essentially trapping most of them in the high-density regions. The neutrons, on the other hand, undergo much smaller scattering effects which consequently allows them to fill all space. The resulting scenario is one which is far removed from homogeneous SBB theory. Not only do we now have density inhomogeneities but also chemical inhomogeneities.

Figs. 5 and 6 show nucleosynthesis yields as a function of time from a typical inhomogeneous calculation [Kaw91] in the proton-rich (high-density) and neutron-rich regions, respectively. Here a density ratio $R = 5000$, and a volume fraction in the high density phase $f_v = 0.1$ is adopted. In order to obtain the final nucleosynthesis yields one must take a mass average of the two regions.

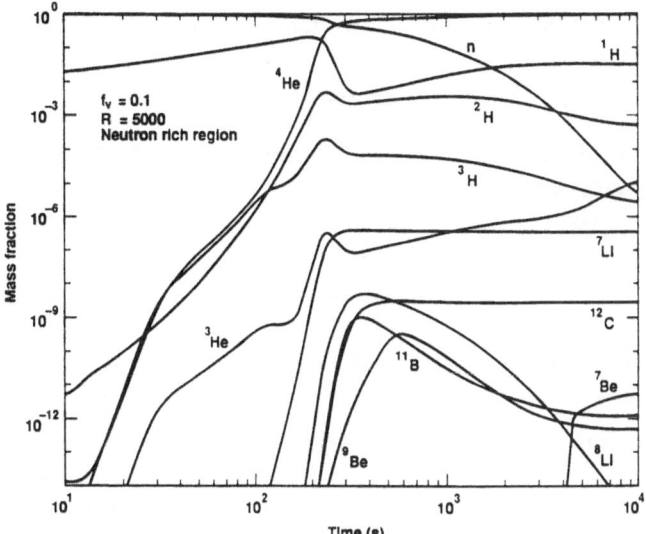

Fig. 6. Evolution of isotopic abundances (mass fractions) in neutron-rich region [Kaw91]

Detailed calculations of the nucleosynthesis arising from such chemical and density inhomogeneous models showed that an $\Omega_b = 1$ universe could be reconciled with the observed abundances of ^{2}H, ^{3}He, and ^{4}He [App87, Alc87, Ful88, Mal88]. This is a remarkable departure from the SBB where the same isotopes constrain the baryon density to be $\Omega_b < 0.15$. One possible difficulty with these early calculations, however, was the apparent large overabundance of ^{7}Li they predicted of at least a factor of 5 to 50, depending on what was adopted as the primordial lithium value (Pop I or Pop II).

[Mal88] showed how the late-time diffusion of neutrons could drastically reduce the ^{7}Li abundance, depending on how readily the neutrons diffused back into the proton-rich regions of the universe. However, as noted by [Mal88], if the neutrons back-diffuse too soon the resulting nucleosynthesis of the other isotopes is also changed. The most important of the other isotopes in this regard is ^{4}He. Several studies have subsequently ruled out large regions of the parameter space of the inhomogeneous models based on the overproduction of ^{4}He arising from the early back-diffusion of neutrons [Kur88]. Because of this effect the case for an $\Omega_b = 1$ universe consistent with the inferred primordial abundance of ^{4}He abundance was weakened.

Complicated diffusion processes, however, could take place near the end of the nucleosynthesis epoch [Alc90] further clouding the issue. In addition, if the primordial lithium abundance is significantly higher than hitherto believed, the baryonic-density limits imposed by the inhomogeneity models will have to be re-considered. What is already clear is that the baryon density limits imposed by the inhomogeneity models are substantially larger than those imposed by the SBB theory. Even adopting the small lithium abundances derived from Pop II stars, values of $\Omega_b \sim 0.3$ remain very feasible. Increased limits on Ω_b

has very important implications and is the most important point to be drawn from the QCD phase transition studies.

It is interesting to note that for a given Ω_b the inhomogeneous models produce substantially less ^{4}He ($3 - 5\%$ by mass) than the SBB theory. If Y_p is found to be less than 23.5% then the inhomogeneous models could adequately account for the discrepancy. Of course, high values for primordial lithium also find a natural solution in the inhomogeneous models.

5. Neutrinos in the Early Universe

Neutrinos play a very important role in the early evolution of the universe. The results from the LEP and SLC e^+e^- colliders have effectively ruled out an extra generation of light neutrino family beyond the three we presently know to exist. This eliminates some of the uncertainty from the SBB model. However, there are several extensions to the standard cosmological and particle physics models which would allow the known neutrinos to perturb primordial nucleosynthesis. We now discuss some of these effects.

5.1 Sterile Neutrinos

If the presence of right-handed (RH) "sterile" neutrinos, ν_i^R, ($(SU(2)_L \otimes U(1)_Y$ singlets) are allowed, then interesting scenarios can arise when mechanisms for their conversion into left-handed (LH) "active" neutrinos, ν_i^L, are introduced. Hereafter when we refer to RH neutrinos we mean both RH neutrinos and LH anti-neutrinos. By "sterile" here we mean that they do not transform under the standard model gauge group.

In the SBB model RH neutrinos play no role. Even if they exist they are assumed to have decoupled at $T > 100$ GeV. The neglect of RH neutrinos in nucleosynthesis calculations is to a large extent justified from consideration of the statistical weight, g, of the relativistic particles present before and after the QCD phase transition. Since the evolution of the universe through phase-transition and particle-annihilation epochs is characterized by constant *comoving* entropy density, the amount of heating of the LH sea relative to the RH sea is given by $T_{LH}/T_{RH} = (g_1/g_2)^{1/3}$, where the subscripts 1 and 2 refer to the times before and after the annihilation epoch, respectively, and g_1 and g_2 are the effective statistical weights of relativistic particles at these times. Prior to confinement the statistical weight in relativistic particles is $g_1 \sim 56.5$; whereas, after the quarks are confined the only nearly relativistic strongly interacting particles will be pions, so that $g_2 \sim 17.25$ and $T_{LH}/T_{RH} \sim 1.5$. This implies that at the time of nucleosynthesis the relative energy densities in a LH and RH neutrino species will be $\rho_{LH}/\rho_{RH} \sim 5$, so that a sterile neutrino species counts as less than $1/5$ of an additional neutrino flavor. Muons and pions should annihilate or drop out of equilibrium at roughly $T \sim 100$ MeV, and if we add their statistical weight to the differential heating of the LH and RH seas then a RH neutrino species would count for only about 0.1 of an additional flavor.

On the other hand, if the decoupling of the RH sea is *after* muon and pion annihilation then the RH components count for almost a full extra neutrino flavor, which would clearly be incompatible with the limit $N_\nu \leq 3.4$ discussed in Sect. 2. Thus, the RH components of the neutrino would have to decouple prior to the QCD epoch, that is for $T > T_c$.

It is important to note here that because this argument is based on added degrees of freedom contributed by the RH neutrinos we must assume that neutrinos are purely Dirac particles. Any Majorana contribution to the neutrino mass term will invalidate the argument.

A particle species, x, is decoupled from the background plasma when

$$\sum_i n_i \sigma \ll H \ , \tag{9}$$

where σ is the cross section of the interaction processes involving x, and H is the Hubble parameter. This latter parameter can re-written as

$$H = \left(\frac{4\pi^3}{45}\right)^{1/2} m_{\mathrm{pl}}^{-1} g_{\mathrm{eff}}^{1/2} T^2, \tag{10}$$

where m_{pl} is the Planck mass, and $g_{\mathrm{eff}} = \sum_{\mathrm{Bose}} g_{\mathrm{Bose}} + \frac{7}{8} \sum_{\mathrm{Fermi}} g_{\mathrm{Fermi}}$.

Several studies have used the necessity that any RH neutrino species be decoupled prior to the QCD phase transition to impose constraints on the mass, electromagnetic, oscillation and weak interaction properties of neutrinos. We discuss the mass constraint in more detail.

Consider a neutrino of mass m_ν and energy E_ν. One can show that the cross section for helicity-flip in any scattering event is given by

$$\sigma_R \sim \sigma_L \left(\frac{m_\nu}{E_\nu}\right)^2 \ , \tag{11}$$

where $\sigma_L \sim G_F^2 T^2$ is the cross section for the normal non-flip scattering event (G_F is the Fermi coupling constant). Demanding that eq. (9) be valid at $T = T_c$ and adopting statistical weights relevant to the epoch prior to quark confinement, leads to the following constraint on the neutrino mass [Ful91]

$$m_\nu \ll 150 \ \mathrm{keV} \left(\frac{100 \ \mathrm{MeV}}{T_c}\right)^{1/2} \ . \tag{12}$$

Assuming $T_c \sim 100$ MeV, we find a cosmological limit of $m_\nu \ll 150$ keV. The present experimental upper limits for the muon and tau neutrino are 250 keV and 35 MeV, respectively [Agu88].

Finally, we note that if the tau neutrino is too massive ($m_\nu \gtrsim 1$ MeV) then it can become non-relativistic during nucleosynthesis [Kol82]. In such circumstances the universe is matter dominated as it passes through nucleosynthesis and the nucleosynthesis yields can be perturbed from their SBB values. A detailed study of this latter point is currently in progress.

5.2 Neutrino Oscillations

Neutrino interactions on matter contribute a new term to the mass matrix which is not present in the vacuum. This extra interaction term can give rise to matter-induced (resonant) neutrino oscillations. In the solar interior the charged current neutrino-electron scattering introduces a term in the mass matrix which singles out electron neutrinos. It is widely believed that this effect may resolve the apparent discrepancy between the neutrino flux predicted by the standard solar model and the value measured by the ^{37}Cl experiment.

In order to analyze resonant neutrino oscillations one must introduce a difference in the *effective* squared mass of the neutrinos, viz., $\Delta_{\text{eff}} \equiv m_2^2(\text{eff}) - m_1^2(\text{eff})$, where $m_i(\text{eff})$ is given by the neutrino vacuum mass plus an effective contribution to the neutrino mass arising from interactions with the plasma. Consider the contribution to Δ_{eff} for the two component system in the primordial plasma. There are contributions from the weak charged currents $\Delta_{\text{eff}}^{W^\pm}$ mediated by $W^\pm$ bosons and also from weak neutral currents $\Delta_{\text{eff}}^{Z^0}$ mediated by Z^0 bosons. These are given by [Sav91]

$$\Delta_{\text{eff}}^{W^\pm} = 2\sqrt{2}G_{\text{F}}\left(n_{e^-}(t) - n_{e^+}(t) - n_{x^-}(t) + n_{x^+}(t)\right)p \quad , \tag{13}$$

$$\Delta_{\text{eff}}^{Z^0} = 2\sqrt{2}G_{\text{F}}\left(n_{\nu_e}(t) - n_{\bar{\nu}_e}(t) - n_{\nu_x}(t) + n_{\bar{\nu}_x}(t)\right)p \quad , \tag{14}$$

where p is the neutrino momentum, $n_i(t)$ are the number densities of charged leptons, (which are independent of neutrino oscillations), and $n_{\nu_i}(t)$, $n_{\bar{\nu}_i}(t)$ are the number densities of ν_i and $\bar{\nu}_i$. The time dependence of the neutrino densities arises not only from the expansion of the universe but also from the oscillations themselves.

It is normally assumed in studies of the early universe that the total lepton number of the universe is very small and effectively zero. In terms of the number density of charged leptons n_i, neutral leptons n_{ν_i} and photons n_γ, the lepton number for each generation is defined to be $L_i = (n_i - n_{\bar{i}} + n_{\nu_i} - n_{\bar{\nu}_i})/n_\gamma$.

Under the assumption that the distribution functions for the neutrinos are Fermi-Dirac in nature, then the lepton numbers can be characterized by a set of degeneracy parameters, $\zeta_i = \mu_i/T$ where μ_i is the chemical potential of each neutrino species, and T is the temperature. Non-zero lepton numbers of the universe can effect nucleosynthesis primarily in two ways (see [Mal91] and refs. therein). First, the excess energy density in a neutrino degenerate sea leads to an increased expansion rate for the universe which subsequently allows less time for the neutrons to decay into protons. Secondly, the non-zero electron neutrino degeneracy can directly effect the equilibrium n/p ratio at weak freeze out.

From eqs. (13) and (14) it can be seen that if there is no initial neutrino degeneracy, then neglecting any higher order corrections we have $\Delta_{\text{eff}}^{Z^0} + \Delta_{\text{eff}}^{W^\pm} = 0$. That is, for most practical purposes the primordial plasma of a non-degenerate universe is "neutral" as far as neutrino oscillations are concerned and no net flavor changing process occurs. In a neutrino degenerate universe, however, resonant neutrino oscillations can become important. The neutral currents make the dominant contributions to the effective mass differences between the neutrino species, that is $\Delta_{\text{eff}}^{Z^0} > \Delta_{\text{eff}}^{W^\pm}$ for $L_i \gtrsim 10^{-9}$. The effect of these neutral current resonant oscillations on the SBB model limits for the leptonic charge of the universe have been investigated [Sav91]. It is found that for a large range of Δ and the vacuum mixing angle, θ_V, significant neutrino flavor oscillations can occur in the early universe subsequently altering the degeneracy of individual neutrino species.

Since resonance conditions can occur prior to the nucleosynthesis epoch, it becomes difficult to determine limits on the original neutrino asymmetry of the universe from studies of primordial nucleosynthesis. This is compounded by two effects. The first of these is the re-heating of the universe arising from the equilibration of the neutrino distribution functions. Second, the resonance may occur both before or after weak decoupling, and in the latter case the neutrino oscillations will again give rise to non-thermal energy distributions. The importance of the time at which neutrino transformations occur is illustrated in Fig. 7. In the presence of neutrino oscillations the best limits on the degeneracy of the universe are not in fact those derived from nucleosynthesis studies but from the age of the universe [Fre83], $|\zeta_i| \leq 140$. These latter constraints are significantly less stringent than those imposed by nucleosynthesis studies without neutrino oscillations.

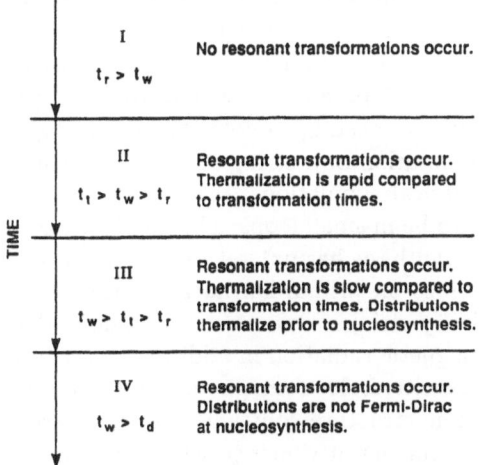

Fig. 7. The modifications of the neutrino distribution functions in four different time epochs. Here t_w, t_r, t_t and t_d are the weak interaction, the resonant, the transformation, and the universal expansion time scales, respectively [Sav91]

Neutrino oscillations in a neutrino degenerate universe should greatly perturb primordial nucleosynthesis. However, the nucleosynthesis yields will be a complicated function of the unknown parameters, θ_V, Δ, the ζ_i's, as well as η_{10}, and detailed calculations have not yet been carried out. It would be worthwhile to investigate the nucleosynthesis yields for a range of these parameters in order to determine the parameter space excluded by the observed primordial abundances.

6. Cosmic Strings

In the simplest scenario ordinary cosmic strings are formed from the breaking of U(1) symmetry by a complex scalar field. The most important cosmic-string parameter is its mass per unit length, μ. This is related to the mass scale at symmetry breaking, ϕ, through $\mu = \phi^2/\hbar$. The mass per unit length required for galaxy formation is believed to be given by $G\mu \sim 10^{-6}$, where G is Newton's constant. A review of cosmic strings and their role in galaxy formation can be found in [Vil85].

The production of cosmic strings in the early universe may also have profound implications for nucleosynthesis. Cosmic strings can undergo relativistic oscillations which subsequently result in the emission of substantial gravitational radiation, the rate of which is given by $L \sim \gamma_g G\mu^2$, where γ_g is a dimensionless number which depends on the shape of the loop (typically, $\gamma_g = 50 - 100$). If the rate of such radiation is large enough a significant fraction of the universal energy density during the radiation dominated era could be in the form of gravitons. The energy density in gravitons is given by

$$\rho(t) \sim \gamma_g^{-1/2}[2\mu^{1/2}G/3t^2] \ln(t/t^*) \; , \qquad (15)$$

where $t^* \sim 10^{-30}$s is the time at which frictional forces become unimportant.

Using the fact that any new radiation source must not contribute more than $\sim 7\%$ of the background radiation at the time of nucleosynthesis, important constraints can be placed on the cosmic string parameters such as the mass per unit length $(G\mu \lesssim \times 10^{-5})$ [Dav85]. It is interesting to note that recent high-resolution simulations [Ben88] of the evolution of a string network differ in a quantitative manner from earlier numerical simulations. Principally, a much larger fraction of the strings are found to be in small loops. The implications of these new results for primordial nucleosynthesis has not yet been worked out.

The effects of superconducting strings on primordial nucleosynthesis can be even more dramatic than for ordinary strings. One reason for this is that superconducting strings emit electromagnetic radiation in addition to gravitational radiation. The rate of radiation emitted electromagnetically is given by $L_{em} = (j/j_c)^2\gamma_{em}\alpha\mu$, where α is the fine structure constant, j is the current carried by the string, and j_c the critical current at which growth is terminated by particle production.

As was the case for gravitational radiation, the additional contribution to the energy density in the form of electromagnetic radiation can perturb the nucleosynthesis yields. The ratio of energy emitted in electromagnetic radiation relative to gravitational radiation is given by (assuming $\gamma_{em} = \gamma_g$)

$$f = (j/j_c)^2\alpha(G\mu)^{-1} \; . \qquad (16)$$

If in fact the superconducting string can carry a very large current, a more profound effect may take place [Mal89]. The moving string and its magnetic field can deflect charged particles away from the immediate vicinity of the string, somewhat like the deflection of the solar wind around the earth. In the rest frame of the string a region completely free of charged particles can develop.

In such a situation, the universe at the onset of nucleosynthesis can consist of proton-rich areas (regions in which a passing string has pushed in additional charge particles), and neutron-rich areas (the wakes of strings into which the neutrons have rapidly diffused). Such a situation is resembles that of the in-homogeneous models discussed in Sect. 3, and the nucleosynthesis yields from both models are very similar.

The calculations of primordial nucleosynthesis in the presence of supercon-ducting cosmic strings carried out to date have not been dynamic. That is, they do not take into account successive passages of the string at any point in space-time, nor do they include diffusion effects. Both these processes will actually occur *during* the nucleosynthesis epoch. Further, it has been pointed out that at relatively late times ($\sim 10^6$s) catastrophic "quenching" of supercon-ducting strings could result in a significant flux of high energy γ-rays capable of photofissioning the light primordial isotopes produced earlier [Hod88]. This clearly is yet another potential source of alteration to the nucleosynthesis yields.

Table 2. Non-Standard Nucleosynthesis Models

Effect	Abundances Relative to SBB
Baryon Inhomogeneities	$\downarrow\uparrow^4$He, $\uparrow^7$Li, $\uparrow$CNO
Particle Decay	$\uparrow^3$He, $\downarrow^4$He, $\uparrow^6$Li
Neutrino Degeneracy	$\downarrow\uparrow^4$He
Neutrino Oscillations	$\downarrow\uparrow^4$He
Neutrino Electromagnetics	$\uparrow^4$He
Massive Neutrinos	$\uparrow^4$He
RH Weak Interactions	$\uparrow^4$He
Cosmic Strings	$\uparrow^4$He (normal); $\downarrow^4$He, $\uparrow^7$Li (SCS)
Anisotropy	$\uparrow^4$He
Time Varying Constants	$\downarrow\uparrow^4$He
Primordial Black Holes	$\uparrow^3$He, $\downarrow^4$He, $\uparrow^6$Li
Mirror Matter	$\uparrow^4$He

7. Conclusions

We have reviewed aspects of some non-standard primordial nucleosynthesis models. In doing so we have investigated the robustness of the standard model by detailing how far one must perturb different aspects of the cosmological model in order to spoil agreement with observation. Also, we have highlighted what constraints one may in turn impose on physics beyond the standard particle-physics model. In Table 2 we summarize the most widely discussed non-standard primordial nucleosynthesis models found in the literature. In column one the type of non-standard model under investigation is listed, and in column two the main effects of this particular model on the nucleosynthesis yields relative to the standard model predictions are given. Clearly the limits on the inferred primordial abundances of the light isotopes allow one to turn the argument around in order to constrain any variant on the SBB model. We have not listed the individual non-standard models in any particular preference or relevance to reality (if any!).

Future observations currently planned or on the next generation of telescopes should allow better discrimination between the different nucleosynthesis models, and verify or rule out extensions to the standard cosmological and particle physics models. Searches for both baryonic and exotic dark matter are already underway or are at the planning phase. Detection of either would clearly have an important impact on our understanding of primordial nucleosynthesis.

Table 2 represents the wealth of information that can be obtained from investigations of primordial nucleosynthesis beyond that of the standard model. We hope that the reader has become convinced that much has, and will continue to be, gained from such studies.

This work was performed under the auspices of the U.S. Department of Energy by the Lawrence Livermore National Laboratory under contract number W-7405-ENG-48. I would like to thank all of my collaborators whose work makes up the body of this review.

References

Agu88	M. Aguilar-Benitez, et al. (Particle Data Group): Phys. Lett. **B204**, 1 (1988)
Alc87	C.R. Alcock, G.M., Fuller, G.J., Mathews: Ap.J. **320**, 439 (1987)
Alc90	C.R. Alcock, D.S. Dearborn, G. M. Fuller, G. J. Mathews, B. S. Meyer: Phys. Rev. Lett. **64**, 2607 (1990)
Ale90	The ALEPH collaboration: Cern preprint PPE/90-104 (1990)
App85	J.H. Applegate, C. Hogan: Phys. Rev. **D30**, 3037 (1985)
App87	J.H. Applegate, C.J. Hogan, R.J. Scherrer: Phys. Rev. **D35**, 1151 (1987) ; J.H. Applegate, C.J., Hogan, R.J. Scherrer: Ap. J. **329**, 572 (1988)
Baa88	D. Baade, P. Magain: Astron. Astrophys. **194**, 237 (1988)
Baa91	D. Baade, S. Cristiani, T. Lanz, R.A. Malaney, K.C. Sahu, G. Vladilo: Astron. Astrophys., submitted (1991)
Ben88	D.P. Bennett, F.R. Bouchet: Phys. Rev. Lett. **60**, 257 (1988)
Boe85	A.M. Boesgaard, G. Steigman: Ann. Rev. Astron. Astrophys. **23**, 319 (1985)
Cau88	G. Caughlan, W.A. Fowler: Atom. Dat. Nucl. Dat. Tab. **40**, 291 (1988)
Cay84	R. Cayrel, G. Cayrel de Strobel, B. Campbell, W. Dappen: Ap. J. **283**, 205 (1984)
Dav85	R.L. Davis: Phys. Lett. **B161**, 285 (1985); D.P. Bennett: Phys. Rev. **D33**, 872 (1986); H.M. Hodges, M.S. Turner: Fermilab preprint 88/115-A (1988)
Del89	C.P. Deliyannis, P. Demarque, S.D. Kawaler, L.M. Krauss, P. Romanelli: Phys. Rev. Lett. **62**, 1583 (1989); C.P. Deliyannis, P. Demarque, S.D. Kawaler: Ap. J. Supp. **73**, 21 (1990)

Fer84	R. Ferlet, M. Dennefeld: Astron. Astrophys. **138**, 303 (1984)
Fre83	K. Freese, E.W. Kolb, M.S. Turner: Ap. J. **148**, 3 (1983)
Ful88	G.M. Fuller, C.R., Alcock, G.J. Mathews: Phys. Rev. **D37**, 1380 (1988)
Ful90	G. Fuller, D. Boyd: preprint (1990)
Ful91	G. Fuller, R.A. Malaney: Phys. Rev. D. (1991) in press
Hob88	L.M. Hobbs, C. Pilachowski: Ap. J. **334**, 734 (1988)
Hod88	H.M. Hodges, J. Silk, M.S. Turner: Fermilab preprint **87/153-A** (1988)
Kaw91	L. Kawano, W.A. Fowler, R.W. Kavanagh, R.A. Malaney: Ap. J. (1991) in press
Kol82	E.W. Kolb, R.J. Scherrer: Phys. Rev. **D25**, 1481 (1982)
Kur88	H. Kurki-Suonio, R.A. Matzner, J.M. Centrella, T. Rothman, J.R. Wilson: Phys. Rev. **D38**, 1091 (1988); H. Kurki-Suonio, R.A. Matzner, K.A. Olive, D.N. Schramm: Ap. J. **353**, 406 (1990)
Mal88	R.A. Malaney, W.A. Fowler: Ap. J. **333**, 14 (1988)
Mal89	R.A. Malaney, M.N. Butler: Phys. Rev. Lett. **62**, 117 (1989)
Mal90	R.A. Malaney, C.R. Alcock: Ap. J. **351**, 31 (1990)
Mal91	R.A. Malaney, G.J. Mathews: Physics Reports, submitted 1991
Mam89	W. Mampe, P. Ageron, C. Bates, J.M. Pendlebury, A. Steyerl: Phys. Rev. Lett. **63**, 2173 (1989)
Mat90	G.J. Mathews, C.R. Alcock, G.M. Fuller: Ap. J. **349**, 449 (1990)
Oli90	K.A. Olive, D.N. Schramm, G. Steigman, T.P. Walker: Phys. Lett. **B236**, 454 (1990)
Olv90	K.A. Olive: Phys. Reps. 190, 181 (1990)
Opa90	The OPAL collaboration: Cern preprint **PPE/90-81** (1990)
Pag89	B.E.J. Pagel, E.A. Simonson: Rev. Mex. Astr. Astrofis. **18**, 153 (1989)
Pag90	B.E.J. Pagel: In *Nobel Symposium 79- The Birth and Early Evolution of our Universe*, ed. by B.S. Skagerstam (Graftavallen, Sweden 1990)
Pei74	M. Peimbert, S. Torres-Peimbert: Ap. J. **193**, 327 (1974)
Sah88	K.C. Sahu, M. Sahu, S.R. Pottasch: Astron. Astrophys. Lett. **207**, L1 (1988)
Sav91	M.J. Savage, R.A. Malaney, G.M. Fuller: Ap. J. (1991) in press
Smi90	V.V. Smith, D.L. Lambert: Ap. J. Lett. **345**, L75 (1989)
Spi82	F. Spite, M. Spite: Astron. Astrophys. **15**, 357 (1982)
Ste89	G. Steigman, J. Gallagher, D.N. Schramm: Comm. Astrophys. **14**, 97 (1989)
Thi91	F.-K. Thielemann, J.H. Applegate, J.J. Cowan, M. Wiescher: In *Nuclei in the Cosmos*, ed. by H. Oberhummer (Springer, Berlin 1991) this volume
Tri87	V. Trimble: Ann. Rev. Astron. Astrophys. **25**, 425 (1987)
Vid87	A. Vidal-Madjar, P. Andreani, S. Cristiani, R. Ferlet, T. Lanz, G. Vladilo: Astron. Astrophys. **177**, L17 (1987)
Vil85	A. Vilenkin: Phys. Rep. **121**, 263 (1985); Ed. by F.S. Accetta, L.M. Krauss: *Cosmic Strings - The Current Status* (World Scientific, Singapore 1990)
Wit84	E. Witten: Phys. Rev. **D30**, 272 (1984)

Production of Heavy Elements in Inhomogeneous Cosmologies

Friedrich-Karl Thielemann[1], *James H. Applegate*[2], *John J. Cowan*[3], *Michael Wiescher*[4]

[1] Harvard-Smithsonian Center for Astrophysics, 60 Garden Street, Cambridge, MA 02138, USA
[2] Department of Astronomy, Columbia University, New York, NY 10027, USA
[3] Department of Physics and Astronomy, University of Oklahoma, Norman, OK 73019, USA
[4] Department of Physics, University of Notre Dame, Notre Dame, IN 46556, USA

1. Introduction

The incorporation of quantum chromodynamics into big bang cosmology predicts that the universe shifted from a quark-gluon plasma to a meson gas when it was roughly $10\,\mu s$ old. If this change is due to a first order phase transition, baryon density inhomogeneities can be produced and survive until nucleosynthesis. At that point they can affect the primordial abundances resulting from element formation in the early universe. In this article we undertake a detailed study of one of the proposed signatures of inhomogeneous nucleosynthesis – the production of heavy elements by neutron capture – and find that heavy elements are produced, but the level of production is probably too low to be observable.

2. The QCD Transition in the Early Universe

The development of quantum chromodynamics (QCD) and its general acceptance as the underlying dynamical theory of the strong interactions has been one of the major advances in physics during the past two decades. Perhaps the most interesting prediction made by QCD is the existence of a new state of matter. Ordinary matter, composed of protons and neutrons, is predicted to dissolve into a quark-gluon plasma when compressed to a density of several times nuclear matter density or when heated to a temperature of order 150 MeV. The incorporation of QCD into the standard model of big bang cosmology predicts that the universe existed in the quark-gluon plasma state until the adiabatic expansion cooled it sufficiently for the plasma to condense into a meson gas. If the shift from the plasma to the meson gas occured in a first order phase transition, then baryon density inhomogeneities will be produced. Under certain conditions, these inhomogeneities can be of large enough amplitude and comoving length scale to survive until the epoch of nucleosynthesis and leave their imprint on the pattern of elements produced in the early universe.

A first order phase transition in the early universe, even if it occurs through a series of near equilibrium configurations, involves a large departure from homogeneity because of phase separation in the two phase region. In addition, a first order phase transition can support a departure from thermal equilibrium if it supercools before nucleating. The significance of these considerations is that departures from homogeneity and thermal equilibrium can be recorded and remembered as inhomogeneities in the distribution of baryons. Under some conditions these inhomogeneities can be of large enough amplitude and length scale to survive until nucleosynthesis and alter the abundances of the elements. Since supercooling and phase separation do not occur in thermal ionization or second order phase transitions, we do not believe that the effects we consider will occur unless the QCD transition in the early universe was a first order phase transition.

Analytical arguments [Yaf82, Bor83, Tom84] and numerical calculations using Monte-Carlo techniques on a spacetime lattice [Fuk89, Uka89] have shown that the transition from color confinement to deconfinement in pure gluon matter, which corresponds to infinite quark masses, occurs through a first order phase transition. In the opposite limit, QCD with massless quarks possesses an exact chiral symmetry which is broken in the low temperature phase and restored in the high temperature phase. The restoration of chiral symmetry in the transition to the plasma state requires a phase transition, but the transition can be either first or second order depending on the number of massless flavors [Pis84]. Nature lies somewhere in between these two extremes. Physical QCD contains nearly massless up and down quarks, and a fairly light strange quark. The presence of finite quark masses means that no exact chiral symmetry exists in physical QCD, so that a phase transition is not required to go to the plasma state. The fact that the quarks are light means that they are dynamically important. Therefore, the calculations that show a first order phase transition in pure gluon matter, while interesting, do not apply to the real world. Monte-Carlo calculations using dynamical but fairly heavy quarks do not show a phase transition, but a smooth transition analogous to the thermal ionization of a gas. A recent calculation [Bro90] which uses quark masses close to the physical values but an unphysically coarse lattice also fails to show a phase transition. Thus all three possibilities for the QCD transition: first order phase transition, second order phase transition, or ionization, remain open. For the remainder of this paper we consider the possible evolution in case that the QCD transition in the early universe was a first order phase transition.

Early work on the cosmological consequences of a first order phase transition in QCD [Suh82, Lod83, Deg84, Kur85, Kaj86] focused on the effect of the two phase region on the expansion of the universe, and on the construction of a general picture of the supercooling, nucleation of new phase, and conversion of material to low temperature phase that would occur. This scenario describes the conversion of a homogeneous quark-gluon plasma to a homogeneous meson gas.

Witten [Wit84] realized that a first order phase transition in QCD, occurring in or near equilibrium could produce drastic changes in the standard model of cosmology. He showed that baryon number would be concentrated in the quark-gluon plasma, if baryon chemical equilibrium could be achieved, and argued that the transport of baryon number at the phase boundary was such that baryon number concentration could still occur, even if complete chemical equilibrium could not be established. This led him to suggest that large amplitude baryon density inhomogeneities could be produced in the QCD transition, and to speculate that, if bulk strange quark matter was the true ground state of strongly interacting matter, then nuggets of strange quark matter could be produced in the early universe and survive today as the dark matter in galaxies and clusters of galaxies. Subsequent work has shown that the baryon number concentrating effect was overestimated in Witten's paper [App85], and that quark nuggets would not survive to the present [Alc85]. Thus, strange quark nuggets are not considered to be attractive dark matter candidates. However, the production of baryon density inhomogeneities is a characteristic feature of a first order transition in QCD.

A major difference between the QCD transition in the early universe and in the laboratory is that gravitation introduces a dimensional parameter into the cosmological problem which is not present in the laboratory. The natural scales of length and time in the early universe are not characteristic of pure QCD, but are determined by a combination of QCD and gravity. The critical temperature and energy density (excluding the photon and lepton contributions) are the same in the early universe as they are in the laboratory. The duration of the phase transition is determined by the expansion rate of the universe and the density of latent heat in the transition. Since the density of latent heat and the energy density are about the same, the duration of the phase transition is comparable to the age of the universe at the onset of the transition, which gives $\Delta t \sim t_c \sim (G\rho)^{-1/2} \sim 10^{-5}$s, where Δt is the duration of the phase transition, t_c is the age of the universe at the onset of the transition, and ρ is the mass density of the universe at the transition. The characteristic length scale of the inhomogeneities is determined by the horizon distance $ct_c \sim 3$km and the amount of supercooling $\Delta T/T_c$ in the transition. Since the rate of nucleation of the low temperature phase is zero at the critical temperature, the universe will continue to expand in the quark-gluon plasma state for a time $\Delta t_N \sim t_c(\Delta T/T_c)$ before rapid nucleation occurs.

Once nucleation occurs, the release of latent heat drives shocks into the quark-gluon plasma, heating it back towards the critical temperature and shutting off further nucleation. Let the typical separation between nucleation sites be L. If Δt_S is the time needed for the shock fronts from different sites to overlap, once rapid nucleation has occurred, the length scale is $L \sim c\Delta t_S$. Detailed calculations [Hog83, Deg84] show that to a very rough approximation $\Delta t_S \sim \Delta t_N$, which gives $L \sim ct_c\Delta T/T_c$. Thus, supercooling of a few percent is equivalent to separation scales between nucleation sites of tens to hundreds of meters.

The universe then expands at temperatures close to T_c, while converting quarks into hadrons at constant entropy. The geometrical structure of possible density inhomogeneities depends on the propagation of the phase boundaries between the hadron and the quark-gluon regions (see e.g. [Pan89]). This in turn is determined by the relative importance of hydrodynamic energy transport versus heat conduction. Heat conduction by neutrinos is the most efficient form, due to the largest mean free path. Under the existing conditions at the time of the phase transition $\lambda_\nu \approx 10$ cm in the quark-gluon region. In case heat conduction by neutrinos dominates the propagation of the phase boundaries, Freese and Adams [Fre90] indicate that the propagation is only spherical until the hadron bubbles reach a size of the order λ_ν. Then instabilities set in and lead to a dendrite or finger like growth. In such a case, even with only one initial nucleation center, the irregular growth would leave irregularly shaped receding quark-gluon (=high baryon density) regions where the phase transitions will occur late within the time interval from the onset to completion of the transition. The average distance l between regions of high baryon density would then be of the order λ_ν. In case hydrodynamic effects dominate the propagation of the phase boundaries, the growing hadron bubbles will stay spherical. The length scale l of the inhomogeneities is then equal to the typical separation between nucleation sites L. Miller and Pantano [Mil89] find that both contributions to the propagation of the phase boundaries – neutrino heat conduction and hydrodynamic energy transport – can be comparable and the outcome is not yet clearly determined. In the following we will analyze the consequencies of the case, where the typical length scale of density inhomogeneities is comparable to the distance between nucleation sites, i.e. of the order of tens to hundreds of meters at the time of the quark-hadron phase transition.

Applegate and Hogan [App85], Bonometto et al. [Bon85], and Sale and Mathews [Sal86] realized that the inhomogeneities could have dramatic consequences for primordial nucleosynthesis. Applegate and Hogan [App85] described a rich set of scenarios for inhomogeneous nucleosynthesis. The most interesting effect they considered is the differential diffusion of neutrons and protons. The neutron, being neutral, has a mean free path roughly 10^6 times that of the proton, because of smaller scattering cross sections with charged particles. Once the weak interactions – transforming neutrons into protons and vice versa via positron and electron captures – fall out of equilibrium, neutrons can diffuse to uniformity over much larger length scales than protons. This has the effect of converting density variations into variations in the local ratio of neutrons to protons. If the density variations are large enough, the neutrons, diffusing into an underdense region, can outnumber the nucleons initially there, producing an environment in which neutrons locally outnumber protons. Nucleosynthesis under neutron rich conditions can give dramatically different results than the standard, proton rich models. The possibilities ranged from ruling out Witten's scenario, to detecting the handiwork of the cosmological QCD transition, to possibly matching the success of the standard model in reproducing the observed abundances of the light elements in a model containing a closure density of baryons. For an in depth treatment of this and other possible scenarios, leading to density inhomogeneities during nucleosynthesis, see [Mal91].

3. The Standard Big Bang

The standard big bang scenario assumes a homogeneous, isotropic expansion in a radiation dominated universe without degenerate or exotic particles. During the expansion, the temperature declines below 1 MeV and eventually slows down the photodisintegration rate of ^{2}H, which is constantly produced via neutron capture on protons. Nucleosynthesis proceeds when a substantial abundance of ^{2}H at $T_{D\gamma} \approx 0.1$ MeV$\approx 1.2 \times 10^9$ K, enables further neutron, proton, and light nuclei capture to form ^{3}H, ^{3}He, ^{4}He, and even heavier nuclei. However, the gaps existing among stable nuclei at mass numbers A=5 and A=8 inhibit the formation of nuclei beyond A=8; therefore the standard big bang can produce only ^{2}H, ^{3}He, ^{4}He, and ^{7}Li in appreciable amounts.

The neutron to proton ratio, which constrains primordial nucleosynthesis, is determined by the conditions during weak decoupling at T$\approx$1 MeV, when the electrons are no longer energetic enough to ensure an equilibrium by the reaction p(e$^-$, ν)n, due to the neutron-proton mass difference of 1.3 MeV, and the positrons needed for the inverse reaction n(e$^+$, $\bar{\nu}$)p are no longer produced by pair creation. Primordial nucleosynthesis conditions are then determined by the particles remaining in thermal equilibrium, initial conditions at time t, and global adiabatic expansion. The initial conditions are set by neutrons, protons, electrons, and photons with densities n_n, n_p, n_e, and $n_\gamma = 2.404/\pi^2(kT/\hbar c)^3$. From charge neutrality it follows that $n_e = n_p \cdot n_p/n_n$ is given by the equilibrium ratio at weak freeze-out.

Therefore, the strength of the standard big bang scenario is that only one free parameter must be specified, the baryon to photon ratio $\eta = n_b/n_\gamma$, to determine all of the primordial abundances, ranging over 10 orders of magnitude (see e.g. [Yan84, Boe85, Kaw88, Oli90]). The parameter η is related to the baryon density $\Omega_b = \rho_b/\rho_c$ in the following way, using all quanitities at present time

$$\rho_b = n_b m_u = \eta_{10} 10^{-10} n_\gamma m_u \propto \eta_{10} T_{\gamma 0}^3 \tag{1a}$$

$$\rho_c = \frac{3H_o^2}{8\pi G} \tag{1b}$$

$$\Omega_b h_{50}^2 = 1.54 \times 10^{-2} (T_{\gamma 0}/2.78K)^3 \eta_{10} \tag{1c}$$

The relation for the critical density follows from the Friedmann equation for the (Hubble) expansion of the universe $H(t)^2 = 8\pi G\rho/3 - k/R^2$ for a flat universe with $k = 0$ and therefore critical density $\rho_c \cdot H_o = h_{50} \times 50$ km s^{-1} Mpc^{-1} and $T_{\gamma 0}$ denote the Hubble constant and the present temperature of the microwave background.

The abundances of individual nuclei depend on η in the following way. A high (baryon) density during the nucleosynthesis phase, i.e. a large η, gives rise to a larger number of capture reactions on ^{2}H and ^{3}He, and consequently leaves less ^{2}H and ^{3}He, but increases the ^{4}He abundance. The behavior of the ^{7}Li abundance is more complex. At low densities ^{7}Li is produced via ^{3}H$(\alpha, \gamma)^7$Li, but is destroyed at higher densities by ^{7}Li$(p,\alpha)^4$He. However, increasing densities lead also to a larger production of ^{7}Be via ^{3}He$(\alpha, \gamma)^7$Be, which is preserved during the nucleosynthesis period and subsequently decays to ^{7}Li. This leads to a predicted minimum in the ^{7}Li/^{1}H ratio of about 10^{-10}, for $2 < \eta_{10} < 4$.

In addition to the η-dependence, the abundances resulting from big bang nucleosynthesis are also dependent upon the number of existing neutrino species and the neutron half life. In the standard scenario the neutron to proton ratio (n/p), resulting from weak decoupling, is always smaller than 1 because of the smaller proton mass. If we assume that all neutrons combine with available protons to form ^{4}He, then the helium mass fractions is given by $X_\alpha = 4Y_\alpha = 4\frac{1}{2}Y_n = 2Y_n = 2X_n$. With $X_n = X_n/(X_n + X_p)$ it follows that $X_\alpha = 2(n/p)/(1+ (n/p))$. Here we made use of the standard notation for the mass fraction X=AY of a nucleus with mass number A and abundance Y, which is useful when the world is more complex than a composition consisting of hydrogen (X), helium (Y) and "metals" (Z). This makes the helium mass fraction X_α a function of the neutron to proton ratio at the time of nucleosynthesis after weak freeze-out at temperature $T_{D\gamma}$. Increasing the number of neutrino species has an effect equivalent to that of a faster expansion, leading to an earlier weak decoupling. This results in a higher (n/p)-ratio and a higher ^{4}He-abundance. A longer beta decay half life of the neutron would have a similar effect.

Primordial abundance information (see [Boe85] and for updates [Pag87, Ban87, Spi86, Hob87, Hob88, Reb88, Oli90, Pag91]) results in the following limits: $(^2$H/H$)\geq 2 \times 10^{-5}$, $(^3$He/H$)\leq 1.2 \times 10^{-5}$, $(^2$H+^{3}He)/H$\leq 11 \times 10^{-5}$, $(^7$Li/H$)\leq2.5\times10^{-10}$, and $0.215 \leq X_\alpha \leq 0.245$. Combining theoretical model predictions with this primordial abundance information, utilizing also the new and very precise neutron half life measurement by Mampe et al. [Mam89] $(\tau_{1/2} = 10.25 \pm 0.015min)$, leads to the following conclusions: $2.6 \leq \eta_{10} \leq 5.0$ and $0.5 \leq N_\nu \leq 3.8$. This can be translated to $0.040 \leq \Omega_b h_{50}^2 \leq 0.077$ and when using a Hubble constant H_o in the range 40-100, this is equivalent to a baryon density in the universe of $0.01 \leq \Omega_b \leq 0.12$.

One major conclusion with regard to the standard big bang model is that, with the inclusion of the recent very precise neutron half life measurement, there remain no nuclear uncertainties which are comparable to those involved in abundance determinations. (The reactions of importance are listed in [Yan84], while Deliyannis et al. [Del89] determined for the first time 1σ abundance uncertainties due to the uncertainties in nuclear cross sections).

4. Inhomogeneous Big Bang Scenarios

4.1 Light Element Production

Density inhomogeneities from the quark-hadron phase transition will remain the same for neutrons and protons before weak decoupling, due to the continuous transformation of protons into neutrons and vice versa by electron and positron capture. The vastly different mean free paths of protons and neutrons create, after weak decoupling, a very proton rich environment in the initially high density regions, while the low density regions are almost entirely filled with diffused neutrons. In order to describe the more complex situation a number of new parameters are introduced, even if we assume that the high and low density regions are of constant density. These parameters are the density ratio R between high and low density regions, the relative volume filling factor f_V of high density zones, the distance l between high density zones (in principle also the geometry of these zones), and η or the total baryon density Ω_b. Using the expression for the ^{4}He abundance $X_\alpha = 4Y_\alpha = 2(n/p)/(1 + (n/p))$ in the proton-rich (high density regions), we expect a smaller He-abundance than in the standard model, due to the efficient neutron diffusion and smaller (n/p)-ratios. However, because of the high densities, ^{2}H and ^{3}He are also underabundant. In the neutron-rich (low density) regions one can derive an analogous expression for the ^{4}He-abundance, assuming now that all protons are incorporated to form ^{4}He, $X_\alpha = 4Y_\alpha = 2(p/n)/(1 + (p/n))$. As a consequence of the very small initial (p/n)-ratio, nucleosynthesis becomes effective after protons are formed by neutron decay, which can lead to $(p/n)=1$ and a large ^{4}He abundance. Also, relatively more ^{2}H and ^{3}He are formed, due to nucleosynthesis at lower densities. Thus, the two regions act in opposite fashions, but the smaller ^{2}H, ^{3}He, and ^{4}He yields in the dominating high density region require an increase of the total (average) density, i.e. η or Ω_b. This led to the hope that, in inhomogeneous scenarios, no non-baryonic matter would be needed to achieve a critical density of the universe, and the primordial abundances of light elements could be obtained with $\Omega = \Omega_b = 1$ [App87, Alc87, Ful88, Aud88, Ter89a,b,c].

Detailed calculations seemed to show that this was possible, with the exception of a problem associated with ^{7}Li, which was overproduced by a factor of 5-100 above its Pop. II value. Back-diffusion of neutrons from the neutron-rich regions after neutron exhaustion in the high density regions, which was originally not considered, was found to be able to reduce this overproduction via ^{7}Be(n, p)^{7}Li(p, α)^{4}He [Mal88]. However, detailed calculations with diffusion showed that this effect was overestimated and that the ^{7}Li problem is not resolved. Late-time dissipation of the high density regions discussed in [Alc90] could cure this problem. However, also the ^{4}He production is too high in $\Omega_b=1$ models. Applegate [App88a] and Applegate, Hogan, Scherrer [App88b] showed analytically that a correct treatment of neutron back diffusion depends crucially on a very fine spatial resolution at the zone boundaries, a resolution which had not been achieved by either of the previous numerical calculations.

Kurki-Sounio et al. [Kur88], Kurki-Sounio and Matzner [Kur89], and Kurki-Sounio et al. [Kur90] performed calculations with such an improved resolution and obtained the result that a ^{4}He overproduction could only be avoided if $\Omega_b < 0.3$. The consensus in the field is that the light element abundances, as they are presently understood, cannot be reconciled with any of the inhomogeneous $\Omega_b = 1$ models, and that these models are ruled out.

Inhomogeneous big bang models did not simplify the topic of primordial nucleosynthesis. The simple standard model with only one free parameter (η or Ω_b), which provided a very good fit to the ^{2}H, ^{3}He, ^{4}He, and ^{7}Li primordial abundances, was replaced by a model which has at least 4 parameters — (R, f_V, l, Ω_b and also the geometry). Nevertheless, if the quark-hadron transition is a first order phase transition, density inhomogeneities will undoubtedly occur. Theoretical estimates of the parameters at the time of the phase transition, are very uncertain. Therefore, a large parameter space had to be allowed for, with constraints being provided by the required fit to primordial abundances. (There are some recent theoretical [Boy89, Mal89] and observational [Rya90] indications that even ^{9}Be might be a product of primordial nucleosynthesis (see also [Pag91]), a fact which could only be explained in an inhomogeneous big bang). One possible final outcome is that the permitted parameter space has to coincide with the standard big bang. Values of l smaller than the proton diffusion length lead to a homogeneous density distribution at nucleosynthesis, i.e. a standard big bang. Very large values of l, larger than the neutron diffusion length, lead to standard type scenarios with varying η and abundances close to the standard model, because an averaging of monotonous functions gives results close to an average η. ^{7}Li however, requiring a value of η which gives its minimum abundance would always be overproduced in the latter case.

4.2 The Formation of Heavy Elements

The barriers that prevent the synthesis of heavy elements in a proton rich environment, such as in standard big bang nucleosynthesis, do not apply in a neutron rich environment. One of the surprises in the first calculations of nucleosynthesis under neutron rich conditions was the large amount of ^{14}C produced [App87, Alc87, Mal87]. Reactions starting with neutron captures on protons ^{1}H$(n, \gamma)^2$H$(n, \gamma)^3$H$(d, n)^4$He$(t, \gamma)^7$Li lead to the production of ^{7}Li and the sequence ^{7}Li$(n, \gamma)^8$Li$(\alpha, n)^{11}$B$(n, \gamma)^{12}$B$(\beta^-)^{12}$C $(n, \gamma)^{13}$C$(n, \gamma)^{14}$C produces carbon. An alternate branching is ^{7}Li$(\alpha, \gamma)^{11}$B. Additional alternative reactions such as ^{7}Li$(t, n)^9$Be$(t, n)^{11}$B, ^{8}Li$(n, \gamma)^9$Li$(\beta^-)^9$Be, ^{8}Li$(d, p)^9$Li, ^{8}Li$(d, n)^9$Be, and ^{8}Li$(d, t)^7$Li might also become important [Boy89, Kaj90a, Bru91]. At three positions (^{7}Li, ^{9}Be, and ^{11}B), proton-induced reactions would prevent the build-up of heavier nuclei in a standard big bang scenario and lead to the destruction of heavy nuclei instead via ^{7}Li$(p, \alpha)^4$He, ^{9}Be$(p, \alpha)^6$Li, or ^{11}B$(p, 2\alpha)^4$He. There are considerable uncertainties associated, because many of these reactions involve unstable nuclei, and also most of the reactions on stable targets in this region of the nuclear chart are not well determined at present [Mal89]. Wiescher, Steininger, and Käppeler [Wie89b] measured the ^{7}Li$(n, \gamma)^8$Li cross section and found a substantial reduction in this rate and the flow to heavier nuclei with respect to previous determinations.

Applegate, Hogan, and Scherrer [App88b] and Applegate [App88a] were able to show that there are possible pathways to heavier nuclei, starting from ^{14}C: e.g. ^{14}C$(\alpha, \gamma)^{18}$O$(n, \gamma)^{19}$O$(\beta^-)^{19}$F$(n, \gamma)^{20}$F$(\beta^-)^{20}$Ne$(n, \gamma)^{21}$Ne$(n, \gamma)^{22}$Ne etc. Wiescher, Görres, and Thielemann [Wie90] demonstrated, however, that the (p, γ) and (n, γ) reactions on ^{14}C are of greater importance for the build-up of heavy elements. Once past the oxygen isotopes, the reactions are dominated by neutron captures and beta decays. The neutron captures are fairly slow until nuclei with mass number A$\sim$60 have been produced, but become rapid afterwards because of larger capture cross sections of heavier nuclei, leading to r-process conditions (for an r-process review see [Cow91]). The neutron captures become sufficiently rapid that an exponential runaway is possible. A nucleus captures neutrons and beta decays until it fissions. Once fission has occured, both daughter nuclei capture neutrons until they have grown to the point of fission. Now four nuclei can capture neutrons and grow. Once they fission, eight nuclei exist. Then sixteen. This runaway fission cycling, first described by Seeger, Fowler, and Clayton [See65], is limited by the neutron supply, which is primarily determined by Ω_b, and by the time required to make a cycle, which is determined by how far the nucleosynthesis path is located from the valley of stability. Applegate [App88a] and Applegate, Hogan, and Scherrer [App88b] used neutron capture cross sections and fission cycle times estimated for the r-process and found that, while the number of fission cycles was quite uncertain, observable, and possibly even excessive, levels of r-process enrichment were possible. Observations of metal-poor halo stars demonstrate the existence of real star-to-star variations in the ratio of the r-process to iron abundances [Gil88]. This nonuniformity in the most metal-deficient stars clearly indicates that a primordial r-process can neither account for their nor the present solar system r-abundances. It could, however, provide a base level of r-process abundances less than, but close to, the lower limit of the observed abundances in stars with metallicities of $\approx 10^{-4} - 10^{-3} \times$ solar.

The results of these analytical estimates inspired us to make a detailed calculation of heavy element nucleosynthesis under neutron rich conditions in the early universe. We begin with pure neutrons just after the weak interactions decouple, and follow the neutron decay and subsequent nucleosynthesis from the production of deuterium to the fission cycling of very heavy elements. The earlier calculations included for light and medium nuclei only reactions for which there existed some information in the literature, involving primarily neutron captures on stable nuclei. With the overwhelming abundance of neutrons in the low density neutron rich regions it is, however, quite feasible that neutron captures on unstable nuclei can bypass these relatively slow beta-decays. For this reason, we have investigated neutron capture reactions on Li, Be, B, C, N, O, F, and Ne isotopes, from stability to the neutron drip line, including the effects of direct capture and all available resonance information from nuclear spectroscopy. The results are presented in Sect. 5.3. For heavier nuclei the results from statistical model calculations [Thi87, Cow91] are applicable.

5. Computational Method

In the following we describe the methods used to calculate the initial conditions, the temperature and density evolution in the high and low density regions during the expansion, the nuclear network utilized to compute the composition changes, and we present the thermonuclear rates involving neutron rich light nuclei, which become important in the inhomogenous big bang applications.

5.1 Inhomogeneous Big Bang in a Two Zone Model

The physical conditions during the expansion of the early universe are set by the Friedmann equation and the first law of thermodynamics

$$\left(\frac{\dot{R}}{R}\right)^2 = H(t)^2 = \frac{8\pi G}{3}\rho \tag{2a}$$

$$\frac{d(\rho R^3)}{dt} + p\frac{dR^3}{dt} = 0 \tag{2b}$$

where ρ denotes the energy density. In the relativistic limit, i.e. $kT >> mc^2$, which is the case for the whole early radiation dominated phase, we have $p = \rho/3$ and, therefore, Eq.(2b) leads to ρR^4=const and consequently Eq.(2a) to $\dot{R} \propto R^{-1}$. The solution is $R(t) \propto t^{1/2}$. On the other hand $\rho \propto gT^4$ with g=3.3626 in the still radiation dominated phase after weak decoupling and Eq.(1a) gives a relation between $1/t^2$ and T^4 with its precise form being [Bör88, Kol90]

$$t = 0.301 g^{-1/2}\frac{m_{Pl}c^2}{(kT)^2}\sec \quad\text{or}\quad T_9 = 13.336/t^{1/2}. \tag{3}$$

With a global $n_b/n_\gamma = \eta$ and $n_\gamma = 2.404/\pi^2(kT/\hbar c)^3$ one can express the baryon number density as a function of temperature

$$n_b = 2.029 \times 10^{28}\eta T_9^3 \text{cm}^{-3} \quad\text{or}\quad \rho_b = n_b m_u = 3.376 \times 10^4 \eta T_9^3 \text{gcm}^{-3}, \tag{4}$$

when we neglect the small effect of binding or neutron-proton mass difference in comparison to the nuclear mass unit. This relation also yields the total neutron and proton densities when introducing the neutron to proton ratio after weak freeze-out and decay until the onset of nucleosynthesis at T=0.1MeV of $n/p=1/7$ or $X_N=0.125$ and $X_p=0.875$ [Kol90].

While these relations determine completely the conditions in the homogeneous standard big bang (unless one wants to follow the weak freeze-out in detail), they only apply globally in inhomogeneous scenarios. In the simplest of all cases, which was followed in our calculation of heavy elements, we assume a constant density in the high and low density regions after the QCD transition. This leads to the number densities n_n^h, n_n^l, n_p^h, and n_p^l in the high and low density regions. In each of the separated regions the n/p ratio is given by the weak freeze-out values. The total baryon density follows Eq.(4), the density contrast between the two regions is $R = n_b^h/n_b^l = \rho_b^h/\rho_b^l$ [not to be confused with the scale factor in Eq.(2)] and their respective volume fractions are f_V and $1 - f_V$. This determines the neutron and proton densities fully, because the mass fractions X_n (or X_p) are the same in each region. The baryon densities are related by $\rho_b = f_V \rho_b^h + (1 - f_V)\rho_b^l$, which leads with the density ratio R between high and low density region to

$$\frac{\rho_b^l}{\rho_b} = \frac{1}{f_V R + 1 - f_V} \quad \text{and} \quad \frac{\rho_b^h}{\rho_b} = \frac{R}{f_V R + 1 - f_V}. \tag{5}$$

The neutrons are assumed to diffuse out uniformly from these initial conditions. This will change the density contrast and the neutron density in both regions. The protons are, however, assumed to remain in place. When utilizing the superscripts 1 and 2 rather than h and l for this new situation the following relations hold: $n_n^1 = n_n^2$ and $n_p^1 = n_p^h$, $n_p^2 = n_p^l$. That translates into $X_n^1 \rho_1 = X_n^2 \rho_2 = X_n \rho_b$ and $X_p^1 \rho_1 = X_p \rho_b^h$, $X_p^2 \rho_2 = X_p \rho_b^l$ and ρ_1 and ρ_2 can be expressed by the global quantities X_n, ρ_b and the parameters R and f_V

$$\rho_1 = X_n^1 \rho_1 + X_p^1 \rho_1 = X_n \rho_b + (1 - X_n)\rho_b^h$$

$$\rho_2 = X_n^2 \rho_2 + X_p^2 \rho_2 = X_n \rho_b + (1 - X_n)\rho_b^l \quad .$$

It is now easy to determine the ratios ρ_1/ρ_b and ρ_2/ρ_b

$$\frac{\rho_1}{\rho_b} = X_n + \frac{(1 - X_n)R}{f_V R + 1 - f_V} \qquad \frac{\rho_2}{\rho_b} = X_n + \frac{(1 - X_n)}{f_V R + 1 - f_V}. \tag{6}$$

Eq.(6) allows us to express the neutron and proton mass fractions in the regions 1 and 2, respectively, as a function of the global mass fractions and the parameters R and f_V

$$X_n^1 = X_n \frac{\rho_b}{\rho_1} = \frac{X_n}{X_n + (1 - X_n)R/(f_V R + 1 - f_V)}$$

$$X_n^2 = X_n \frac{\rho_b}{\rho_2} = \frac{X_n}{X_n + (1 - X_n)/(f_V R + 1 - f_V)} \tag{7}$$

$$X_p^1 = X_p \frac{\rho_b^h}{\rho_1} = \frac{(1 - X_n)R/(f_V R + 1 - f_V)}{X_n + (1 - X_n)R/(f_V R + 1 - f_V)}$$

$$X_p^2 = X_p \frac{\rho_b^l}{\rho_2} = \frac{(1 - X_n)/(f_V R + 1 - f_V)}{X_n + (1 - X_n)/(f_V R + 1 - f_V)}. \tag{8}$$

These relations were already derived by Fuller, Mathews, Alcock [Ful88], when one replaces the densities ρ in our formulation by the corresponding values for Ω_b. Now the temperature relation Eq.(3), the density relations for ρ_b Eq.(4) and ρ_1 and ρ_2 Eq.(6), as well as the determination of $X_{n,p}^{1,2}$ as a function of the global freeze-out value X_n and the values for η, f_V, and R in Eqs.(7) and (8), lead to a uniquely defined initial value problem. One can recognize that for the low density region 2 all quantities depend only on the product $f_V R$ for large values of this product (f_V itself can only range between 0 and 1). The scenario described above is an oversimplification, not following the weak freeze-out in detail, assuming two regions with uniform densities, complete neutron diffusion while proton diffusion is treated as negligible (therefore the length scale l does not appear), and neglecting neutron diffusion during nucleosynthesis. The last point will be treated approximately in a forthcoming paper [Thi91].

5.2 The Nuclear Network

5.2.1 The Charged Particle Network: Our reaction network consisted of two parts, one part for light and intermediate nuclei, the second part being an r-porcess code. For light and intermediate nuclei from neutron and protons to krypton (Z=36), from stability to the neutron drip line, we made use of a general nuclear network (of 655 nuclei) which includes neutron, charged particle, and photon induced reaction as well as weak reactions. The nuclear abundances are defined by $Y_i = n_i/(\rho N_A)$. For a nucleus with atomic weight A_i, $A_i Y_i$ represents the mass fraction of this nucleus, therefore $\sum A_i Y_i = 1$. In terms of nuclear abundances Y_i, a reaction network is described by the following set of differential equations

$$\dot{Y}_i = \sum_j N_j^i \lambda_j Y_j + \sum_{j,k} N_{j,k}^i \rho N_A < j,k > Y_j Y_k$$

$$+ \sum_{j,k,l} N_{j,k,l}^i \rho^2 N_A^2 < j,k,l > Y_j Y_k Y_l. \tag{9}$$

The reactions listed on the right hand side of the equation belong to the three categories of reactions: (1) decays, photodisintegrations, electron and positron captures, (2) two-particle reactions, and (3) three-particle reactions like the triple-alpha process, which can be interpreted as successive captures with an intermediate unstable target (see e.g. [Nom85]). The individual N^i's are given by: $N_j^i = N_i$, $N_{j,k}^i = N_i/(N_j!N_k!)$, and $N_{j,k,l}^i = N_i/(N_j!N_k!N_l!)$. The N_i's can be positive or negative numbers and specify how many particles of species i are created or destroyed in a reaction. The denominators, including factorials, avoid double counting of the number of reactions when identical particles react with each other (for example in the $^{12}C + ^{12}C$ or the triple-alpha reaction; for details see [Fow67]). This set of differential equations is solved with a fully implicit treatment. Then the stiff differential equations can be rewritten (see e.g. [Pre86], §15.6) as difference equations of the form $\Delta Y_i/\Delta t = f_i(Y_j(t + \Delta t))$, where $Y_i(t + \Delta t) = Y_i(t) + \Delta Y_i$. In this treatment, all quantities on the right hand side are evaluated at time $t + \Delta t$. This results in a set of non-linear equations for the new abundances $Y_i(t + \Delta t)$, which can be solved using a multi-dimensional Newton-Raphson iteration procedure.

The individual rates in Eq.(9) λ_j, $N_A < j,k >$, and $N_A^2 < j,k,l >$ were taken from the following sources. Beta decay half lives if not known experimentally from [Kla84], electron and positron captures from [Ful82, Ful85], experimental nuclear rates for light nuclei from the most recent compilation [Cau88], experimental neutron capture cross sections from [Bao87], rates for unstable light nuclei are given in [Mal88, Mal89], and in [Wie86, Wie87, Wie88a, b, Wie89a, b, Wie90]. For neutron rich light unstable nuclei from Li to Ne we utilized experimentally available information about resonances and bound states to determine mainly neutron capture cross sections. This is an area of the nuclear chart, previously not investigated for thermonuclear reaction rates because no practical application was recognized. Details of these determinations are discussed in Sect. 5.3. For the vast number of medium and heavy nuclei which exhibit a high density of excited states at capture energies, Hauser-Feshbach (statistical model) calculations are applicable , e.g. [Thi87]. For a detailed discussion of the methods involved and neutron capture cross sections for heavy unstable nuclei see also Sect. 3.4 and the appendix in [Cow91].

5.2.2 The r-Process Network: The second part was an r-process code that determines the abundances of the heavy nuclei. This network extends up to Z=114 and contains all nuclei from the so-called valley of beta-stability to the neutron-drip line. Using the mass formula of Truran, Cameron and Hilf [Tru70] to determine the drip line resulted in a total of 6,033 nuclei in this r-process network. The individual nuclear masses, the neutron capture and beta-decay Q values, and the shell and pairing energies were also calculated with the mass formula of [Tru70]. The neutron capture rates were calculated with statistical model methods and the beta decay rates were taken from the gross theory or from [Kla84], where experimental values were not available.

The numerical methods employed in the solution of the r-process rate equations involve the straightforward application of implicit differencing techniques. Consider the following representative segment of the nuclear reaction network:

$(Z+1,A-2)$ $(Z+1, A-1)$ $(Z+1,A)$

$\qquad\qquad\quad$ $(Z,A-1)$ (Z,A) $(Z,A+1)$

$\qquad\qquad\qquad\qquad$ $(Z-1,A)$ $(Z-1,A+1)$ $(Z-1,A+2)$.

We include all possible neutron capture and photodisintegration reactions, as well as beta decays leading to the emission of zero, one, two, or three delayed neutrons. The rate of change of the abundance of nucleus (Z, A) is then given by

$$
\begin{aligned}
\dot{Y}(Z,A) = {} & n_{\mathrm{n}} Y(Z, A - 1)\sigma_{A-1} + Y(Z, A + 1)\lambda_{A+1} \\
& - Y(Z,A)(n_{\mathrm{n}}\sigma_A + \lambda_A + \lambda_\beta^A + \lambda_{\beta n}^A + \lambda_{\beta 2n}^A + \lambda_{\beta 3n}^A) \\
& + Y(Z-1,A)\lambda_\beta^{Z-1,A} + Y(Z-1,A+1)\lambda_{\beta n}^{Z-1,A+1} \\
& + Y(Z-1,A+2)\lambda_{\beta 2n}^{Z-1,A+2} + Y(Z-1,A+3)\lambda_{\beta 3n}^{Z-1,A+3}, \quad (10)
\end{aligned}
$$

where $Y(Z,A)$ is the abundance of nucleus (Z,A), n_{n} is the neutron number density, σ_{A-1} is the thermally averaged (n,γ) reaction rate $(\langle\sigma v\rangle)$ of nucleus

$(Z, A-1)$, λ_{A+1} is the photodisintegration rate (γ, n) for nucleus $(Z, A+1)$, λ_β^A is the beta-decay rate of nucleus (Z, A), $\lambda_{\beta n}^A$ is the rate of beta-decay followed by the emission of one delayed neutron, $\lambda_{\beta 2n}^A$ is the rate of beta-decay followed by the emission of two neutrons, and $\lambda_{\beta 3n}^A$ is the rate of beta-decay followed by the emission of three neutrons.

In a fully implicit treatment, the stiff differential equations in Eq.(10) can be rewritten (see e.g. [Pre86], §15.6) as difference equations of the form $\Delta Y(Z, A)/\Delta t = f(Y(Z_i, A_i, t + \Delta t))$, where $Y(Z, A, t + \Delta t) = Y(Z, A, t) + \Delta Y(Z, A)$. In this treatment, all quantities on the right hand side are evaluated at time $t + \Delta t$. If the neutron abundance $(Y_N = n_N/[\rho N_A])$ were treated in the same manner, this would result in a set of non-linear equations for the new abundances $Y(Z, A, t + \Delta t)$, which can be solved using a multi-dimensional Newton-Raphson iteration procedure. If we assume that n_N is constant (over a time step Δt), Eq. (10) leads to a set of linear equations for the ΔYs, which can be solved with methods for tridiagonal matrices in case a reaction flow proceeds only from small Z's to large Z's (see [Cow91] for details). This is essential and allows the advantages associated with the use of implicit techniques, even for a full r-process network containing 6033 nuclei.

For the calculations in this paper we also introduced (beta delayed) fission of heavy nuclei, as calculated by Thielemann, Metzinger, Klapdor [Thi83] with the Howard and Möller [How80] fission barriers *and* masses. Such a choice is presently the most realistic and consistant treatment (see [Cow91]). In this case nuclear fission cycles material from large Z's back to smaller Z's and a fission term has to be added in Eq.(10), proportional to $Y(Z_f, A_f)$ with $Z_f \approx 2Z$ and $A_f \approx 2A$. This cannot be treated implicitly within the method described above. Therefore, that part is implemented in an explicit way, limiting time steps in the r-process network to half lives of beta delayed fission of appropriate nuclei.

5.2.3 Network Coupling: These two networks are coupled together such that they both run simultaneously at each time step, and the number of neutrons produced and captured is transmitted back and forth between them. The neutron abundance Y_n is a variable in the charged particle network where the dominant abundances are retained which affect the neutrons. Then this neutron abundance is used to obtain the neutron number density $n_n = \rho N_A Y_n$ in the r-process network and is assumed to be constant over the (same) time step. The flux between both networks is determined by treating the Kr isotopic chain as a part of both networks. In each time step neutron captures and beta decays will lead to a (possibly artificially high) build-up of Kr in the charged particle network, due to the inhibition of beta decays of Kr isotopes and neutron captures by the most neutron rich Kr isotope. The Kr abundances resulting from this application of the charged particle network are fed into the r-process code and given the chance to follow the correct path of neutron captures and beta decays for that time step, thus forming heavier nuclei. The next timestep is started in the charged particle network with these Kr isotopic abundances from the r-process code. The neutrons abundance change during the previous time step due to nuclei in the r-process network can be written as

$$\Delta Y_{\mathrm{n}} = \dot{Y}_{\mathrm{n}} \Delta t$$
$$= \sum_{Z,A} Y(Z,A)(-n_{\mathrm{n}} \sigma_A + \lambda_A + \lambda^A_{\beta \mathrm{n}} + 2\lambda^A_{\beta 2\mathrm{n}} + 3\lambda^A_{\beta 3\mathrm{n}})\Delta t, \qquad (11)$$

where the same symbols are used as in Eq. (10). This is used as a correction to the neutron abundance in the charged-particle network to insure an accurate accounting of the number of neutrons produced and absorbed and to insure mass conservation. The total abundances of all nuclei in both networks are normalized to $\sum_i A_i Y_i = 1$.

There exists one source of error in this treatment. In case of small photo-disintegration rates when only neutron captures are important, the Kr isotopes are exposed to neutron capture twice in one time step, once in the charged particle and once in the r-process network. When one is close to an $(n, \gamma) - (\gamma, n)$-equilibrium, photodisintegrations stabilize the abundances in each isotopic chain (i.e. also for Kr) and the position of the abundance peak remains fixed. Thus, any additional neutron capture is cancelled by an additional photodisintegration and no error is introduced. Also the error mentioned above is small with respect to the total process. If any, only small local disturbances are expected.

5.3 Important Reaction Rates and Their Determination

An overview over reactions, important in the standard big bang, is given in [Yan84] and [Del89]. In an inhomogeneous big bang the break-out to heavier nuclei has to be considered. We do not want to repeat the whole list of reactions from Sect. 4.2 and also from Table 1 by Malaney and Fowler [Mal89] or Table 2 by Malaney [Mal90]. Almost all of the reactions starting at ^{8}Li, which lead to the formation of heavier nuclei, are poorly determined and precise measurements are highly desirable. Only then reliable nucleosynthesis calculations are possible. The most (?) important reaction ^{8}Li$(\alpha, n)^{11}$B might actually be accessable by simple theoretical means, a statistical model calculation. The main requirement which determines whether the statistical model is applicable is a high density of states (resonances) at the projectile binding energy in the compound nucleus, so that an averaging over resonances becomes meaningful. The critical minimum number quoted is 10 MeV^{-1} [Woo78, Thi87]. In our case the alpha binding energy in ^{12}B is 10MeV and the level density at this energy fulfills the requirement quoted. We display our results in Table 1 in comparison to the rate given by Malaney and Fowler [Mal89]. The calculation was performed with the statistical model code SMOKER ([Thi87]).

The two columns represent the results obtained by a detailed statistical model calculation with realistic optical potentials and inclusion of experimentally known excited states (SMOKER), or by employing the approximation formula of [Woo75]. The difference of 50-100% underlines actually a reasonable agreement. Such calculations are only expected to be correct within a factor of

Table 1. $^8\mathrm{Li}(\alpha, \mathrm{n})^{11}\mathrm{B}$

$$N_A < \sigma v >$$

T_9	SMOKER	Malaney and Fowler (1989)
0.10	8.17−05	2.24−04
0.20	4.84−01	7.60−01
0.30	2.44+01	3.70+01
0.40	2.78+02	4.17+02
0.50	1.54+03	2.29+03
0.60	5.60+03	8.27+03
0.70	1.55+04	2.28+04
0.80	3.56+04	5.24+04
0.90	7.12+04	1.05+05
1.00	1.28+05	1.90+05
2.00	3.14+06	4.98+06
3.00	1.24+07	2.12+07
4.00	2.72+07	4.91+07
4.50	3.61+07	6.63+07
5.00	4.57+07	8.49+07
6.00	6.64+07	1.25+08
7.00	8.81+07	1.65+08
8.00	1.10+08	2.04+08
9.00	1.32+08	2.41+08
10.00	1.54+08	2.74+08

2, when using global parameters not finetuned to a specific nuclear reaction. However, an experimental verification of this most important reaction is highly desirable. The first reports by Paradellis et al. [Par90], utilizing the inverse reaction, takes only into account the transition to the $^{11}\mathrm{B}$ ground state. A statistical model calculation, just for that specific transition, is planned, in order to compare the results.

It can also be seen from Sect. 4.2, that (n, γ) reactions on light nuclei in the range Li to Ne are of key significance in inhomogeneous big bang scenarios. We know that even the ones involving stable targets in this region of the nuclear chart bear large uncertainties [Bao87, Bee89], due to their minute cross sections. The high neutron concentrations in low density regions might also make neutron captures feasible on light unstable, neutron rich, targets. Presently only a few rates for these nuclei are available and they stem from old estimates [Wag67, Wag69, Wag73]. Level densities at the relatively small neutron separation energies of neutron rich compound nuclei are large enough to justify the use of statistical model calculations for heavy and perhaps also for medium nuclei. This is absolutely not the case for nuclei lighter than Ne.

For this reason we calculated neutron capture rates for neutron-rich nuclei from Li to Ne, based on experimental information. The reaction rates are determined by resonant contributions from resonant capture into neutron unbound levels near the neutron threshhold of the compound nucleus and by direct capture into low lying bound states of the final nucleus $< \sigma v >=< \sigma v >_{\mathrm{res}} + < \sigma v >_{\mathrm{dc}}$. The resonant reaction rate is estimated only for known states, therefore, this contribution may be larger than quoted here if the actual level density is higher that currently known. However, for most of the rections the Q-values are very small and only a few resonances are expected to contribute. The resonant reaction rate is mainly determined by the resonance energy $E_r(\text{MeV})$ and the resonance strength $(\omega\gamma)_r(\text{MeV})$,

$$< \sigma v >_{\text{res}} = 2.56 \times 10^{-13}(\omega\gamma)_r A^{-3/2} T_9^{-3/2} \exp(11.605 E_r/T_9). \qquad (12)$$

A is the reduced mass in atomic mass units. The resonance energy is calculated from the known excitation energies of the unbound levels, the resonance strength was estimated as described in [Wie86].

In case of small level densities, direct capture is the dominant reaction mechanism, determining the reaction rates. The cross section $\sigma(E)$ is calculated treating this process as an extra nuclear channel electromagnetic transition between the continuum wave function of the target-projectile system and the final bound state in a Woods-Saxon potential. Details are described in [Wie89b]. The reaction rate for s-wave neutron capture is constant because of the $1/v$ cross section dependence. Here we have taken into account that the neutron direct capture on neutron-rich light isotopes with a closed p-shell populates predominantly low excited states with a predominant sd-configuration. Therefore, the E1 direct capture to these states is characterized by p-wave neutron capture only. With the approximation for the p-wave neutron penetrability $P_1 = \sqrt{\mu} E^{3/2}$ the cross section can be expressed by $\sigma(E) = S_n \sqrt{\mu E}$, S_n being the S-factor for p-wave neutrons. Then the reaction rate has the form

$$< \sigma v >_{\text{dc,p}} = 1.787 \times 10^{-16} S_n [\text{barn}/\text{MeV}^{1/2}] T_9 \text{cm}^3/\text{s}. \qquad (13)$$

Using this formalism for p-wave capture and the formalism for s-wave capture as described in [Fow67], we calculated the total direct capture contribution. The sum of direct capture and resonant terms results in the expressions given in Table 2. This is an enhanced and revised version of Table 2 in [Thi90]. If the the neutron rich nuclei ^{9}Li, ^{10}Be, and ^{13}B are produced, further (α, n)-reactions can become important. We are currently investigating these rates.

6. Results

The aim of the present investigation was to explore the full consequences of inhomogeneous big bang scenarios to the production of heavy elements in the neutron rich low density regions. Full network calculations as described in Sect. 5.2 were performed, utilizing the temperature relation Eq. (3), the density relations Eqs. (4) and (6), and the initial neutron and proton mass fractions from Eqs.(7) and (8) for the low density neutron rich region 2. A value of η had to be chosen. We made use of the relation between Ω_b and η in Eq. (1c). While the background temperature $T_{\gamma 0}$ is well known, the Hubble constant is unfortunately not, which leaves some uncertainty. With a Hubble constant of

Table 2. Neutron-Induced Reactions on Light Unstable Targets

Reaction	$N_A < \sigma v >$
$^{7}Li(n,\gamma)^{8}Li$	3.144e+3+4.26e+3/t932*exp(-2.576/t9)
$^{8}Li(n,\gamma)^{9}Li$	3.260e+3+6.328e+4/t932*exp(-2.866/t9)
$^{9}Be(n,\gamma)^{10}Be$	1.01e+3+1.01e+4/t932*exp(-6.487/t9) +5.41e+4/t932*exp(-8.471/t9)
$^{10}Be(n,\gamma)^{11}Be$	5.96e+2+6.66e+2/t932*exp(-14.85/t9)
$^{11}Be(n,\gamma)^{12}Be$	3.56e+2
$^{11}B(n,\gamma)^{12}B$	7.29e+2+3.86e+3/t932*exp(-0.241/t9) +3.30e+4/t932*exp(-4.990/t9)
$^{12}B(n,\gamma)^{13}B$	1.700e+3+9.548e+3/t932*exp(-1.625/t9) +1.562e+3/t932*exp(-2.669/t9) +1.163e+4/t932*exp(-5.919/t9)
$^{13}B(n,\gamma)^{14}B$	1.02e+1+4.950e+1*t9+4.940e+4/t932*exp(-4.76/t9)
$^{11}C(n,\gamma)^{12}C$	3.18e+4+3.30e+3/t932*exp(-0.917/t9)
$^{11}C(n,\alpha)^{8}Be$	1.66e+8+1.60e+8/t932*exp(-0.917/t9)
$^{11}C(n,p)^{11}B$	1.99e+8+2.47e+6/t932*exp(-0.917/t9)
$^{13}C(n,\gamma)^{14}C$	1.82e+2+3.083e+4/t932*exp(-1.636/t9)
$^{14}C(n,\gamma)^{15}C$	3.24e+3*t9+2.05e+3/t932*exp(-21.14/t9)
$^{15}C(n,\gamma)^{16}C$	5.27e+2*t9+3.28e+4/t932*exp(-21.56/t9)
$^{16}C(n,\gamma)^{17}C$	3.66e+2*t9
$^{13}N(n,\gamma)^{14}N$	1.13e+6+2.65e-2/t932*exp(-2.99/t9)
$^{13}N(n,p)^{13}C$	6.22e+8+6.63e+5/t932*exp(-2.99/t9)
$^{15}N(n,\gamma)^{16}N$	4.64e+3*t9+2.04e+3/t932*exp(-9.98/t9)
$^{16}N(n,\gamma)^{17}N$	1.18e+2+5.75e+2*t9+4.756e+3/t932*exp(-2.29/t9)
$^{17}N(n,\gamma)^{18}N$	2.91e+2*t9
$^{18}N(n,\gamma)^{19}N$	3.23e+2*t9
$^{19}N(n,\gamma)^{20}N$	1.54e+3*t9
$^{17}O(n,\gamma)^{18}O$	7.15e+1+4.03e+4/t932*exp(-0.937/t9) +2.85e+4/t932*exp(-1.961/t9) +4.70e+3/t932*exp(-2.762/t9)
$^{17}O(n,\alpha)^{14}C$	3.11e+4+2.28e+5/t932*exp(-0.937/t9) +3.52e+7/t932*exp(-1.961/t9) +3.69e+6/t932*exp(-2.762/t9)
$^{18}O(n,\gamma)^{19}O$	2.12e+1+2.55e+3/t932*exp(-1.769/t9)
$^{19}O(n,\gamma)^{20}O$	7.32e+2*t9+9.70e+3/t932*exp(-0.209/t9) +5.86e+4/t932*exp(-1.741/t9)
$^{20}O(n,\gamma)^{21}O$	1.22e+3*t9
$^{21}O(n,\gamma)^{22}O$	1.74e+1*t9
$^{21}F(n,\gamma)^{22}F$	2.12e+2*t9+8.42e+3/t932*exp(-0.464/t9) 1.82e+3/t932*exp(-3.667/t9)
$^{22}F(n,\gamma)^{23}F$	1.96e+2*t9
$^{23}F(n,\gamma)^{24}F$	3.52e+3*t9
$^{24}F(n,\gamma)^{25}F$	1.20e+3*t9
$^{25}F(n,\gamma)^{26}F$	3.08e+1*t9
$^{23}Ne(n,\gamma)^{24}Ne$	9.53e+2+4.69e+2*t9
$^{24}Ne(n,\gamma)^{25}Ne$	4.93e+3*t9
$^{25}Ne(n,\gamma)^{26}Ne$	7.81e+2*t9
$^{26}Ne(n,\gamma)^{27}Ne$	5.81e+2*t9

$H_o=62$, i.e. $h_{50}=1.24$, we obtain $\eta_{10}=100$ or $\eta=10^{-8}$ if $\Omega_b=1$. Three calculations have been performed, employing Ω_b values of 1.0, 0.5, and 0.1. Earlier calculations explored a range of $5\leq f_V R\leq 10^4$ [Mat90] and found the best fit with primordial abundances of light elements for a combination of $f_V R\approx 10$. This leads to $\rho_2/\rho_b\approx 1/5$. Even for $f_V R$ going to infinity this ratio does not change much, being bound by $\rho_2/\rho_b=X_n=0.125=1/8$. For our calculations we chose $\rho_2/\rho_b=1/6$. The neutron mass fraction in region 2 is bound by $X_n=1$ for infinite $f_V R$. We will use this extreme value, i.e. $X_p=0$, in order to achieve the highest possible neutron density and the most favorable environment for the formation of heavy elements.

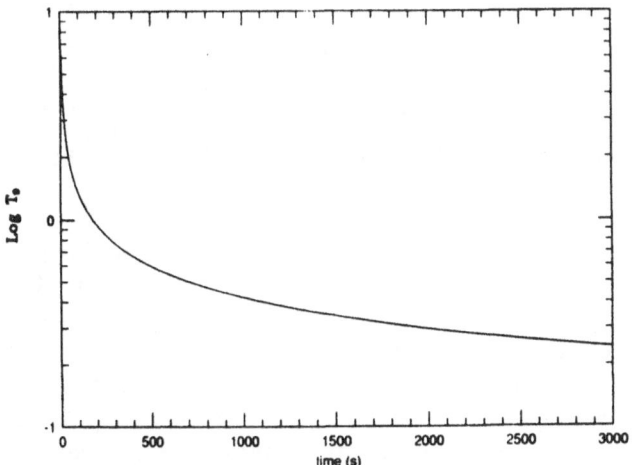

Fig. 1a. Global temperature as a function of time in big bang conditions. Nucleosynthesis and formation of ^{2}H, ^{3}He, ^{4}He, ^{7}Li, and heavier nuclei sets in after the photodisintegration of ^{2}H ceases at T$\approx$0.1 MeV$\approx$1.16$\times$10^9 K

In the following we will discuss the details of the calculations, taking the most favorable case for the formation of heavy elements $\Omega_b=1$. Figs.1a and b show the temperature T_9 and neutron number density n_n as a function of time. The onset of nucleosynthesis takes place after about 200s at T$\approx$0.1MeV$\approx$10^9 K when ^{2}H ceases to photodisintegrate. This is followed by a sudden drop in the neutron density over about 100s during which most of the primordial nucleosynthesis and the formation of ^{4}He occurs. This can also be seen in Figs. 2ab. In Fig. 2a only n, p and ^{4}He mass fractions are displayed. This is equivalent to Fig. 2a in [App88a]. We see that in the neutron rich region almost all nucleons end in ^{4}He.

The more interesting part, the formation of heavy elements, is shown in Fig.2b, where the dashed line identifies the total mass fraction in Ne and all heavier elements. This calculation used only the network of 655 nuclei from neutrons and protons up to Kr and does not include an r-process network, thus neglecting the fission cycling effect discussed in Sect. 2. After 1000 s essentially all nuclei.beyond Ne have been converted to Kr by neutron capture and beta

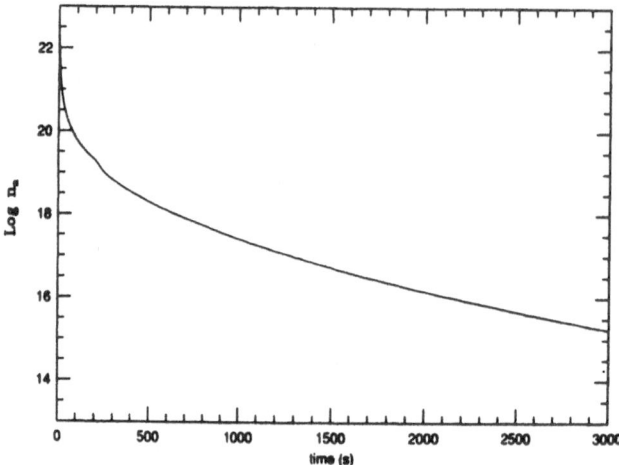

Fig. 1b. Neutron number density n_n in low density (neutron rich) regions of an inhomogeneous big bang with $\Omega_b = 1$ and $\rho_2/\rho_b=1/6$. The drop after 200s corresponds to the onset of nucleosynthesis. For the first 500s the number densities are close to r-process conditions

decay, where abundances pile up because no draining to heavier nuclei is permitted. The timescale for the build-up of nuclei up to Kr is governed by small neutron capture cross sections and long beta decay half lives of lighter nuclei. The total mass fraction of 10^{-13} without fission cycling is very close to the prediction by Applegate [App88a] in his Eq. (3.13). [Mat88, Kaj89, Kaj90b], utilizing a network up to Si, obtained similar results but with somewhat larger abundances. This reflects the use of larger neutron capture rates, in particular when they employed statistical model rates for neutron rich light species – lighter than Ne – with small Q-values.

The details of the flow to heavier nuclei are given in Figs. 3a-e. The time derivative of each individual abundance i in Eq. (9) is the sum of possible forward and reverse reactions, linking nucleus i with another nucleus j. These can be (n, γ) and the corresponding (γ, n) reactions, (α, n) and (n, α) etc. That means we can write Eq.(9) as

$$\dot{Y}_i = \sum_j f_{ij} = \sum_j (\dot{Y}_{i,j} - \dot{Y}_{j,i}), \tag{14}$$

where the net fluxes f_{ij} are expressed by the differences between terms of the abundance derivatives of i and j which describe the appropriate reaction from i to j or vice versa. Figs. 3a-e show the time integrated fluxes $\int f_{ij} dt$ as arrows, fixed at the point of origin i and pointing towards endpoint j. This means that the angle of the arrow defines the endpoint, which is located along the extension of the arrow. We also added different reactions with the same origin and endpoint, e.g. β^+ and (n,p). The maximum flux found in any reaction is normalized to 1. In order to visualize minor fluxes, being orders of magnitude smaller, a logarithmic normalization was introduced for the length of the vector $l_{ij} = 1/(1 + [log_{10}(|f_{ij}/f_{max}|)]^{3/2})$.

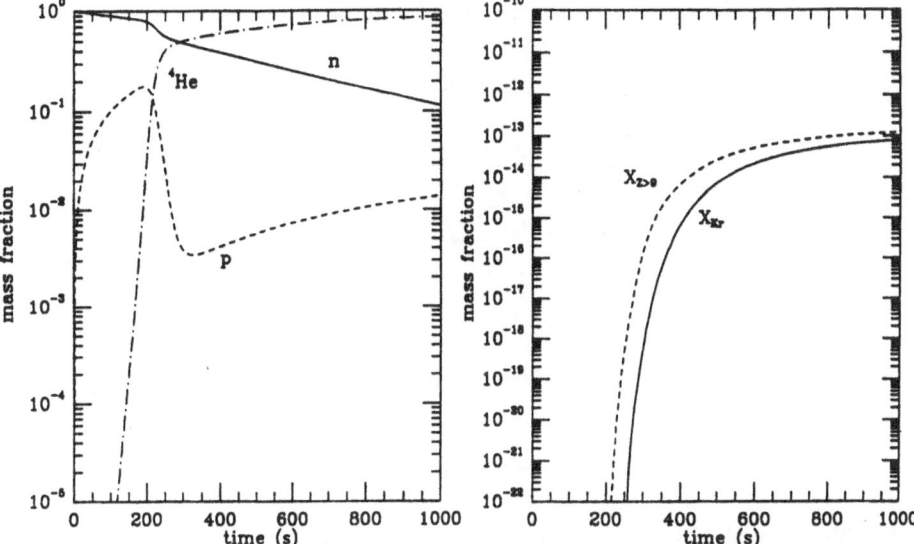

Fig. 2a. Light nuclei production for the same conditions as shown in Figs.1ab. Only neutrons, protons and ^{4}He are displayed. Compare with Fig.2 in [App88a]

Fig. 2b. Heavy nuclei production, including only nuclei up to Kr in this calculation and not allowing for further processing. The shift between the two curves, which display the mass fraction of all Kr isotopes (the end of the network), or all nuclei heavier than Ne, indicates that about 50s are needed to process nuclei from Ne to Kr via neutron captures and beta decay. In total, a mass fraction of 10^{-13} of heavy elements is produced

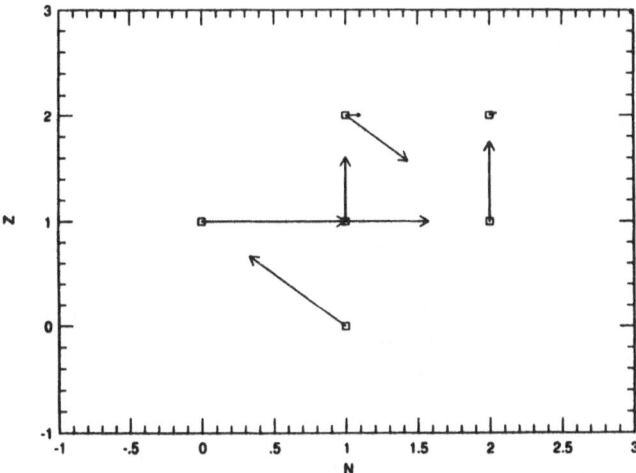

Fig. 3a. Reaction flow pattern for the formation of ^{4}He. Plotted are the time integrated fluxes f_{ij} as defined in Eq.(14), summed over all reactions which originate at nucleus i and end in nucleus j (see text for the normalization). The arrows start at nucleus i and point in the direction of nucleus j. Only very minor reaction flows go beyond ^{4}He

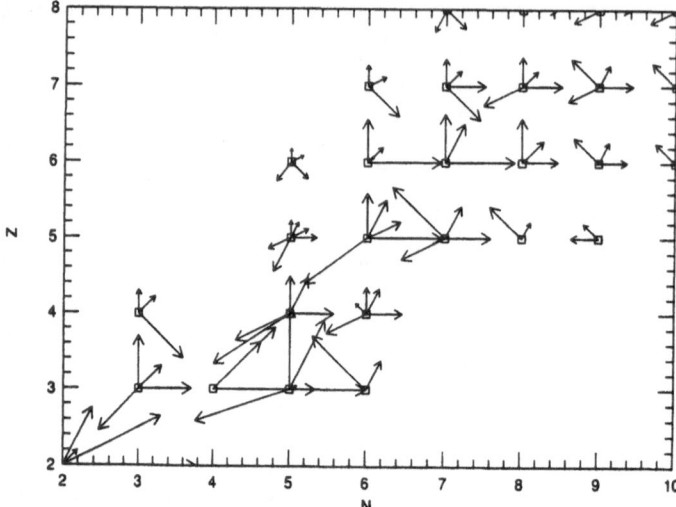

Fig. 3b. ^{4}He(t, γ)^{7}Li and ^{7}Li(n, γ)^{8}Li(α, n)^{11}B or ^{7}Li(α, γ)^{11}B are the major reactions forming heavier nuclei. Large losses back to light elements are due to ^{8}Li(β^-)2^{4}He, ^{9}Be(p, d)2^{4}He, and ^{11}B(p, α)2^{4}He. Beyond B, reactions are dominated by neutron and proton captures as well as beta decays

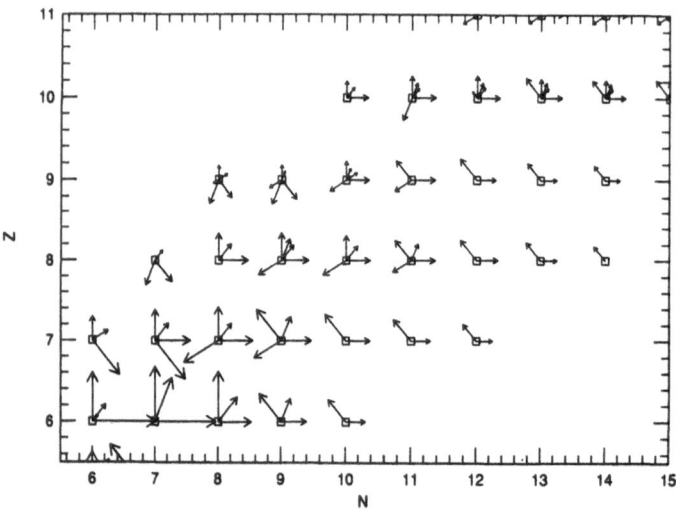

Fig. 3c. With increasing charge, the proton (and even more so the alpha) captures decrease and neutron captures and beta decays become the major reactions. ^{14}C has a small neutron capture Q-value of 1.22MeV. Large photodisintegration rates cause a small net (n, γ) flow. Therefore, the proton capture has to carry the major flux, which is further depleted by (p, α) reactions in the N and O chains

In Fig. 3a the maximum flux is passing through ^{1}H(n, γ)^{2}H. Only the small fluxes originating in ^{4}He, already enhanced by logarithmic display, lead to the formation of heavier nuclei. Fig. 3b starts from there. All fluxes are scaled and the scaling factor can be determined by comparing the ^{4}He(t, γ)^{7}Li vectors in Figs. 3a and b. The major reactions towards heavier nuclei are

^{7}Li$(n,\gamma)^8$Li$(\alpha,n)^{11}$B or ^{7}Li$(\alpha,\gamma)^{11}$B. Large losses back to light elements are due to ^{8}Li$(\beta^-)2^4$He, ^{9}Be$(p,d)2^4$He, and ^{11}B$(p,\alpha)2^4$He. Beyond B, reactions are dominated by neutron and proton captures as well as beta decays. With increasing charge, the proton (and even more so the alpha) captures decrease and neutron captures and beta decays become the major reactions. ^{14}C, having a very long half life, which makes it stable during big bang nucleosynthesis, has a small neutron capture Q-value of 1.22 MeV. This triggers large photodisintegration reactions and a small net (n,γ) flow. Therefore, the proton capture has to carry the major flux, which is further depleted by (p,α) reactions in the N and O chains. Only beyond F, (p,α) reactions do not form major obstacles anymore and the flow is dominated by (n,γ) reactions and beta decays. The major reaction flow towards heavier nuclei passes through ^{20}O, ^{21}F, ^{23}Ne, ^{27}Na, ^{29}Mg, ^{31}Al, ^{34}Si, ^{37}P, ^{41}S, ^{44}Cl, ^{46}Ar, ^{49}K, ^{52}Ca, ^{53}Sc, ^{54}Ti, ^{57}V, ^{62}Cr, ^{65}Mn, ^{66}Fe, etc. This is already 10 mass units from stability and will deviate even further for heavier nuclei. At that distance from stability even small arrows indicating beta-delayed neutron emission become visible.

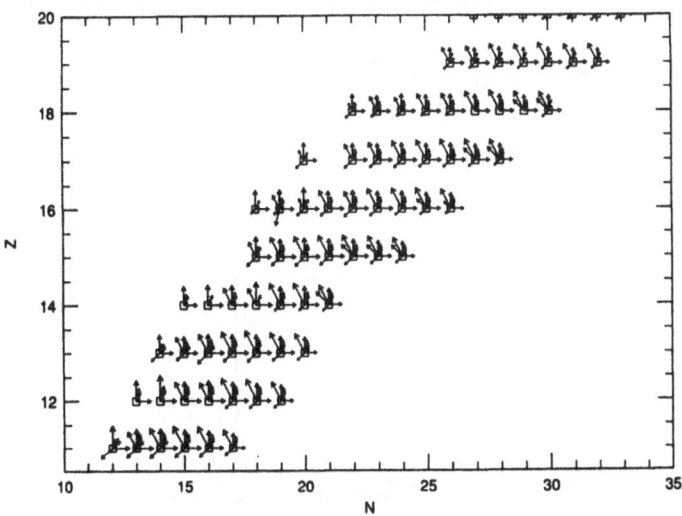

Fig. 3d. A flow pattern dominated by (n,γ) reactions and beta decays starts to form. The major reaction flow towards heavier nuclei passes through ^{20}O, ^{21}F, ^{23}Ne, ^{27}Na, ^{29}Mg, ^{31}Al, ^{34}Si, ^{37}P, ^{41}S, ^{44}Cl, ^{46}Ar, and ^{49}K

Figs. 4a-c display the abundances as a function of A from A=1 to A=90 for the Ω_b=1.0, 0.5, and 0.1. One notices the prominent peak at ^{14}C. The change in Ω_b and therefore ρ_b and $n_n = X_n\rho_b$ (even when $X_n\approx1$) leads with decreasing Ω_b to a neutron capture path closer to stability, where longer half lives are encountered and heavier nuclei are formed with a smaller reaction flux. The shift by one mass unit can already make a tremendous difference when two neighoring nuclei have very different beta decay half lives. When changing Ω_b from 1 to 0.5 the abundance at A=14 drops by an order of magnitude and another factor of 10 up to A=80. Changing it to 0.1 (by a factor of 5 rather than 2) enhances the effect and leads to a drop by more than two orders of

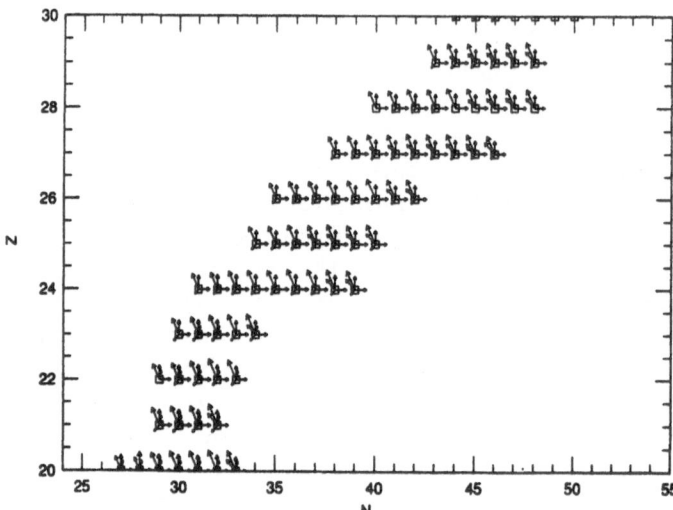

Fig. 3e. The main reaction path proceeds via ^{52}Ca, ^{53}Sc, ^{54}Ti, ^{57}V, ^{62}Cr, ^{65}Mn, ^{66}Fe, etc., deviating more and more from stability. This distance is widening towards heavier nuclei

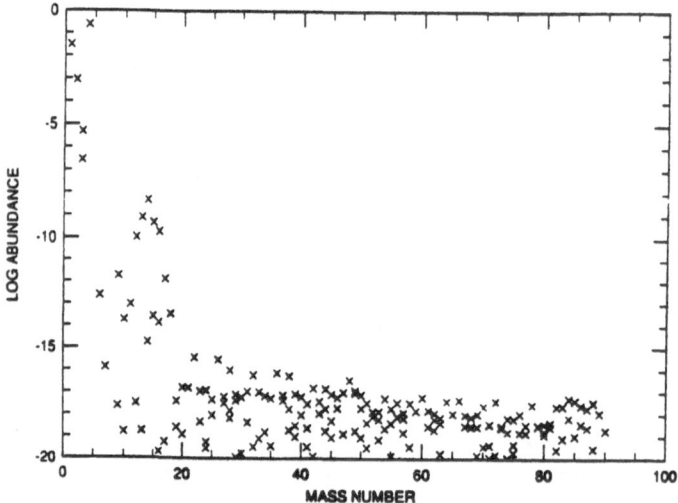

Fig. 4a. Abundances as a function of A from A=1 to A=90 for $\Omega_b \dot{=} 1$. One notices the prominent peak at ^{14}C, caused by the necessity of charged particle reactions to transport the reaction flow beyond A=14

magnitude at A=14 and more than another factor of 100 up to A=80. Thus, values of Ω_b <0.3, permitted by constraints to primordial abundances of light elements, leave only very small yields of intermediate elements.

For neutron densities in excess of $10^{19} - 10^{20}$ cm^{-3}, the line in the nuclear chart beyond which beta decay dominates over neutron capture would be located 15-20 mass units away from stability for heavy nuclei. Typical beta decay half lives of $10^{-2} - 10^{-3}$ s are then encountered, with a few exceptions of the order 10^{-1} s. Along an r-process path at this distance from stability, Fe-nuclei

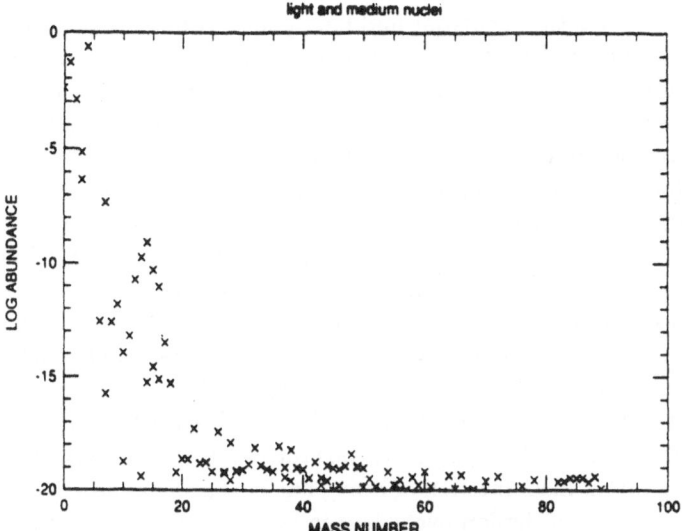

Fig. 4b. The same as Fig. 4a for $\Omega_b=0.5$. This change in Ω_b leads to a neutron capture path closer to stability. In comparison to the $\Omega_b=1$ results the abundances are smaller by an order of magnitude at A=14 and another factor of 10 up to A=80

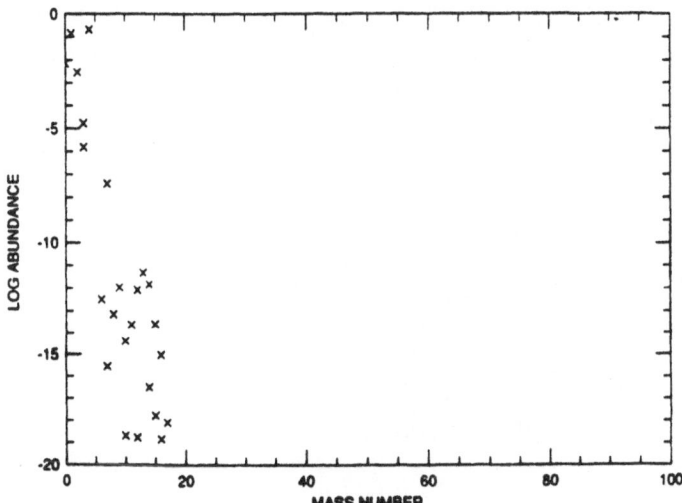

Fig. 4c. The same as Fig. 4b for $\Omega_b=0.1$. The change in Ω_b by a factor of 5 leads to a drop by more than two orders of magnitude at A=14 and more than another factor of 100 up to A=80

can be transformed to U within a few seconds. The transformation from Ne to Fe took much longer, however. The actual neutron number density found in the calculations (shown in Fig. 1b) are of the order 10^{19} cm^{-3} between 200 and 300 s, when most of the light nuclei are produced.

Fission in the r-process path can have a strong influence on the total abundances of heavy nuclei as a consequence of "fission cycling", whereby each of the fission fragments can capture neutrons and finally form heavy nuclei which fission again [See65]. This is of particular importance in environments with a long duration of high neutron densities, and was therefore suggested as relevant to primordial nucleosynthesis in neutron rich zones of an inhomogeneous big bang [App88a]. In contrast to the operation of the r-process in explosive stellar environments, confined to a few seconds, this process in the neutron rich regions associated with an inhomogeneous big bang is only limited by the neutron half-life and can go on for an extended period of time. One of the remarkable features of an r-process with fission cycling is that the production of heavy nuclei is not limited to the r-process flow (neutron captures and beta-decays) coming from light nuclei, but requires only a small amount of fissionable nuclei to be produced initially (a mass fraction of 10^{-13} in Fig. 2b).

The total mass fraction of heavy nuclei is doubled with each fission cycle and can thus be written as $X_r = 2^n X_{seed}$. Here, n is the number of fission cycles and X_{seed} denotes the initial mass fraction of heavy nuclei. The main parameter that determines the number of fission cycles is the rate of the r-process flux, which is a function of the location of the r-process path with respect to the stability line, and thus dictates the typical beta decay half lives. The location of the r-process path is a function of n_n and T, coming closer to stability for decreasing n_n (or increasing T). In Table 3 we list the results of test calculations which were performed to determine the fission cycle times τ_{cycl} for T=7 × 10^8 K as a function of n_n. We notice a tremendous change in τ_{cycl} with decreasing n_n and thus do not expect much cycling once n_n dropped down to 10^{16} cm^{-3} at 2000 s (see Fig. 1b).

Table 3. Fission cycle times τ_{cycl}

n_n(cm^{-3})	τ_{cycl} (sec)
10^{20}	2.84×10^0
10^{19}	1.16×10^1
10^{18}	3.46×10^1
10^{17}	3.73×10^2
10^{16}	2.23×10^3

Fig. 5a and b show the results for the low density (neutron rich) region of an inhomogeneous big bang with Ω_b=1 at t=400 s and 3000 s, when performing calculations with a full r-process network beyond Kr. In Fig. 5b one notices that fission cycling was indeed effective, because the final mass fraction of heavy nuclei is 1.5×10^{-11}; this is a factor of 150 larger than the mass fraction of 10^{-13}, produced without the r-process network and fission (see Fig. 2b). The results of these calculations differ, however, from the initial estimates of Applegate. This is due to the fact that, for the first time, the reactions for nuclei above Ne and Si are included; in previous studies, their effect was only estimated. The amount of

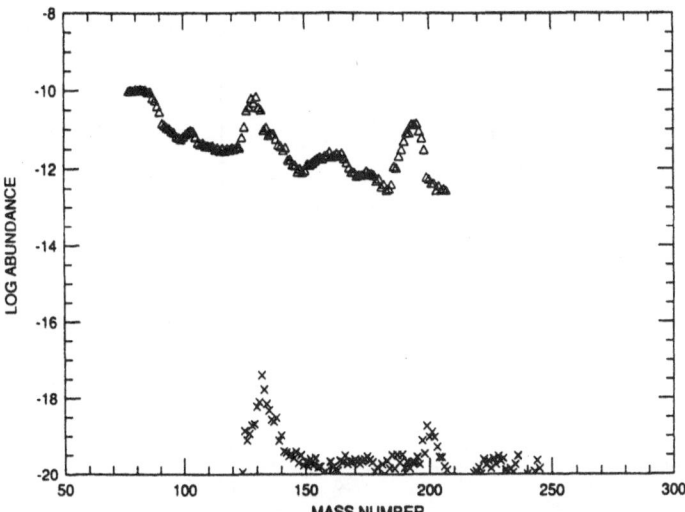

Fig. 5a. Heavy element abundances due to r-process production. Solar r-process abundances are indicated by triangles. This is compared with calculations for the low density region of an inhomogeneous big bang with $\Omega_b=1$ and $\rho_2/\rho_b=1/6$ at $t=400$ s. The major r-process features and peaks start to form at approximately the correct positions

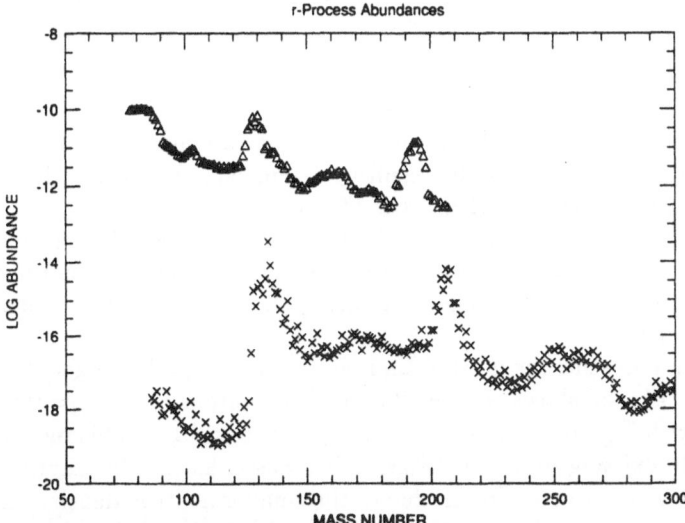

Fig. 5b. Composition for $\Omega_b=1$ at 3000 s (final). Fission cycling was effective by producing a final mass fraction of heavy nuclei of 1.5×10^{-11}, a factor of 150 larger than the mass fraction obtained without the r-process network and fission. The abundances are more than a factor of 1000 smaller than solar, which has to be diluted by another factor of 6 to obtain global abundances after mixing with the high density regions. This is about a factor of 10 smaller than the lowest r-abundances in low metallicity stars. The peaks are shifted close to s-process positions

fission cycling was overestimated in earlier work. For an initial neutron density of 10^{19} gcm^{-3}, the r-process path closely approximates a path reproducing the

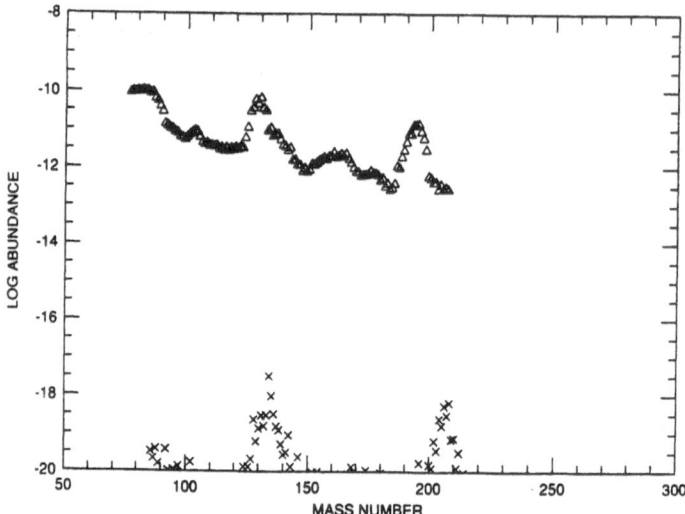

Fig. 5c. Composition for $\Omega_b=0.5$, being on average by a factor of 10^4 smaller than the results for $\Omega_b=1$. The difference at A=80 (see Figs.4ab) was a factor of 100, i.e. the smaller n_n, resulting from a decreae in Ω_b, led to an r-process path closer to stability with enhanced beta decay half lives and fission cycling times. This more realistic case for Ω_b would place primordial r-abundances more than 4 orders of magnitude below observed r-abundances in low metallcity stars, and therefore render such an effect unobservable. Neutron back diffusion into high density regions would reduce these abundances further. In addition, another drop by 4 orders of magnitude is expected for $\Omega_b=0.25$

solar r-process abundance pattern. However, during the subsequent expansion the neutron density decreases as well, resulting in an r-process path closer to stability. When an average is taken over the total duration of big bang nucleosynthesis, one obtains a cycle time of 50-70 s rather than the shorter initial estimates. Since the abundances are proportional to 2^n, with n denoting the number of fission cycles, the change to longer cycle times leads to smaller abundance predictions.

This calculation represents the most optimistic case with $\Omega_b=1$ and neutron diffusion neglected. Fig. 5c shows the results for a calculation with $\Omega_b = 0.5$. When comparing with Fig.5b we see a difference by a factor of 10^4. This means that the factor 100, existing at A=80 (Figs. 4ab), was enhanced by another 10^2 loss towards heavier r-process nuclei, due to the smaller neutron density. In addition, it is clear from Figs. 5bc that these calculated primordial abundances do not even follow an r-process pattern. The small neutron densities at late time led to a shift of the path very close to stability and abundance maxima close to s-process peaks, rather than the r-process peaks. This can be recognized in Fig. 6 which shows the r-process path of the $\Omega_b = 1$ calculation at 500, 1500 and 3000 s.

Based on the calculation for $\Omega_b = 0.5$, the constraints from primordial abundances of light elements – which leads to $\Omega_b < 0.3$ – and the fact that the inclusion of neutron back diffusion into high density regions will further reduce the neutron densities, it appears questionable whether an r-process abundance

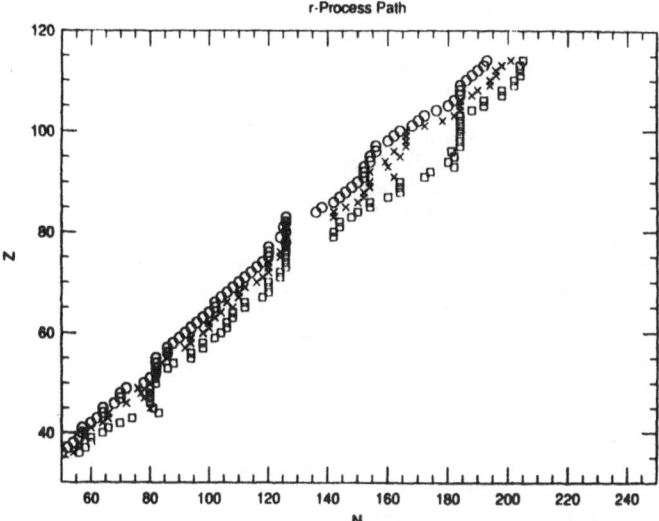

Fig. 6. Positions of maximum abundances in each isotopic chain – r-process path – as a function of time: squares 500s, crosses 1000s, circles 3000s. At the magic number N=126 the nucleus ^{208}Pb, an s-process nucleus, is encountered. That explains the tendency that the abundance peaks move towards higher mass numbers with time and coincide almost with s-process peaks, rather than r-process peaks, at the end of the primordial nucleosynthesis period

level, that would be observationally detectable, was produced in an inhomogeneous big bang.

Acknowledgements

We want to thank F. Adams, S. Bonometto, J. Görres, C. Hogan, T. Kajino, H. Kurki-Suonio, R. Malaney, G. Mathews, B. Meyer, O. Pantano, K. Sato, and R. Scherrer for stimulating discussions. This research was supported in part by NSF grants AST 89-13799 and PHY 88-03035, and NASA grant NGR 22-007-272. JHA is an Alfred P. Sloan Foundation Fellow. JJC thanks the University of Oklahoma Research Council for partial support. The computations were performed at the National Center for Supercomputer Applications at the University of Illinois (AST 890009N).

References

Alc85   C.R. Alcock, E. Farhi: Phys. Rev. **D32**, 1273 (1985)
Alc87   C.R. Alcock, G.M. Fuller, G.J. Mathews, G.J.: Ap. J. **320**, 439 (1987)
App88a   J.H. Applegate: Phys. Rep. **163**, 141 (1988)
App85   J.H. Applegate, C.J. Hogan: Phys. Rev. **D31**, 3037 (1985)
Alc90   C.R. Alcock, D.S. Dearborn, G.M. Fuller, G.J. Mathews, B. Meyer: Phys. Rev. Lett. **64**, 2607 (1990)

176   F.-K. Thielemann et al.

App87   J.H. Applegate, C.J. Hogan, R.J. Scherrer: Phys. Rev. **D35**, 1151 (1987)
App88b   J.H. Applegate, C.J. Hogan, R.J. Scherrer: Ap. J. **329**, 572 (1988)
Aud88   J. Audouze, P. Delbourgo-Salvador, H. Reeves, P. Saleti: In *The Origin and Distribution of the Elements*, ed. by G.J. Mathews, (World Scientific, Singapore 1988), p. 556
Ban87   T.M. Bania, R.T. Rood, T.L. Wilson: Ap. J. **323**, 30 (1987)
Bao87   Z.Y. Bao, F. Käppeler: At. Nucl. Data Tables **36**, 411 (1987)
Bee89   H. Beer, G. Rupp, F. Voss, F. Käppeler: In *Proc. 5th Workshop on Nuclear Astrophysics*, ed. by W. Hillebrandt, E. Müller (Max-Planck-Institut MPA/P1, Garching 1989), p.6
Bör88   G. Börner: *The Early Universe*, (Springer-Verlag, Berlin 1988)
Boe85   A. Boesgaard, G. Steigman: Ann. Rev. Astron. Astrophys. **23**, 319 (1985)
Bor83   C. Borgs, E. Seiler: Nucl. Phys. **B215** [FS7], 125 (1983)
Boy89   R.N. Boyd, T. Kajino: Ap. J. Lett. **336**, L55 (1989)
Bon85   S.A. Bonometto, P.A. Marchetti, S. Matarrese: Phys. Lett. **B157**, 216 (1985)
Bro88   F.R. Brown et al.: Phys. Rev. Lett. **65**, 2491 (1988)
Bru91   C.R. Brune, R.W. Kavanagh, S.E. Kellog, T.R. Wamg: Phys. Rev. **C43**, 875 (1991)
Cau88   G.R. Caughlan, W.A. Fowler: At. Nucl. Data Tables **40**, 283 (1988)
Cow91   J.J. Cowan, F.-K. Thielemann, J.W. Truran: Phys. Rep., in press
Deg84   T. Degrand, K. Kajantie: Astrophys. Space. Sci. **10**, 203 (1984)
Del89   C.P. Deliyannis et al.: Phys. Rev. Lett. **62**, 1583 (1989)
Fow67   W.A. Fowler, G.E. Caughlan, B.A. Zimmerman: Ann. Rev. Astron. Astrophys. **5**, 525 (1967)
Fre90   K. Freese, F.C. Adams: Phys. Rev. **D41**, 2449 (1990)
Fuk89   H. Fukugita: Nucl. Phys. B (Proc. Suppl.) **9**, 291 (1989)
Ful82   G.M. Fuller, W.A. Fowler, M. Newman: Ap. J. Suppl. **48**, 279 (1982)
Ful85   G.M. Fuller, W.A. Fowler, M. Newman: Ap. J. **293**, 1 (1985)
Ful88   G.M. Fuller, G.J. Mathews, C.R. Alcock: Phys. Rev. **D37**, 1380 (1988)
Gil88   K.K. Gilroy, C. Sneden, C.A. Pilachowski, J.J. Cowan: Ap. J. **327**, 298 (1988)
Hob87   L.M. Hobbs, D.K. Duncan: Ap. J. **317**, 786 (1987)
Hob88   L.M. Hobbs, C. Pilachowski: Ap. J. Lett. **326**, L23 (1988)
Hog83   C. Hogan: Phys. Lett. **B133**, 172 (1983)
How80   W.M. Howard, P. Möller: Atomic Data Nuclear Data Tables **25**, 219 (1980)
Kaj90a   T. Kajino, R.N. Boyd: Ap. J. **359**, 267 (1990)
Kaj86   K. Kajantie, H. Kurki-Sounio: Phys. Rev. **D34**, 1719 (1986)
Kaj89   T. Kajino, G.J. Mathews, G.M. Fuller: In *Heavy Ion Physics and Nuclear Astrophysical Problems*, ed. by S. Kubono, M. Ishihara, T. Nomura (World Scientific, Singapore 1989), p.51
Kaj90b   T. Kajino, G.J. Mathews, G.M. Fuller: Ap. J. **364**, 7 (1990)
Kaw88   L. Kawano, D.N. Schramm, G. Steigman: Ap. J. **327**, 750 (1988)
Kla84   H.V. Klapdor, J. Metzinger, T. Oda: At. Nucl. Data Tables **31**, 81 (1984)
Kol90   E.W. Kolb, M.S. Turner: *The Early Universe*, (Addison-Wesley 1990)
Kur85   H. Kurki-Suonio: Nucl. Phys. **B255**, 231 (1985)
Kur89   H. Kurki-Suonio, R.A. Matzner: Phys. Rev. **D39**, 1046 (1989)
Kur90   H. Kurki-Suonio, R.A. Matzner, K.A. Olive, D.N. Schramm: Ap. J. **353**, 406 (1990)
Lod83   J. Lodenquai, V. Dixit: Phys. Lett. **B124**, 199 (1983)
Mal90   R.A. Malaney: In *Primordial Nucleosynthesis*, ed. by W. Thompson, B. Carney (World Scientific, Singapore 1990), p.49
Mal87   R.A. Malaney, W.A. Fowler: In *Origin and Distribution of the Elements*, ed. by G.J. Mathews (World Scientific, Singapore 1987), p.76
Mal88   R.A. Malaney, W.A. Fowler: Ap. J. **333**, 14 (1988)
Mal89   R.A. Malaney, W.A. Fowler: Ap. J. Lett. **345**, L5 (1989)
Mal91   R.A. Malaney, G.J. Mathews: Phys. Rep., in press
Mam89   W. Mampe et al.: Phys. Rev. Lett. **63**, 593 (1989)
Mat88   G.J. Mathews, G.M. Fuller, C.R. Alcock, T. Kajino: In *Dark Matter*, ed. by J. Audouze, Tran Thanh Van (Editions Frontiers, France 1988), p.319
Mil89   J.C. Miller, O. Pantano: Phys. Rev. **D40**, 1789 (1989)
Nom85   K. Nomoto, F.-K. Thielemann, S. Myaji: Astron. Astrophys. **149**, 239 (1985)

Oli90   K. Olive, D.N. Schramm, G. Steigman, T. Walker: Phys. Lett. **B236**, 454 (1990)
Pag87   B.E.J. Pagel: In *A Unified View of the Macro- and Micro-Cosmos*, ed. by A. De
            Rujula, D.V. Nanopoulos, P.A. Shaver (World Scientific, Singapore 1987), p.399
Pag91   B.E.J. Pagel: Physica Scripta, in press
Pan89   O. Pantano: Phys. Lett. **B224**, 195 (1989)
Par90   T. Paradellis et al.: Z. Phys. **A337**, 211 (1990)
Pis84   R.D. Pisarski, F. Wilczek: Phys. Rev. **D29**, 338 (1984)
Pre86   W.H. Press, B.P. Flannery, S.A. Teukolsky, W.T. Vetterling: *Numerical Recipes*
            (Cambridge Univ. Press, Cambridge 1986)
Reb88   R. Rebolo, P. Molaro, J.E. Beckman: Astron. Astrophys. **192**, 192 (1988)
Rya90   S.G. Ryan, M.S. Bessell, R.S. Sutherland, J.E. Norris: Ap. J. Lett. **348**, L57
            (1990)
Sal86   K.E. Sale, G.J. Mathews: Ap. J. Lett. **309**, L1 (1986)
See65   P.A. Seeger, W.A. Fowler, D.D. Clayton: Ap. J. Suppl. **97**, 121 (1965)
Spi86   F. Spite, M. Spite: Astron. Astrophys. **163**, 340 (1986)
Suh82   E. Suhonen: Phys. Lett. **B119**, 81 (1982)
Ter89a   N. Terasawa, K. Sato: Phys. Rev. **D39**, 2893 (1989)
Ter89b   N. Terasawa, K. Sato: Prog. Theor. Phys. **81**, 254 (1989)
Ter89c   N. Terasawa, K. Sato: Prog. Theor. Phys. **81**, 1085 (1989)
Thi91   F.-K. Thielemann, J. Applegate, J.J. Cowan, M. Wiescher: To be submitted to
            Ap. J.
Thi87   F.-K. Thielemann, M. Arnould, J.W. Truran: In *Advances in Nuclear Astro-
            physics*, ed. by E. Vangioni-Flam et al. (éditions frontières, Gif sur Yvette 1987),
            p.525
Thi83   F.-K. Thielemann, J. Metzinger, H.V. Klapdor: Z. Phys. **A309**, 301 (1983)
Thi90   F.-K. Thielemann, M. Wiescher: In *Primordial Nucleosynthesis*, ed. by W.
            Thompson, B. Carney, (World Scientific, Singapore 1990), p.92
Tom84   E.T. Tomboulis, L.G. Yaffe: Phys. Rev. Lett. **52**, 2115 (1984)
Tru70   J.W. Truran, A.G.W. Cameron, E. Hilf: CERN **70-30**
Uka89   A. Ukawa: Nucl. Phys. B (Proc. Suppl.) **10A**, 66 (1989)
Wag73   R.V. Wagoner: Ap. J. **179**, 343 (1973)
Wag67   R.V. Wagoner, W.A. Fowler, F. Hoyle: Ap. J. **148**, 3 (1967)
Wag69   R.V. Wagoner: Ap. J. Suppl. **18**, 247 (1969)
Wie86   M. Wiescher, J. Görres, F.-K. Thielemann, H. Ritter: Astron. Astrophys. **160**,
            56 (1986)
Wie87   M. Wiescher, V. Harms, J. Görres, F.-K. Thielemann, L.J. Rybarcyk: Ap. J. **316**,
            162 (1987)
Wap88   A.H. Wapstra, G. Audi, R. Hoekstra: At. Nucl. Data Tables **39**, 281 (1988)
Wie88a   M. Wiescher, J. Görres: Z. Phys. **A329**, 121 (1988)
Wie88b   M. Wiescher, J. Görres, F.-K. Thielemann: Ap. J. **326**, 384 (1988)
Wie89a   M. Wiescher, J. Görres, S. Graaf, L. Buchmann, F.-K. Thielemann: Ap. J., in
            press
Wie89b   M. Wiescher, R. Steininger, F. Käppeler: Ap. J. **344**, 464 (1989)
Wie90   M. Wiescher, J. Görres, F.-K. Thielemann: Ap. J. **363**, 340 (1990)
Wit84   E. Witten: Phys. Rev. **D30**, 272 (1984)
Woo75   S.E. Woosley, W.A. Fowler, J.A. Holmes, B.A. Zimmerman: Caltech Rep. OAP-
            422 (1975)
Woo78   S.E. Woosley, W.A. Fowler, J.A. Holmes, B.A. Zimmerman: At. Nucl. Data Ta-
            bles **22**, 371 (1978)
Yaf82   L.G. Yaffe, B. Svetitsky: Phys. Rev. **D26**, 963 (1982)
Yan84   J. Yang, M.S. Turner, G. Steigman, D.N. Schramm, K.A. Olive: Ap. J. **281**, 493
            (1984)

The s-Process: Branchings and Chronometers

F. Käppeler

Kernforschungszentrum Karlsruhe, Institut für Kernphysik, Postfach 3640, D-7500 Karlsruhe, Germany

1. Introduction

This contribution presents a discussion of the various potential s-process chronometers. In principle, this means that one compares the observed abundance of an unstable isotope or of its daughter nucleus with the original abundance by which they were produced in the slow neutron capture process (s-process). The identification and discussion of potential chronometers requires appropriate models for the s-process, which serve a twofold purpose: they define the relevant time scales and they must allow to determine the original abundances of the respective unstable isotopes. The concepts for investigating nuclear chronometers are outlined in Sect. 2, together with some critical comments on the principal uncertainties inherently related to this subject. In Sect. 3, a brief summary of the current s-process concepts is presented, followed by a discussion of the related time scales in Sect. 4. Quantitative studies of these chronometers require the reliable description of their original s-process abundances. Since the abundances of nearly all potential chronometers are affected by branchings in the s-process capture path, this aspect needs to be addressed in some detail. In addition, the branchings themselves can be considered as chronometers, describing the shortest possible time scale in the s-process. In Sect. 5, these ideas are illustrated at the examples of ^{176}Lu (a cosmic clock with thermal defects) and of the chronometric pair ^{187}Re/^{187}Os (that yields information on a very short and possibly on a cosmic time scale as well).

2. Chronometers – Models and Reality

An ideal nuclear chronometer is expected to work unambiguously. It, therefore, should consist of an unstable isotope that is produced in a well defined proportion with its daughter. After production, it should be quickly confined in a closed system where it decays freely. Then, the age of the system can be expressed in terms of the mother/daughter ratio. The example of ^{14}C dating represents such an ideal chronometer in good approximation, but even there are problems with the $^{14}C/^{12}C$ ratio changing as a function of time or with contaminations by 'modern' carbon [Hed86].

Nuclear chronometers can be distinguished depending on whether the abundance of the mother or of the daughter nucleus is followed. If - as in the case of ^{14}C - the disappearance of the mother is considered, one obtains

$$N_m^o = N_m^* \cdot \exp[-\lambda \Delta t] \tag{1a},$$

where N_m^* and N_m^o denote the abundances of the mother isotope after production and at the time of observation, respectively. If the daughter belongs to a less abundant element than the mother, it may be more favorable to follow the appearance of the daughter,

$$N_d^o = N_d^m \cdot \{\exp(\lambda \Delta t) - 1\} \tag{1b}.$$

Provided that one deals with an ideal chronometer, the time interval of interest, Δt, can easily be determined from these expressions.

In reality, however, practically none of the s-process chronometers falls in this category. Instead, the interpretation of nuclear chronometers is hampered in various respects:

(i) The initial abundance of the mother is neither constant nor well defined in time. It is characterized by at least two different time scales, the 10^7 years of s-processing during the asymptotic giant branch (AGB) phase of a particular star, and by the roughly 10^{10} years of stellar nucleosynthesis during galactic evolution.

(ii) The mother isotope is most often not confined in a closed system. The s-process material is mixed with matter of very different composition already at the stellar site and again during and after its ejection into the interstellar medium (ISM) by stellar winds, in a planetary nebula, or in a supernova.

(iii) Even the decay rate, λ, turns out to be subject to considerable uncertainties. Under the extreme conditions of the stellar plasma, the decay rate is no longer a constant of nature as we are used to it in the laboratory. Rather, the decay can be accelerated by orders of magnitude due to the action of high temperatures and densities [Tak87]. Examples of this type will be presented in Sect. 5.

Clearly, these difficulties do not allow for unbiased interpretations, but require complex analyses including a comprehensive description of s-process nucleosynthesis as a whole. On the other hand, it is this complexity that makes the study of nuclear chronometers a fascinating field of research. In particular, it offers the possibility for testing the various s-process models for internal consistency.

3. s-Process Concepts

3.1 The Classical Approach

The s-process is commonly accepted to occur during the helium burning stages
of stellar evolution. The classical picture [Bur57] assumes a small fraction of the
observed ^{56}Fe abundance as a seed for this process. Given an appropriate distribution of neutron exposures [See65], the classical s-process approach turned
out to be a very successful tool for reproducing the observed s-abundances
within typical uncertainties of 3% [Käp90]. It is one of the characteristics of
the s-process that the abundances are anti-correlated with the respective stellar neutron capture cross sections. Isotopes with large cross sections are easily
transformed into heavier nuclei and do not develop large abundances, whereas
small cross sections cause the build-up of large abundances before the neutron
capture flow can proceed further.

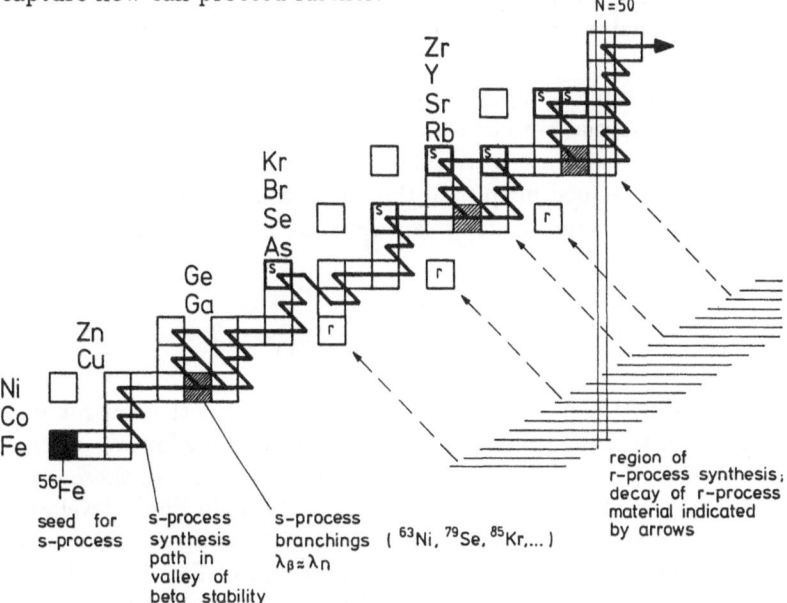

Fig. 1. The s-process neutron capture chain from Fe to Zr (solid line) Sufficiently long-lived
isotopes, which may represent potential chronometers, are indicated by shaded boxes. For
details see text

In general, the observed abundances [And89] are a mixture of contributions
from the s-process and from the rapid neutron capture process (r-process). The
latter is associated with stellar explosions and is characterized by extremely
high neutron densities. Accordingly, the synthesis path is shifted from the valley
of beta stability to the neutron rich side by about 15 to 20 mass units as
indicated in Fig. 1. Only after termination of the r-process, the synthesized
material falls back to the valley of beta stability via beta decay chains that end
at the first stable isotopes, respectively. The dashed arrows in Fig. 1 point to

those nuclei, that are not reached by the *s*-process because of their short-lived neighbors, and which can, therefore, be considered as being of pure *r*-process origin. In turn, their isobars - indicated by heavy boxes - are completely shielded against any *r*-process contributions and can, therefore, be ascribed to the *s*-process. (To a small extent, these *s*-nuclei may receive an additional abundance contribution from the so-called *p*-process that is responsible for the existence of the few neutron poor isotopes that lie off the *s*-process path. These minor corrections are not essential in this context and will not be discussed further).

The two ensembles of pure *s*- and *r*-process nuclei are the cornerstones of any quantitative model for either one of these processes. As mentioned above, the classical approach succeeds to reproduce the *s*-only isotopes within 3%. In addition, it allows to decompose the *s*- and *r*-process contributions for the mixed isotopes by subtracting the *s*-process yields from the observed abundances. It turns out that these *r*-process residuals are in perfect agreement with the abundance distribution of the *r*-only nuclei. This success in reproducing the observed *s*- and *r*-abundance patterns suggests that the classical approach can be considered as a reliable basis for the discussion of the potential *s*-process chronometers.

In principle, all unstable isotopes in the neutron capture chain of Fig. 1 can be understood as a potential *s*-process chronometer, provided that it defines a time scale that corresponds to a physically significant quantity. The shortest time scale that can be defined in spite of the schematic nature of the classical approach is the inverse of the neutron capture rate, i.e. the neutron capture time for an isotope A

$$\tau_{n(A)} = \frac{1}{\lambda_{n(A)}} = \frac{1}{n_n \cdot v_T \cdot \sigma(A)} \tag{2}$$

that is characterized by the neutron flux $n_n \cdot v_T$, where n_n is the neutron density and v_T the mean thermal velocity. Potential chronometers for this shortest time scale are all isotopes with half-lives comparable to τ_n . In all these cases, the *s*-process path, which is indicated in Fig. 1 by the solid line, develops a branching as a consequence of the competition between beta decay and neutron capture. The corresponding examples in the mass range of Fig. 1 are the isotopes ^{63}Ni, ^{79}Se, and ^{85}Kr, which are indicated by shaded boxes. The strength of a branching can be expressed by the branching ratio,

$$f_\beta = \frac{\lambda_\beta}{\lambda_\beta + \lambda_n} \tag{3}$$

where λ_n and λ_β are the rates for neutron capture and beta decay. The branching ratio is preserved in the abundance pattern of the involved isotopes. By analyzing this pattern in terms of an *s*-process model, it is possible to extract the actual neutron capture rate, and, hence, the mean neutron density at the *s*-process site. This holds as long as the beta decay rate is not affected by temperature. The systematic study of all suited branchings (see Sect. 4.1) yields, indeed, consistent solutions for the mean neutron density with an average of

$n_n = (3.4 \pm 1.1) \cdot 10^8$ cm^{-3} [Käp90]. This result shows that typical neutron capture times are between 0.2 and 4 years, depending on the respective stellar capture cross sections.

The only other time scale that follows naturally from the classical approach is that related to the age of the s-process elements in general. This aspect will be addressed in Sects. 4.2 and 5.1.

3.2 Stellar s-Process Models

That the s-process occurs during the helium burning stages of stellar evolution is accepted since the discovery of the element technetium in the atmosphere of S stars, a subclass of Red Giants [Mer52]. Since all technetium isotopes are unstable with half-lives much shorter than the age of these stars, this was clear evidence for the production of the observed Tc in just these stars. The neutrons for the s-process are produced in the helium burning zones predominantly by (α,n)-reactions on the isotopes ^{13}C and ^{22}Ne.

Two different s-processes can be inferred from the observed abundances, one that has produced the s-abundances from zirconium to bismuth, and a second one that has been more efficient between iron and yttrium. In terms of the classical approach, these two processes are called the main and the weak s-process component. Here, the discussion will be restricted to stellar models of the main component, since practically all potential chronometers are found at mass numbers $A > 90$.

The main s-process component is characterized by an exponential distribution of neutron exposures. This property that was empirically found by [See65], was explained by [Ulr73] as the natural consequence of the s-process occurring in repeated He shell burning episodes. While subsequent episodes move outward towards larger stellar radii, the corresponding s-processing zones exhibit a fractional overlap of about 40 to 60%. Therefore, rather large fractions of s-processed material experience only low exposures during one or two irradiation episodes, whereas the probability for larger exposures becomes ever smaller. The resulting distribution of exposures is indeed an exponential as suggested empirically by the classical approach.

The situation of helium shell burning is sketched in Fig. 2. Nuclear energy is generated by hydrogen burning in the CNO cycle at the bottom of a deeply convective envelope. The so-produced helium accumulates in a thin shell around the inert carbon-oxygen core. When the helium shell has reached a critical thickness, the bottom temperature becomes high enough for helium burning to ignite. The enormous energy production during the helium burning episode causes an expansion and cooling of the envelope. As a result, hydrogen burning is extinguished. When the helium fuel is exhausted the envelope shrinks back and heats up until hydrogen burning starts again to provide the stellar energy and to produce the fuel for the next helium burning instability. This cycle is repeated many times, with the helium shell moving outward in radius. The helium burning zone of each instability overlaps with that of the preceding

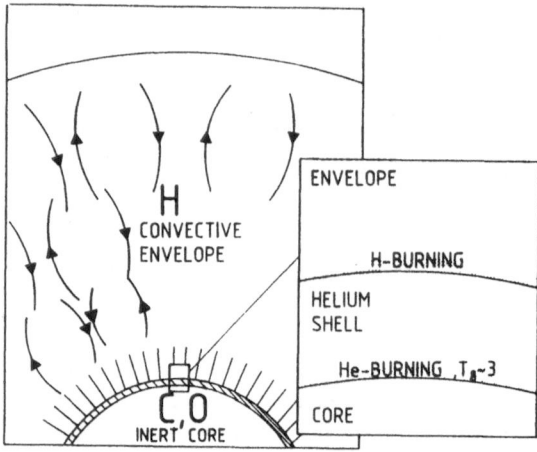

Fig. 2. Schematic diagram of helium shell burning during the Red Giant stage of stellar evolution. Hydogen burning at the bottom of the deeply convective envelope produces helium that accumulates in a thin layer around the inert carbon/oxygen core. When the bottom of the helium shell reaches a critical temperature, helium burning is ignited and moves outward towards the He-H-interface. Subsequent He-burning episodes do overlap each other as indicated in Fig. 3

one, thus producing the exponential distribution of exposures characteristic for the observed s-process abundances.

During the helium burning instabilities, the expansion and shrinkage of the envelope provides for a possibility of mixing freshly synthesized material to the stellar atmosphere, thereby explaining the actually observed s-process enhancements in AGB stars. This scenario for the s-process defines three more time scales that characterize the dynamics of helium shell burning. These are the time interval between and the typical duration of the helium shell burning episode, as well as the mean transport time for the s-processed material to the stellar surface.

The first stellar model of this scenario that succeeded in a quantitative description of s-process abundances was worked out for intermediate mass stars in the range between 3 and 8 M⊙ [Wei66, Sch67, Ibe76, Ibe83]. In general, the operation of the $^{22}Ne(\alpha,n)^{25}Mg$ reaction as a neutron source in these stars was shown to produce the s-process abundances in the mass range $90 < A < 206$ in solar proportions [Tru77]. However, the very high neutron densities predicted for these stars resulted in systematic discrepancies for the abundance patterns in s-process branchings [Des80, How86], leading to severe underabundances of the bypassed s-only nuclei.

The problems with the $^{22}Ne(\alpha,n)^{25}Mg$ source in intermediate mass stars and the fact that significant s-process enhancements are only observed in low mass AGB stars (LMS) of low metallicity have focused the theoretical efforts on these LMS stars, particularly since Iben and Renzini [Ibe82a, Ibe82b, Ibe83] had shown that semiconvective mixing of protons into the helium burning shell could produce enough ^{13}C to make the $^{13}C(\alpha,n)^{16}O$ reaction the dominant neutron source. This alternative for the site of the s-process was investigated

by a number of recent studies [Ibe82a, Ibe82b, Boo88, Hol88, Hol89]. In fact,
it was shown to reproduce the observed abundances in the solar system and in
evolved stars, and it also satisfies the main constraints imposed by the chemical
evolution of the Galaxy [Gal88, Gal89, Käp90].

In this model, there are two neutron bursts in each thermal instability. The
main burst is caused by the $^{13}C(\alpha,n)^{16}O$ reaction at temperatures of $\sim 1.5 \cdot 10^8$
K, when a zone enriched in ^{13}C is engulfed by the convective He shell. A
second, smaller burst follows at the end of the instability, when the bottom
temperature of the He shell rises sharply to $\sim 3 \cdot 10^8$ K. In a way, this second
burst adjusts the s-process branchings, which otherwise would reflect the higher
neutron density and lower temperature during the main burst. These features
of thermal instabilities during helium shell burning in LMS are sketched in
Figs. 3 and 4.

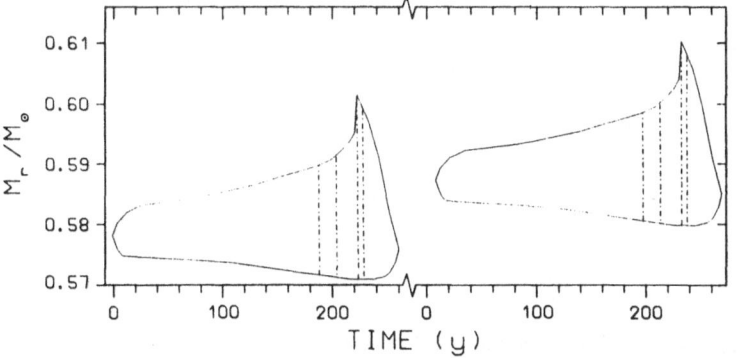

Fig. 3. The position and extension of the helium burning shell in a LMS for two subsequent
thermal instabilities as a function of time. The two instabilities are separated by about $2 \cdot 10^5$
yr. Note the radial overlap of the two zones that causes the characteristic exponential distribution of neutron exposures. The dashed-dotted lines indicate the periods when neutrons
are produced via the $^{13}C(\alpha,n)^{16}O$ and $^{22}Ne(\alpha,n)^{25}Mg$ reactions, respectively (see Fig. 4)

Fig. 3 shows the extension of the strongly convective helium burning zone
inside a LMS as a function of time. The ordinate represents the stellar mass
fraction and indicates the position and size of the instability inside the star.
Helium burning starts rather violently and grows then at an almost constant
rate. The main neutron burst occurs after ~ 190 yr when the He burning front
reaches a zone enriched in ^{13}C. This ^{13}C originates from proton captures on
^{12}C to ^{13}N (and subsequent β^+ decays), when protons from the envelope are
mixed into the helium shell by semiconvection during the preceding instability.
The mean neutron density in this main burst is determined by the speed of
the He burning front and by the extension of the ^{13}C rich zone. At this point,
the mean neutron density and, hence, the neutron capture time introduced in
the classical approach turns out to be related to the dynamics of helium shell
burning in LMS stars! As illustrated in Fig. 4, this main neutron burst lasts
for about 20 yr. It is followed by a neutron-free interval of ~ 15 yr until the
temperature rises sufficiently for the $^{22}Ne(\alpha,n)^{25}Mg$ reaction to take place. The

corresponding second neutron burst lasts for about 2.5 yr and consumes part of the ^{22}Ne that was produced by double α-captures on the ^{14}N left over from the CNO cycle.

The time scales defined by this s-process model range from about 0.2 to 10^5 yr: from the neutron capture time and the duration of the two neutron bursts, over the duration of a thermal instability, the transport time of processed matter to the stellar surface, to the interval between thermal instabilities. After identification of these time scales, the next question is, of course, whether there are suited chronometers to study these phenomena.

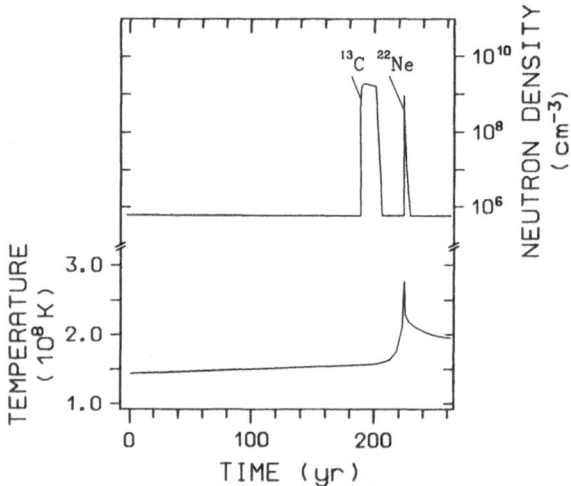

Fig. 4. Neutron production during helium shell burning episodes in LMS: A main neutron burst is produced when the He burning front hits a zone enriched in ^{13}C, and a second burst occurs when the temperature rises at the end of the instability so that the ^{22}Ne(α,n)^{25}Mg reaction becomes activated. Typical time scales are 20 and 3 yr for the neutron bursts, 225 yr for the duration of the instability, and about 10^5 yr between subsequent instabilities

4. s-Process Time Scales

4.1 Helium Shell Burning

Short time scales can best be investigated via appropriate s-process branchings. This holds mostly for the neutron capture time, but in two cases also for the duration of the neutron bursts. The longer transport time of freshly synthesized material to the stellar surface requires quantitative abundance determinations of unstable isotopes by astronomical methods.

4.1.1 Mean Neutron Density : Branchings that are suited for determining the mean neutron density are expected to satisfy the following criteria:

- The neutron capture time of the unstable branch point isotope should be short compared to the duration of the neutron bursts in Fig. 4 (see Sect. 4.1.2).
- The half-life of the unstable branch point isotope must be insensitive to the prevailing temperatures.
- Uncertainties introduced by the input data should still allow for a meaningful analysis.

Under these conditions, the best candidates are the branchings at $A = 147/148$ and at $A = 185/186$ with the branch point isotopes ^{147}Nd, ^{147}Pm, ^{148}Pm, and ^{185}W, ^{186}Re, respectively. Other possible examples were the branchings at ^{95}Zr, ^{169}Er/^{170}Tm, and ^{191}Os/^{192}Ir, but these do not satisfy all of the above criteria. Analysis of the two relevant branchings by means of the classical approach yields a mean neutron density $n_n = (3.4 \pm 1.1) \cdot 10^8$ cm^{-3}. On average, this neutron density is smaller than predicted by the LMS model (Fig. 4). This suggests that the ^{13}C may be diluted in a larger part of the He shell as a consequence of more efficient mixing.

4.1.2 Duration of s-Process Pulses : The abundance patterns of branchings can be affected by the duration of the neutron bursts of Fig. 4 only if the neutron capture time of the branch point isotope is compatible with the length of a pulse. Since the branch point isotope decays completely between subsequent helium burning episodes, in such cases, the neutron capture flow never reaches equilibrium before the end of the neutron irradiation. Branch points with sufficiently small cross sections are ^{85}Kr and ^{95}Zr with correspondingly large capture times of 3.8 and 6.5 years. While these times are still too short to make the branchings sensitive to the first neutron burst, they are just right for the shorter burst at the end of the helium burning episode. However, both branchings are difficult to analyze because of significant abundance contributions from other processes, e.g. from the weak s-process component and from the r-process. Therefore, a satisfactory solution for the duration of the s-process pulses is still pending.

4.1.3 Transport to Stellar Surface: It has been mentioned in Sect. 3.2 that technetium is observed in the atmosphere of S giants. By improved observational techniques and atmospheric models, technetium abundances can now be determined quantitatively [Smi83, Smi86, Smi88]. These direct observations are all the more exciting since the decay rate of ^{99}Tc, which is the technetium isotope produced in the s-process, is drastically enhanced at high temperatures [Cos81, Sch83, Tak87]. The terrestrial half-life of $2.1 \cdot 10^5$ yr persists up to $1 \cdot 10^8$ K but is then steeply reduced to a few years at typical s-process temperatures of $3 \cdot 10^8$ K.

This behavior implies that ^{99}Tc is exposed to high temperatures only for a short time, in agreement with the LMS model sketched above. As soon as it is mixed into cooler zones with temperatures below 10^8 K, it decays with its terrestrial half-life. That it is actually observed on the surface means that the transport time from the s-process zone must be of the order of the terrestrial half-life or shorter. A second chronometer for the transport time is provided by ^{93}Zr ($t_{1/2} = 1.5 \cdot 10^6$ yr), which is complementary to ^{99}Tc since its half-life is almost independent of temperature. While ^{93}Zr is not easily detected in astronomical observations due to the presence of the stable Zr isotopes, its decay can be followed via the daughter ^{93}Nb which is the only stable Nb isotope. Since it is not produced in the s-process, its appearance is a sensitive indicator for the decay of ^{93}Zr.

The ^{99}Tc and ^{93}Zr abundances have been followed during their transport to the surface for the case of intermediate mass stars [Mat86, Tak86]. It was found that these isotopes represent indeed a useful clock for the time scales related to the third dredge-up phase of thermally pusing stars. These investigations need to be extended to the more successful LMS model [Bus90], and definitely require improved field studies of ^{99}Tc and ^{93}Zr/^{93}Nb abundances in AGB stars.

4.2 Remixing to the Interstellar Medium
    and the Age of the Galaxy

The s-process occurring in AGB stars leads naturally to the related questions:

- How was the s-processed material remixed to the interstellar medium, when and how was it incorporated into the protosolar cloud?
- When during galactic evolution did the s-process start, and at which rate did it change in time?

The first question stands for the problem at which stage and by which mechanisms remixing to the ISM takes place. A possible ansatz to investigate these problems are the anomalous isotopic ratios discovered in meteorites. The second question aims at the galactic age and hence at the age of the universe itself.

4.2.1 Isotopic Anomalies: Though element abundances exhibit large differences in samples of different origin due to chemical processing, isotopic ratios are fairly constant or are at most slightly modified by physical effects, e.g. by diffusion. This fact reflects the very homogeneous mixture of the protosolar cloud. The only known exceptions are anomalous isotope patterns that were discovered in small inclusions of certain meteorites. It was found that these anomalies reflect the admixture of material from specific nucleosynthesis sites to the average abundance distribution. For example, s-process anomalies have been reported for xenon and krypton [Sri78, Ala80, Ott88] and for neodymium [Lug83a, Lug83b].

Some of these anomalies are the result of in situ decay of now extinct radionuclides with half-lives in the 10^6 to 10^8 yr range, e.g. ^{26}Al, ^{107}Pd, ^{129}I, and ^{244}Pu. If these radionuclides were alive during formation of solid bodies in the early solar system, they must have been synthesized shortly before. One possible explanation for their origin were a local supernova, which produced these isotopes and also triggered the collapse of the protosolar cloud. In this scenario, the isotopic anomalies produced by the decay of radionuclides could establish a chronometry of the early solar system.

However, this interpretation was questioned by Clayton [Cla79], who pointed out that presolar grains could have formed much earlier and could have been ingested by the protosolar cloud as part of the ISM. In that picture, isotopic anomalies of radiogenic origin would not carry information on the early solar system but rather on the condensation time scale of the grains (for details see reviews by [Cla79, Beg80, Was82], for example).

The s-process aspect of that story is intriguing insofar as it may reveal, which mechanism is most efficient in remixing s-processed material into the ISM: stellar winds during or the ejection of a planetary nebula at the end of the AGB phase, or the final supernova explosion. For that purpose it would be necessary to identify an isotopic anomaly associated with an unstable s-isotope that could serve as an appropriate clock for the relevant time scale of 10^6 to 10^7 yr.

Yokoi et al. [Yok85] have shown that ^{205}Pb ($t_{1/2} = 1.4 \cdot 10^7$ yr) would be such an example. This chronometer was long overlooked, because the decay of ^{205}Pb is strongly enhanced at stellar temperatures. Therefore, it was not expected to be produced in significant amounts during the s-process. However, the small difference in atomic mass between ^{205}Pb and ^{205}Tl causes the latter isotope to become unstable against bound state beta decay in the stellar plasma. In this way, an equilibrium abundance between the $A = 205$ isobars is established and the so produced ^{205}Pb was found to survive the transport to the cooler outer zones. The final ^{205}Pb/^{204}Pb ratios were estimated to range between 10^{-5} and unity. Hence, isotopic anomalies in excess of the upper limit of $9 \cdot 10^{-5}$ that was obtained by Huey and Kohman [Hue72] may well exist. A renewed search for extinct ^{205}Pb is, therefore, required to exploit this chronometer further.

4.2.2 The Cosmic Time Scale:
In first attempts, nucleocosmochronolgy was restricted to relate the observed abundances of long-lived radionuclides to their original abundances predicted by model calculations. While in these first estimates the rate of galactic nucleosynthesis was simply assumed to have decreased exponentially [Fow72], the analysis of cosmic clocks turned out to be more complicated when galactic chemical evolution was considered as an additional constraint [Tin80, Cla84, Cla88]. In all these studies, the rate of nucleosynthesis had to be parameterized. It was hoped that the combined analysis of several long-lived radionuclides may yield a satisfactory fit of these parameters, provided that the respective half-lives were sufficiently different.

Studies of this type concentrated on the r-process chronometers ^{232}Th, ^{235}U, and ^{238}U. Depending on the assumptions for the rate of nucleosynthesis and on the calculated r-process yields, galactic ages between 10 and 20 billion years were reported [Fow78, Fow86, Thi86, Cow87]. This range is in agreement with, but no more accurate than, current estimates derived from the Hubble time [San84, Huc87, Bra87] or from the age of globular clusters [San82, Van83, Van85].

Among the s-process chronometers, ^{176}Lu is the only long-lived isotope that seemed to be suited as a cosmic clock [Aud72, Arn73]. However, it was shown recently [Kla91, Les91] that the decay rate of ^{176}Lu is so strongly enhanced during the s-process, that this isotope can hardly ever be interpreted as a reliable clock (see Sect. 5.1). The remaining long-lived species in the chart of nuclides, ^{40}K [Bee87], ^{87}Rb [Bee84a] and ^{187}Re [Yok83, Arn84, Win86, Käp91] are produced in at least two different processes. For this reason, they are difficult to interpret, apart from the problems related to the enhancement of the stellar decay rate of the latter two isotopes [Tak87]. The case of ^{187}Re will be discussed in more detail in Sect. 5.2.

An important progress in nucleocosmochronology has been achieved recently by the direct determination of long-lived radionuclides in old stars [But87, But88]. By combining the time scales set by nuclear clocks with stellar ages, such analyses are no longer restricted to solar material, but can directly be related to stellar and galactic evolution. The systematic study of ^{232}Th abundances in G-dwarfs of different galactic ages relative to the elements Ba, Nd, and Eu showed no evidence for a variation of the thorium abundance relative to the other elements, a result that is difficult to reconcile with galactic evolution models [Cla87, Cla88, Mat88]. Such investigations will certainly produce new insights in the history of our galaxy as observational uncertainties are further reduced.

4.3 Summary of Potential Chronometers

All potential s-process chronometers are summarized in Table 1 according to the various phenomena they characterize. Despite of their importance for a fascinating field of research, it is to be noted that practically all problems in this table are still waiting for a satisfactory solution. This situation is partly due to insufficient nuclear physics data, but mainly because the required abundance information is not yet accurate or detailed enough. Most of the progress in this field will, therefore, be triggered by new and better abundance determinations. This concerns improved astronomical observations of the surface composition in AGB stars as well as an extended and more comprehensive search for isotopic anomalies in meteorites.

5. Selected Examples

In this section, the influence of temperature and of branchings on the analysis of s-process chronometers is discussed for two examples. The stellar decay rate of ^{176}Lu is so extremely temperature-dependent that it can no longer be considered as a meaningful cosmic clock. The branchings at ^{185}W and ^{186}Re can be used to derive an estimate for the neutron density. They also illustrate the influence of s-process branchings on the analysis of chronometers, an aspect that concerns 90% of all cases listed in Table 1.

5.1 ^{176}Lu – a Defect Cosmic Clock

For many years, ^{176}Lu was considered as a promising s-process chronometer with respect to galactic history [Aud72, Arn73]. The s-process flow in the Yb-Lu-Hf region is sketched in Fig. 5; it follows the solid line by subsequent neutron captures and beta decays. At $A = 176$, abundance contributions from the r-process (indicated by arrows) are accumulated at ^{176}Yb, leaving the s-process as the only production mechanism for the isobars ^{176}Lu and ^{176}Hf (apart from minor p-process contributions). If the original s-process abundance of ^{176}Lu can be determined, e.g. by means of the classical approach [Käp89], the only complication in interpreting ^{176}Lu as an s-process chronometer seemed to be the existence of an isomeric state, which causes part of the s-process flow to bypass the long-lived ground state. This effect is considered in Fig. 5 by showing ground state and isomer separately. As indicated by the strength of the lines, neutron capture on ^{175}Lu leads more frequently to the isomer, leaving only a small probability for producing the long-lived ground state.

Table 1. Potential s-process chronometers and related phenomena

Dynamics of He shell burning		
(i)	neutron capture time, neutron density at s-process site:	$0.1 \leq \tau_n \leq 4$yr	branchings at ^{147}Nd, ^{147,148}Pm; ^{169}Er, ^{170}Tm; ^{185}W, ^{186}Re. ^{191}Os, ^{192}Ir.
(ii)	s-process pulses:	$\Delta \tau \sim 2.5$yr	branchings at ^{85}Kr and ^{95}Zr.
(iii)	transport to stellar surface:	$\tau < 10^5$ yr	decay of ^{99}Tc, ^{93}Zr/^{93}Nb.
Stellar evolution, remixing to ISM, star formation		
(i)	duration of AGB phase:	$10^5 \leq \tau \leq 10^6$ yr	decay of ^{99}Tc, ^{93}Zr/^{93}Nb.
(ii)	mechanism for ejection of s-processed matter	$10^5 \leq \tau \leq 10^7$ yr	decay of ^{107}Pd, ^{205}Pb.
Cosmology		
(i)	age of s-process	$\tau \sim 1.5 \cdot 10^{10}$ yr	decay of ^{176}Lu, ^{87}Rb.
(ii)	age of r-process	$\tau \sim 1.5 \cdot 10^{10}$ yr	^{187}Re/^{187}Os, ^{207}Pb.

Since the isomer decays exclusively to ^{176}Hf with a half-life of only 3.68 h, the original s-process abundance of ^{176}Hf is determined by the fraction of neutron captures feeding the isomer. The idea that ground state and isomer could be treated like two different nuclei appeared plausible, because direct

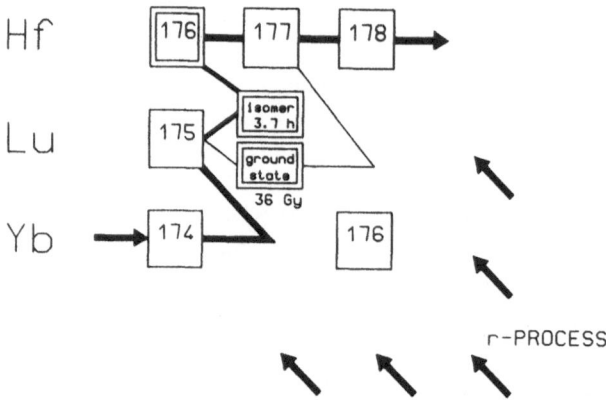

Fig. 5. The s-process neutron capture chain between Yb and Hf. Note, that ^{176}Lu and ^{176}Hf are shielded against the r-process

transitions are completely inhibited by selection rules. However, this picture was questioned by an obvious discrepancy: The partial capture cross section for populating the isomer in ^{176}Lu was found too large (e.g. [Zha91]) resulting in a ^{176g}Lu production that was not sufficient to account even for the observed abundance. On the other hand, Beer et al. [Bee84b] found from a comparison with the ^{176}Hf abundance that the s-process flow through ^{176g}Lu was considerably more efficient than expected from this result.

Therefore, additional feeding mechanisms for the long-lived ground state had to be identified. The most important one was that higher lying states may be excited in the hot stellar photon bath [War80, Bee81], which then decay back either to the isomer or to the ground state. In this way, the initial population of these two states can be efficiently changed, and even thermal equilibrium can be reached. Hence, the s-process abundances of ^{176}Lu and ^{176}Hf are no longer determined by their cross sections alone, but by the competition between their stellar beta decay and neutron capture rates, resulting in an abundance ratio that is sensitive to temperature and neutron density during the s-process.

The quantitative discussion of this mechanism became possible only after the level scheme of ^{176}Lu had been carefully established in two recent investigations [Kla91, Les91]. These studies identified an $I^\pi = 5^-$ state at 838.6 keV to be the most efficient mediating level for the coupling of ground state and isomer. The inset of Fig. 6 illustrates these thermally induced transitions for the case that ^{176g}Lu is exposed to high temperatures. As a consequence, the half-life of ^{176}Lu is drastically reduced above $1.5 \cdot 10^8$ K, i.e. at typical s-process conditions. The shaded band in Fig. 6 is obtained with the experimental limits for the life time of the mediating level at 838.6 keV, whereas the dashed-dotted an the dashed lines indicate the solutions for more conservative assumptions on the role of K-forbidden transitions. In particular, note the extremely sharp decrease of $t_{1/2}$ over 10 orders of magnitude!

The s-process branching that results from the reduced half-life of ^{176}Lu at stellar temperatures can now be described quantitatively for the first time.

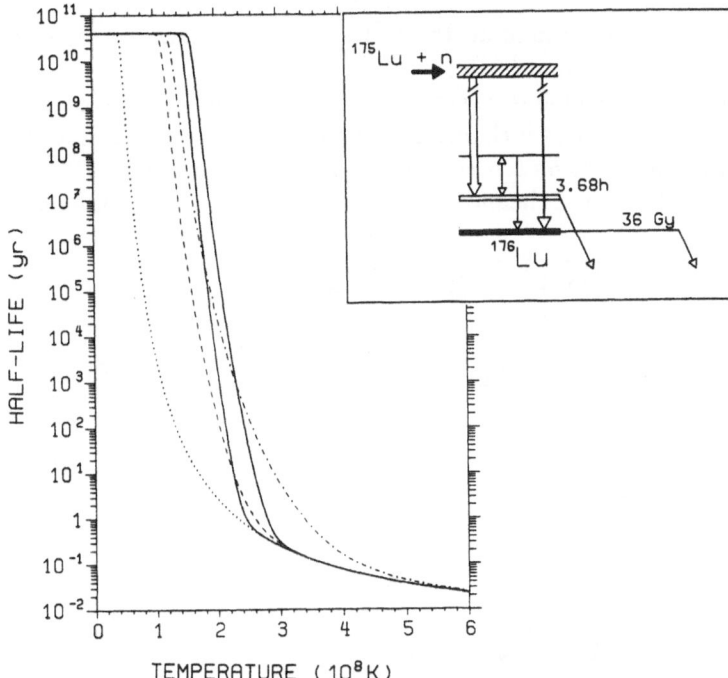

Fig. 6. The half-life of ^{176}Lu as a function of temperature

Nevertheless, the remaining uncertainties are by far too large to restore ^{176}Lu as a cosmic clock: the chronometer remains obscured by thermal effects. Instead, it can rather be considered as a suitable s-process thermometer, yielding a temperature range between $2.4 \cdot 10^8$ and $3.6 \cdot 10^8$ K [Kla91], in good agreement with similar estimates from other s-process branchings and with current stellar models [Käp90].

5.2 The Branchings at ^{185}W and ^{186}Re – Neutron Density and r-Process Age

The mass region between tungsten and osmium has received considerable attention, mostly due to the possible use of ^{187}Re as a chronometer for the r-process. Fig. 7 illustrates the s-process flow through the W-Re-Os isotopes and the corresponding contributions from the r-process beta decay chains. Obviously, ^{186}Os and ^{187}Os are shielded against the r-process by their isobars, and hence are produced as pure s-process nuclei. However, ^{187}Os has accumulated an additional abundance contribution from the decay of ^{187}Re, which is predominantly produced in the r-process. The s-process part of the ^{187}Os abundance can be reliably determined as the σN values of ^{186}Os and ^{187}Os are known to be equal according to the 'local approximation'. The excess in the ^{187}Os abundance can then be ascribed to the decay of ^{187}Re (e.g. [Sym81, Bro81, Win82]).

The quantitative interpretation of the chronometric pair ^{187}Re–^{187}Os is, however, complicated

- by the *s*-process contributions to the ^{187}Re abundance that results from the branchings at ^{185}W and ^{186}Re,
- by the problem of the stellar decay of ^{187}Re that can even be reversed under stellar conditions [Tak87], which means that the *s*-process corrections could be sensitively dependent on temperature and neutron density, and
- by problems related to the chemical evolution of the galaxy [Yok83, Arn84].

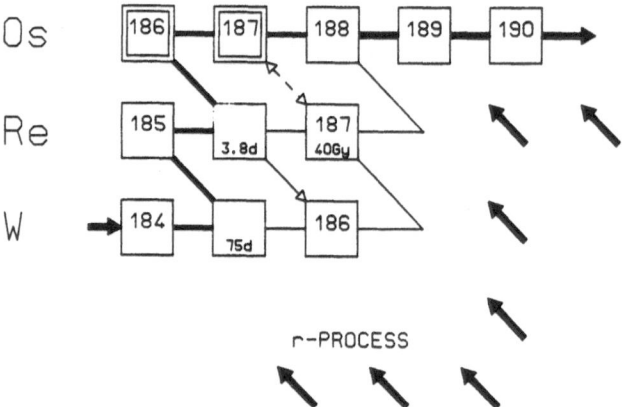

Fig. 7. The *s*-process flow in the W-Re-Os region. The *s*-only isotopes ^{186}Os and ^{187}Os are indicated by heavy boxes. Note, that ^{187}Os has received an additional contribution from the decay of ^{187}Re. This decay can even be reversed under stellar conditions [Tak87]

Apart from their importance for cosmochronometry, the isotopes ^{186,187}Os are of interest as normalization points for the *s*-process branchings at ^{185}W and ^{186}Re, which are indicated in Fig. 7. These branchings define the small *s*-process contribution to the ^{187}Re abundance, but are mainly useful for estimating the mean neutron density during the *s*-process. It is important in this respect that the beta decay rates of the branch point isotopes ^{185}W and ^{186}Re are almost unaffected at typical *s*-process temperatures [Tak87].

The uncertainties of the stellar capture rates, which hampered previous analyses of these branchings [Arn84, Bee84a, Bee84b, Bee88] have been considerably reduced [Käp91], leading to a mean neutron density,

$$n_{\rm n} = (4.1\pm^{1.2}_{1.1}) \cdot 10^8 {\rm cm}^{-3}.$$

This value is in good agreement with the results obtained from similar branchings (sections 3.1 and 4.1.1), but significantly smaller than the mean neutron density expected from stellar models, e.g. that for LMS outlined in Sect. 3.2. The discrepancy indicates that the parameterization of, for example, convective mixing, may have to be reconsidered in stellar models.

An important byproduct of the branching analysis for deriving the mean neutron density are the *s*-process contributions to the Re and Os abundances, which are required for correcting the ^{187}Re/^{187}Os clock. Indeed, one finds that ^{187}Re is not of pure *r*-process origin. If the *s*-contribution to the ^{187}Re abundance is taken into account, the age estimates based on that chronometer are

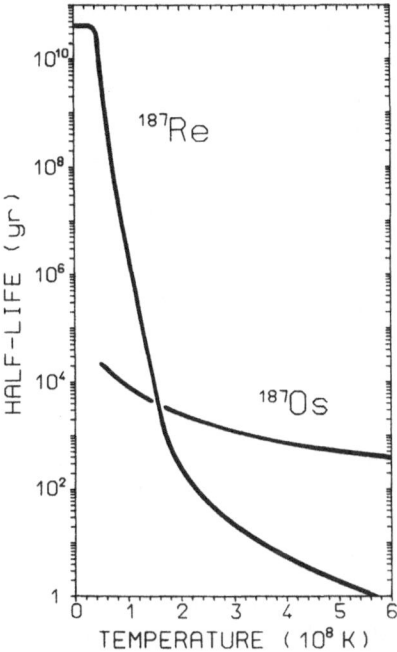

Fig. 8. The half-lives of ^{187}Re and ^{187}Os as a function of temperature

reduced by $\sim 6\%$. In view of the remaining uncertainties related to stellar models or to galactic evolution, these s-process corrections do not appear too dramatic.

The temperature sensitivity of the decay rates is another aspect to consider here. The half-lives of ^{187}Os and ^{187}Re are plotted in Fig. 8 for an assumed electron density of $1 \cdot 10^{27}$ cm^{-3} [Tak87]. It is important to note from this figure, that both isotopes still exhibit rather long half-lives at a typical s-process temperature (1242 yr and 21.3 yr at $3 \cdot 10^{8}$ K, respectively). Since their beta decay is much slower than the corresponding neutron capture times of ~ 0.25 yr, these chronometric isotopes are practically stable during the s-process, and are not expected to develop unusual s-process abundance ratios.

The validity of this qualitative argument was checked by means of the s-process model for LMS in order to study the influence of the dynamical helium shell burning environment on the s-process contributions to the ^{187}Re and ^{187}Os abundances. Using the profiles for neutron density and temperature of the LMS model outlined in Sect. 3.2, the abundance ratio $N_s(^{187}\text{Re})/N_s(^{187}\text{Os})$ was followed with a reaction network comprising 72 nuclei between ^{178}Hf and ^{209}Bi [Käp91]. Fig. 9 illustrates the result of this calculation: The dashed line corresponds to the second neutron burst (Fig. 4) that is characterized by steep gradients for temperature and neutron density. The s-process abundance ratio of the chronometric pair exhibits a certain decrease as the neutron burst develops, but stabilizes then close to the value obtained with the classical approach.

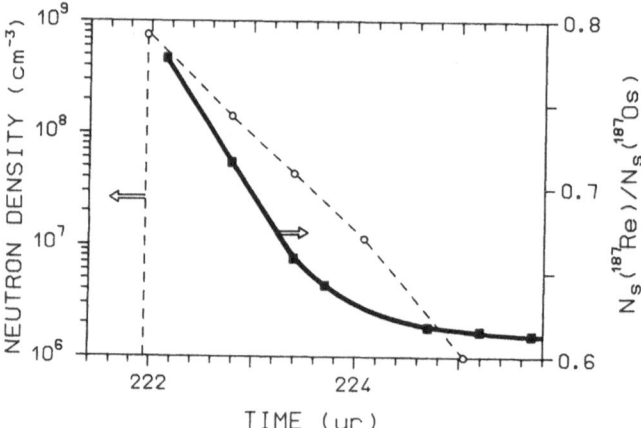

Fig. 9. The *s*-process abundance ratio ^{187}Re/^{187}Os (right scale) as it develops in the second neutron burst during a helium shell burning instability in LMS. The corresponding neutron density profile is given by the dashed line (left scale)

From this discussion, two of the above concerns, the *s*-corrections to the observed abundances as well as the temperature dependence of the beta decay rates of ^{187}Re and ^{187}Os, were found to be no strong limitations for this chronometer; in other words, there seem to be no stringent problems as far as the *s*-process environment is concerned. The third aspect of galactic evolution remains to be solved, but can now be addressed with renewed hope for the Re/Os clock.

6. Conclusions

The essentials of this report on the current status of *s*-process chronometers can be summarized in three points:

- The characteristic time scales span a wide range, from the dynamics of helium shell burning scenarios to galactic ages. Quantitative studies of the related chronometers may yield solutions to a variety of open questions that can hardly be achieved otherwise.
- The investigation of *s*-process chronometers requires the combination of results from very different fields, e.g. from nuclear physics, stellar models, abundance determinations by astronomical observations, and isotopic anomalies. Therefore, the success of such studies depends strongly on the interaction of specialized groups.
- Satisfactory answers do not yet exist for any of the potential chronometers! In view of their importance for so many aspects of astrophysics, the investigation of nuclear chronometers represents a promising and rewarding field of research.

References

Ala80   L. Alaerts, R.S. Lewis, J.-I. Matsuda, E. Anders: Geochim Cosmochim. Acta **44**, 189 (1980)

And89   E. Anders, N. Grevesse: Geochim. Cosmochim. Acta **53**, 197 (1989)

Arn73   M. Arnould: Astron. Astrophys. **22**, 311 (1973)

Arn84   M. Arnould, K. Takahashi, K. Yokoi: Astron. Astrophys. **137**, 51 (1984)

Aud72   J. Audouze, W.A. Fowler, D.N. Schramm: Nature **238**, 8 (1972)

Bee81   H. Beer, F. Käppeler, K. Wisshak, R.A. Ward: Ap. J. Suppl. **46**, 295 (1981)

Bee84a   H. Beer, G. Walter: Space Sci. Rev. **100**, 243 (1984)

Bee84b   H. Beer, G. Walter, R.L. Macklin, P.J. Patchett: Phys. Rev. **C30**, 464 (1984)

Bee87   H. Beer, R.-D. Penzhorn: Astron. Astrophys. **174**, 323 (1987)

Bee88   H. Beer, R.L. Macklin: Ap. J. **331**, 1047 (1988)

Beg80   F. Begemann: Rep. Prog. Phys. **43**, 1309 (1980)

Boo88   A.I. Boothroyd, I.-J. Sackmann: Ap. J. **328**, 632, 641, 653, 671 (1988)

Bra87   D. Branch: Ap. J. Letters **320**, L23 (1987)

Bro81   J.C. Browne, B.L. Berman: Phys. Rev. **C23**, 1434 (1981)

Bur57   E.M. Burbidge, G.R. Burbidge, W.A. Fowler, F. Hoyle: Rev.Mod.Phys. **29**, 547 (1957)

Bus90   M. Busso, R. Gallino, G. Picchio, C.M. Raiteri: In *Nuclei in the Cosmos*, ed. by H. Oberhummer and W. Hillebrandt (Max-Planck Institut für Physik und Astrophysik, Garching 1990), 233

But87   H.R. Butcher: Nature **328**, 127 (1987)

But88   H.R. Butcher: The ESO Messenger **51**, 1 (1988)

Cla79   D.D. Clayton: Space Sci.Rev. **24**, 147 (1979)

Cla84   D.D. Clayton: Ap.J. **285**, 411 (1984)

Cla87   D.D. Clayton: Nature **329**, 397 (1987)

Cla88   D.D. Clayton: Mon.Not.R.astr.Soc. **234**, 1 (1988)

Cos81   K. Cosner, J.W. Truran: Astrophys.Space Sci. **78**, 85 (1981)

Cow87   J.J. Cowan, F.-K. Thielemann, J.W. Truran: Ap.J. **323**, 543 (1987)

Des80   K.H. Despain: Ap.J.Letters **236**, L165 (1980)

Fow72   W.A. Fowler: In *Cosmology, Fusion and Other Matters*, ed. by F. Reines (Colorado Associated University Press, Boulder 1972), 67

Fow78   W.A. Fowler: In *21st Welch Foundation Conf. on Chemical Research*, ed. by W.D. Milligan (Robert A. Welch Foundation, Houston 1978) 61.

Fow86   W.A. Fowler, C.C. Meisl: In *Cosmogonical Processes*, ed. by W.D. Arnett et al. (VNU, Singapore 1986) 83

Gal88   R. Gallino, M. Busso, G. Picchio, C.M. Raiteri, A. Renzini: Ap.J. Letters **334**, L45 (1988)

Gal89   R. Gallino: In *The Evolution of Peculiar Red Giants*, IAU Symposium No.106, ed. by H.L. Johnson and B. Zuckerman (Cambridge University Press, Cambridge 1989) p. 176

Hed86   R.E.M. Hedges, J.A.J. Gowlett: Scientific American **254/1**, 82 (1986)

Hol88   D.E. Hollowell, I. Iben Jr.: Ap.J. Letters **333**, L25 (1988)

Hol89   D.E. Hollowell: Ap. J. **340**, 966 (1989)

How86   W.M. Howard, G.J. Mathews, K. Takahashi, R.A. Ward: Ap.J. **309**, 633 (1986)

Huc87   J.P. Huchra: In *13th Texas Symposium on Relativistic Astrophysics*, ed. by M.P. Ulmer (World Scientific, Singapore 1987) p. 1

Hue72   J.M. Huey, T.P. Kohman: Earth Planet. Sci. **16**, 401 (1972)

Ibe76   I. Iben Jr.: Ap. J. **208**, 165 (1976)

Ibe82a   I. Iben Jr., A. Renzini: Ap.J. Letters **259**, L79 (1982)

Ibe82b   I. Iben Jr., A. Renzini: Ap. J. Letters **263**, L23 (1982)

Ibe83   I. Iben Jr., A. Renzini: Ann. Rev. Astron. Astrophys. **21**, 271 (1983)

Käp89   F. Käppeler, H. Beer, K. Wisshak: Rep. Prog. Phys.**52**, 945 (1989)

Käp90   F. Käppeler, R. Gallino, M. Busso, G. Picchio, G., C.M. Raiteri: Ap. J. **354**, 630 (1990)

Käp91   F. Käppeler, S. Jaag, Z.Y. Bao, G. Reffo: Ap. J.**366**, 605 (1991)

Kla91   N. Klay, F. Käppeler, H. Beer, G. Schatz: Phys. Rev. (1991), in press

Les91   K.T. Lesko, E.B. Norman, R.-M. Larimer, B. Sur: Phys. Rev. (1991), in press
Lug83a   G.W. Lugmair, T. Shimamura, R.S. Lewis, E. Anders: Science **222**, 1015 (1983)
Lug83b   G.W. Lugmair, T. Shimamura, R.S. Lewis, E. Anders: Lunar Planet. Sci. **14**, 448 (1983)
Mat86   G.J. Mathews, K. Takahashi, R.A. Ward, W.M. Howard: Ap.J. **302**, 410 (1986)
Mat88   G.J. Mathews, D.N. Schramm: Ap. J. Letters **324**, L67 (1988)
Mer52   P.W. Merrill: Science **115**, 484 (1952)
Ott88   U. Ott, F. Begemann, J. Yang, S. Epstein: Nature **332**, 700 (1988)
San82   A. Sandage: Ap. J. **252**, 553 (1982)
San84   A. Sandage, G.A. Tammann: Nature **307**, 326 (1984)
Sch83   G. Schatz: Astron. Astrophys. **122**, 327 (1983)
Sch67   M. Schwarzschild, R. Härm: Ap.J. **150**, 961 (1967)
See65   P.A. Seeger, W.A. Fowler, D.D. Clayton: Ap.J.Suppl. **11**, 121 (1965)
Smi83   V.V. Smith, G. Wallerstein: Ap. J. **273**, 742 (1983)
Smi86   V.V. Smith, D.L. Lambert: Ap. J. **311**, 843 (1986)
Smi88   V.V. Smith, D.L. Lambert: Ap. J. **333**, 219 (1988)
Sri78   B. Srinivasan, E. Anders: Science **201**, 51 (1978)
Sym81   E.M.D. Symbalisty, D.N. Schramm: Rep. Prog. Phys. **44**, 293 (1981)
Tak86   K. Takahashi, G.J. Mathews, R.A. Ward, S.A. Becker: In *Nucleosynthesis and its Implications on Nuclear and Particle Physics*, ed. by J. Audouze and N. Mathieu (Dordrecht, Reidel, 1986), 285
Tak87   K. Takahashi, K. Yokoi: Atomic Data and Nuclear Data Tables **36**, 375 (1987)
Thi86   F.-K. Thielemann, J.W. Truran: In *Galaxy Distances and Deviations from Univesral Expansion*, ed. by B. F. Madore and R. B. Tully (Reidel, Dordrecht 1986) p. 185
Tin80   B.A. Tinsley: Fund. Cosmic Phys. **5**, 411 (1980)
Tru77   J.W. Truran, I. Iben Jr.: Ap. J. **216**, 797 (1977)
Ulr73   R.K. Ulrich: In *Explosive Nucleosynthesis*, ed. by D.N. Schramm and W.D. Arnett (Austin, University of Texas, 1973), 139
Van83   D.A. VandenBerg: Ap.J. Suppl. **51**, 29 (1983)
Van85   D.A. VandenBerg: Ap.J. Suppl. **58**, 711 (1985)
War80   R.A. Ward: 1980 (private communication)
Was82   G.J. Wasserburg, D.A. Papanastassiou: In *Essays in Nuclear Astrophysics*, ed. by C.A. Barnes et al. (Cambridge University Press, Cambridge 1982) p. 77
Wei66   A. Weigert: Z. Astrophys. **64**, 395 (1966)
Win82   R.R. Winters, R.L. Macklin: Phys. Rev. **C25**, 208 (1982)
Win86   R.R. Winters, R.F. Carlton, J.A. Harvey, N.W. Hill: Phys.Rev. **C34**, 840 (1986)
Yok83   K. Yokoi, K. Takahashi, M. Arnould: Astron. Astrophys. **117**, 65 (1983)
Yok85   K. Yokoi, K. Takahashi, M. Arnould: Astron. Astrophys. **145**, 339 (1985)
Zha91   W.R. Zhao, F. Käppeler: Phys. Rev. (1991), in press

Beta Decay Far from Stability and Double Beta Decay, and Consequences for Astrophysics

H. V. Klapdor-Kleingrothaus

Max-Planck-Institut für Kernphysik, D-6900 Heidelberg, Germany

1. Introduction

The recognition of the close connection between the laws of microphysics (nuclear and particle physics) and macrophysics (astrophysics and cosmology) is one of the most important discoveries of this century (see Fig. 1 and [Gro90, Kla88a, Kla90c, Kol90]). The barriers between these "classical" disciplines are becoming in some sense more and more meaningless. Nuclear physics, e.g., may give – as non-accelerator particle physics – important contributions to beyond standard model physics (double beta decay, proton decay, ...). One of the best examples for this situation is given by the neutrino (Fig. 2).

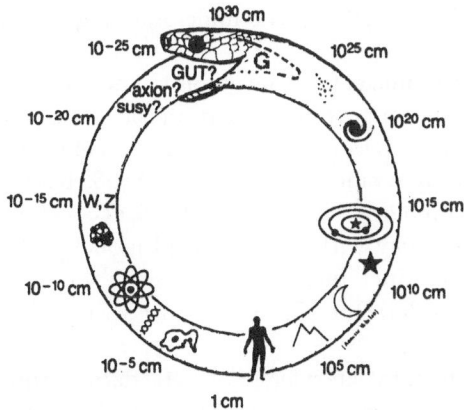

Fig. 1. Illustration of the interconnection between different physical disciplines (after Glashow, see [Gro90])

The neutrino plays – by its nature (Majorana or Dirac particle) and its mass – a key role for the structure of modern particle physics theories (GUTs, SUSYs, SUGRAs, ...), which is at least as important as the role it played at the time of Pauli and Fermi for the understanding of the weak interaction. It is

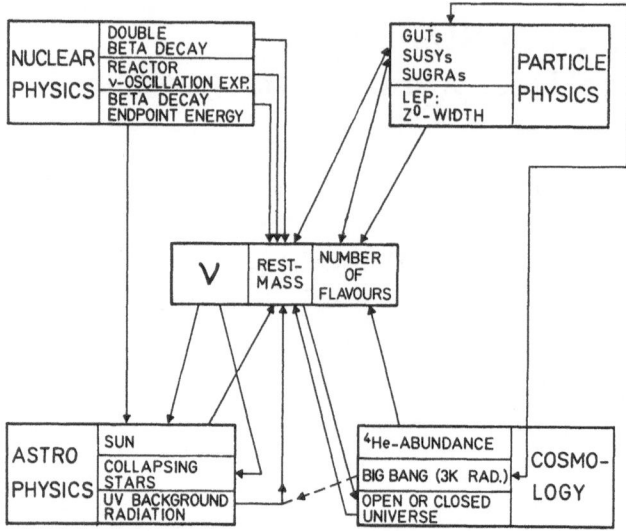

Fig. 2. The neutrino and its role in micro- and macrophysics

at the same time candidate for non-baryonic dark matter in the universe, and thus important for cosmology, it plays a decisive role for the energy transport in supernovæ, and solar neutrino observation might help to answer the beyond standard model question of neutrino-oscillations. On the other hand the presently sharpest limits on the electron neutrino mass are coming from nuclear physics experiments: double beta decay, tritium decay and reactor neutrino oscillation experiments. Prerequisite for a reliable calculation of nuclear double beta decay matrix elements required for a deduction of the neutrino mass from double beta decay experiments is on the other hand the understanding of the distribution of nuclear beta strength.

Turning to macrophysics beta decay of nuclei far from stability and electron capture by hot nuclei are some of the most important nuclear physics input for star explosion calculations and in particular for studies of the synthesis of heavy elements in the universe and of its age by cosmochronometers [Gro90, Kla83, Kla86a], which again is connected by cosmology with the cosmological constant of general relativity [Kla86b] and in this way with grand unified models. The nuclear beta strength distribution determines further the efficiency of solar neutrino detectors, such as the Gallium detector [Gro86b] – just to mention some of these interdisciplinary connections.

The present situation of our experimental knowledge of beta decay properties far from stability and of the needs for astrophysical applications are summarized in Fig. 3. Shown is in a Z vs. N plane the range of stable and unstable nuclides realized, or realizable in nature.

Border lines of the latter are given by the neutron and proton drip lines (denoted by $S_n = 0$ and $S_p = 0$, respectively). The hatched "peninsula" corresponds to nuclei whose β-decay properties are known experimentally. The dashed "island" right to the upper end of the peninsula denotes the range of nuclides for which β-delayed fission is expected. The black points denote nu-

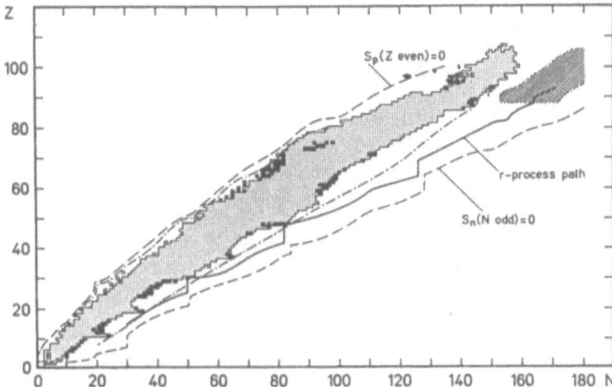

Fig. 3. The range of stable and unstable nuclei realized or realizable in nature (in a Z versus N plane). For explanation see text

clides which have been discovered after completion of the β half-life calculations of Klapdor, Metzinger and Oda (KMO) published in 1984 [Kla82a, Kla84a]. The dash-dotted line denotes the border line which may be optimistically expected to be reached experimentally in future with the new SIS accelerator in Darmstadt. The nuclides of interest for the synthesis of heavy elements by the r-process are those between the line of β stability and the line denoted by "r-process path". This latter line indicates the element distribution built up for short time during supernova explosions. By the successive β decay of this distribution the r-process is thought to build up the observed cosmic r-element distribution.

β-decay half-lives, β-delayed neutron emission rates and for very heavy nuclides beta-delayed fission (Fig. 4), therefore, have decisive influence on the r-process. They determine its time scale and yield severe constraints on possible astrophysical sites for this process [Kla81ab, Kla83, Kla86a, Thi83].

Fig. 3 shows that experimental progress in investigating new neutron-rich isotopes is rather slow - on the scale of astrophysical needs. Therefore, although in two points, with ^{80}Zn and ^{130}Cd, probable r-process waiting point nuclei could be reached [Kra86, Lun86, Gil86], most nuclides participating in the r-process for the foreseeable future are not accessible in terrestrial laboratories. Consequently, for studies of cosmic element synthesis theoretical predictions are required and remain indispensable. The improved understanding of nuclear beta decay reached in recent years [Kla83, Kla85, Kla86a] which had allowed for the set of predictions published in the 1984 At. Data Nucl. Data Tables [Kla82a, Kla84a] and which allows for an improved second-generation set discussed later on, is, therefore, of great importance, also for other fields of physics.

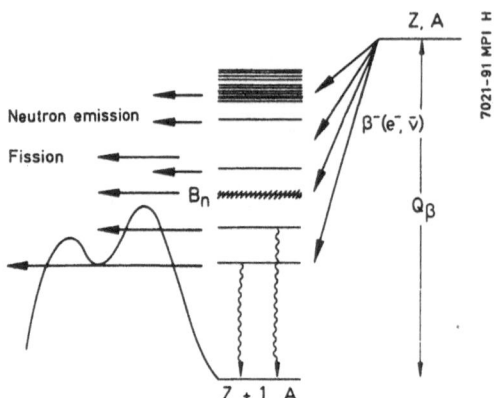

Fig. 4. According to the energy of the level excited by beta decay in the daughter nucleus, the latter decays via gamma emission, beta-delayed neutron emission or beta-delayed fission. B_n denotes the neutron separation energy. The figure also shows the double-humped fission barrier which usually occurs in heavy nuclei

2. Calculations of Beta Decay Properties of Nuclei Far from Stability

2.1 Half-lives

There exist at present three calculations for all neutron-rich nuclides far from stability out to the neutron drip line

(1) the gross-theory calculations by [Tak73], recently modified [Tac88]
(2) microscopic calculations using a Tamm-Dancoff approximation by [Kla82a, Kla84a] (KMO)
(3) second-generation microscopic calculations using the proton-neutron quasiparticle RPA by [Ben88, Sta89, Sta90a] (SBMK).

A discussion also of the first two approaches is given in [Kla83, Kla86a, Kla89a]. In this paper we shall concentrate on the second-generation calculations.

This new approach [Sta89, Sta90a] uses the proton-neutron quasiparticle RPA with a GT residual interaction. The QRPA takes in addition to a BCS treatment (which has been realized in some simplified way in [Kla82a, Kla84a]) spin-isospin g.s. correlations into account (Fig. 5).

For our application an extension of the QRPA formalism discussed in [Hal67] had to be made for GT decay of odd-odd nuclei, for which formulae did not exist in the literature. This has been published recently [Mut89a].

In nuclei with an odd number of protons and/or neutrons there occur two types of transitions: one is a phonon transition, where the odd quasiparticle(s) act(s) as a spectator. The other is a transition of the odd particle(s), which is (are) described as a phonon-correlated one-quasiparticle state or two-quasiparticle state. In that case we treat phonon correlations in first order perturbation [Mut89a].

initial state

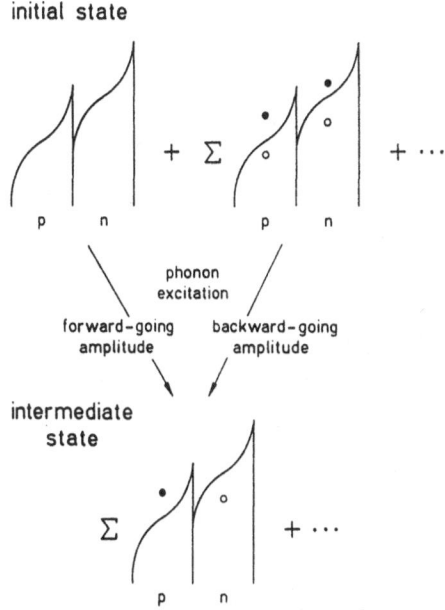

Fig. 5. A phonon excitation from the ground state (QRPA vacuum) to a one-phonon state in the proton-neutron QRPA. The dominant contribution of the forward-going amplitude term (from the zero-quasiparticle state to a proton-neutron pair state) interferes with the backward-going amplitude term (from four-quasiparticle states to the pair state). In the case of $\beta\beta$ decay, this phonon excitation corresponds to the transition from the initial (final) 0^+ state to an intermediate state

In deriving the QRPA formulae for β^--decay we can neglect the nonseparable particle-particle force, which plays, however, a decisive role in calculations of 2ν double beta decay (see below). We further use instead of a realistic NN interaction used in $\beta\beta$-decay (G matrix of the Paris or some other potential [Mut88a,b, Mut89b , Sta90c,d]), a schematic GT interaction as residual interaction in the RPA. These assumptions simplify the QRPA matrix equation to an algebraic equation, because the particle-hole interaction has separable matrix elements.

Some results of the new calculations are shown in Figs. 6–8. The mean deviation of the ratio $t_{1/2}^{\mathrm{cal}}/t_{1/2}^{\mathrm{exp}}$ from 1 is smaller than in the other existing calculations. The description of light nuclei (Fig. 8) is seen to be able to compete with that given by the large-scale shell-model calculations by [Wil83].

The dependence of the calculated half-lives on the parameters involved: GT interaction strength, deformation of the Nilsson potential, strength of the pairing force and β-decay Q values (mass formulae) has been investigated in [Sta89, Sta90a], where also the results for about 6000 nuclides are given.

Concluding, the second-generation calculations yield a set of β^--decay half-life predictions improved over those from other approaches. This is in particular the case for light nuclei.

Figs. 9,10 show that this approach also works remarkably well for β^+ decay. In general on the β^+ side the particle-particle force has to be included. Calcu-

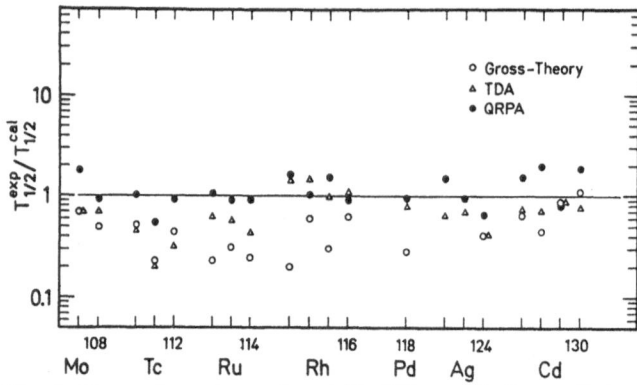

Fig. 6. Comparison of experimental half-lives with the predictions of the revised gross theory [Tac88], the TDA model [Kla82a, Kla84a] and the QRPA [Sta89, Sta90a]

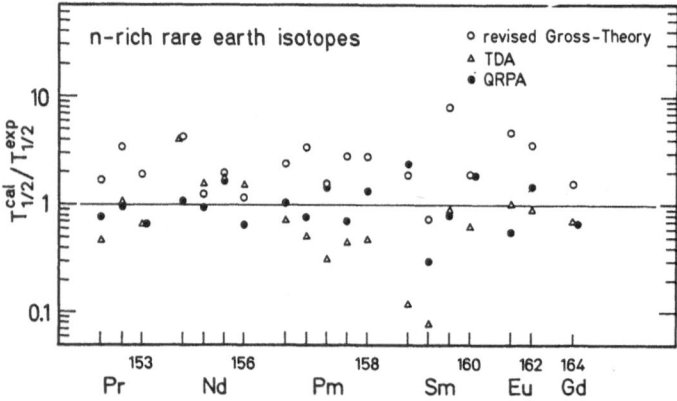

Fig. 7. Comparison of experimental half-lives [Gre88] with the predictions of the revised gross theory [Tac88], the TDA [Kla82a, Kla84a] and the QRPA-model [Sta89, Sta90a]

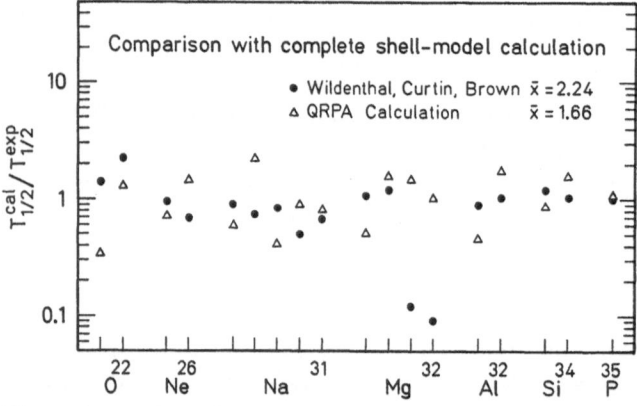

Fig. 8. Comparison of experimental β^--decay half-lives of s-d shell nuclei with the second-generation microscopic approach and the calculations of [Wil83] (from [Sta89, Sta90a])

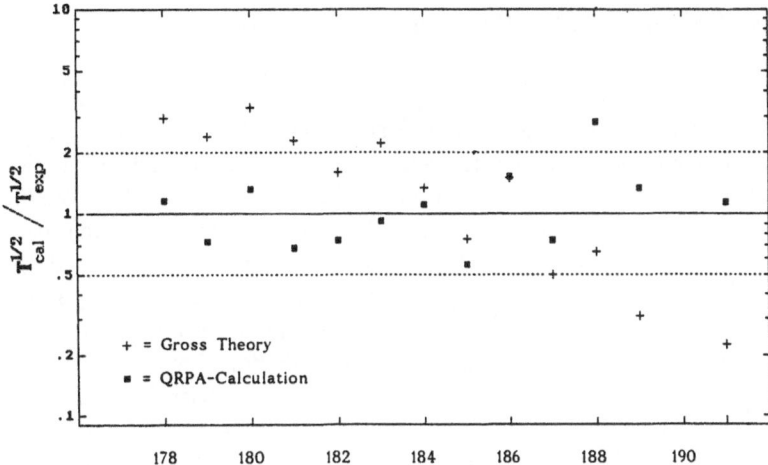

Fig. 9. Comparison of measured β^+-decay half-lives with calculations by the second-generation microscopic approach (from [Hir91ab])

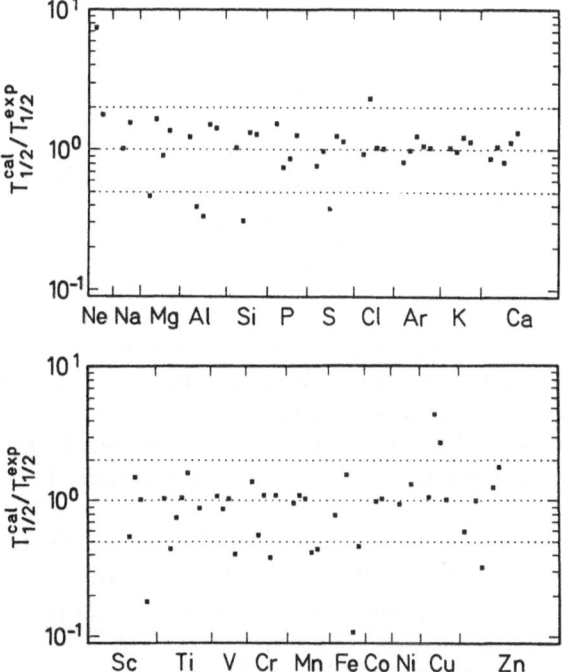

Fig. 10a,b. Comparison of experimentally known β^+ half-lives in the mass range $A = 18 - 63$ with calculations by the second-generation microscopic approach (from [Hir91ab])

lations of this type for all β^+-emitters of the nuclidic chart (~ 2000 nuclides) have been performed recently [Sta90b, Hir91ab].

2.2 β-Delayed Fission and β-Delayed Neutron Emission

Berlovich and Novikov were the first to point out the possible existence of a
large area of neutron-rich nuclides, where fission may follow the β-decay to an
excited state [Ber69]. In the following years this island of β^--delayed fission
(βdf) was confirmed by subsequent studies [Kla83, Thi83, Wen75, Wen76].
We performed earlier [Thi83] a systematic calculation of βdf (and βdn) using
matrix elements calculated in a TDA approach and discussed in detail the
consequences for the r-process and age of the galaxy. However, [Hof86] pointed
out somewhat later that the βdf-values obtained by [Thi83] seem to be too large
compared to thermonuclear explosion yields. We therefore have performed a
new calculation of βdf [Sta91] in the region of n-rich nuclei with $75 < Z < 100$
and $140 < N < 184$ using the strength functions of [Sta89], and applying
essentially the same model as used by [Thi83].

The calculation of β-delayed fission probabilities involves essentially two
inputs. One is the β-strength distribution which can reliably be predicted within
the framework of the pnQRPA. The second – at present maybe more delicate
– input which is required are nuclear masses, especially fission barrier heights
of nuclei far off stability. The predictions of fission barrier heights by Howard
and Möller [How80] are applied in the present work. Q_β-values and neutron
separation energies S_n and S_{2n} are chosen according to the mass formulae of
[Gro76, Hil76], respectively. The latter was shown to give reliable results for
the half-lives of neutron-rich isotopes far away from the line of β-stability. This
choice is discussed in detail in [Sta89, Sta90a]. We apply these two sets of
masses also for the sake of consistency, since the β-strength distributions of
[Sta89, Sta90a] are computed with the deformations given by these two mass
formulae.

The results of our calculations are displayed in Fig. 11 for the masses from
[Gro76, Hil76]. The β-delayed fission rates P_f are given in % for the fissioning
daughter nucleus. Only P_f-values greater than 1% are considered.

It is obvious that delayed fission has to be considered in the decay back
to stability either in a nuclear explosion or in the very heavy region of the
r-process.

In the region around $Z = 94$ we predict P_f-values of 100%, independently
of the underlying mass formula and in agreement with [Kla83, Thi83]. This
area acts as a sink to the r-process and prevents the synthesis of superheavy
elements in nature.

The recently observed discrepancy between calculated and measured abundances in nuclear explosions with a ^{238}U target, traced back by [Hof86] to
overestimated β-delayed fission rates by [Thi83], seems to be removed in the
present work [Sta90b, Sta91].

Preliminary calculations of fission barrier heights of neutron-rich heavy elements within the Yukawa-plus-exponential [How87] model seem to indicate
higher fission barriers than in the work of [How80]. Another difference is that
the ground-state shapes may differ considerably. Both features may affect the
probabilities of β-delayed fission [Sta91]. Despite some promising features of

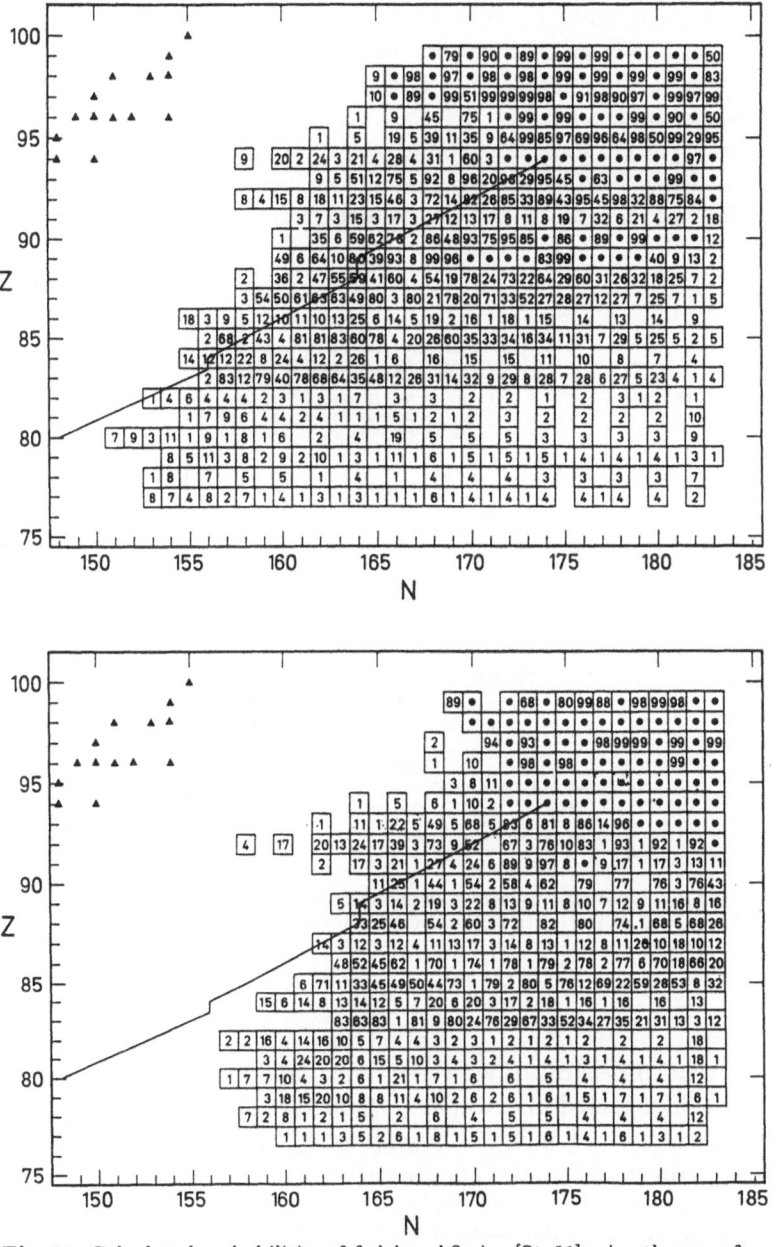

Fig. 11. Calculated probabilities of β-delayed fission [Sta91] using the mass formula of [Hil76] (a), and of [Gro76] (b)

the new mass formulae, there are still large uncertainties in the predictions of nuclear masses of neutron-rich nuclides far from stability [Gra90, Sta89, Sta90a].

In consequence, the new estimations of fission barriers by [How87] must be considered as uncertain as those of [How80]. This leads to the conclusion that, at present, there seems to be no alternative to the extensive work of [How80]. But one should keep in mind that new developments in this field are in progress and that new results on fission barriers might affect the fission rates discussed here and presented in [Sta91].

We find [Sta91] as the most important conclusion for astrophysics, that the area of 100% β-delayed fission remains, even if we increase the fission barriers by about 3 MeV, to an extent, that the r-process still will be stopped by βdf (see Figs. 12, 13).

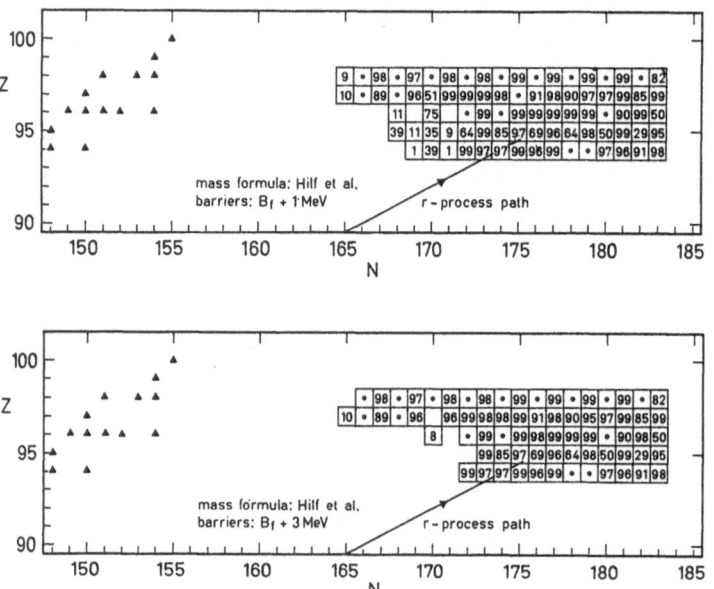

Fig. 12. Calculated β-delayed fission probabilities for the isotopic chains with $94 \leq Z \leq 98$, using the mass formula of [Hil76], with the fission barriers of [How80], but raised by 1 and 3 MeV, respectively (from [Sta91])

Fig. 14 a,b compares our second-generation calculations of beta-delayed neutron emission probabilities [Hir91] with experiment and thus gives a feeling for the reliability to be expected in extrapolations further out from stability. The experimental P_n-values used in Fig. 14 (from [Lun89]) can be regarded as a status report on experimental P_n-values in the region of the fission products. Error bars in Fig. 14 indicate only the experimental uncertainty. In (a) the neutron separation energies S_n were determined from the most recent compilation of experimental nuclear masses [Wap88], in (b) the S_n were taken from the mass formulae of [Gro76, Hil76, Moe81].

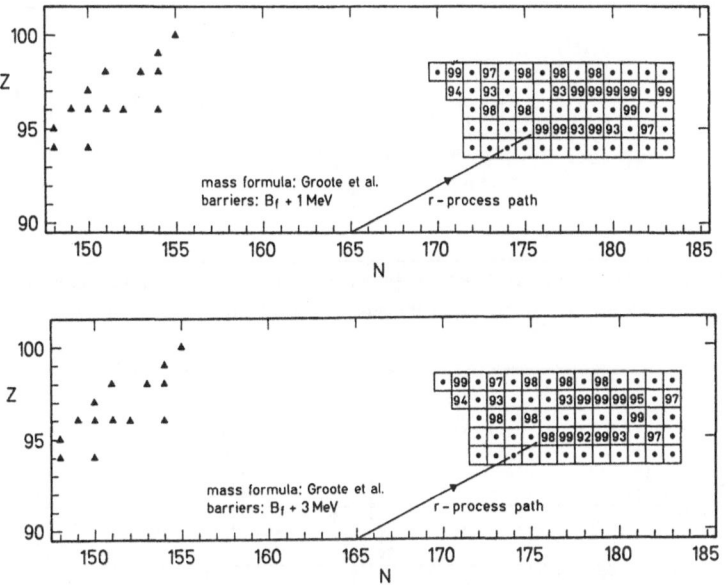

Fig. 13. Same as Fig. 12, but with the mass formula from [Gro76] (from [Sta91])

3. Beta Decay Far from Stability and the Astrophysical r-Process

3.1 Synthesis of Heavy Elements in the Universe

The first element synthesis occurred some 10^2 s after the big bang. At this time the temperature of the universe had sunk to $0.9 \cdot 10^9$ K, and because then deuterons once formed could no longer be disintegrated, the existing neutrons were "boiled" into helium, which since then has formed roughly 25% of the mass of the universe. Only some 700 000 years later, with the decoupling of matter and radiation, the Jeans condition was fulfilled (i.e. the gravitational energy exceeded the molecular thermal energy), at which time the formation of galaxies and stars and thus the synthesis of all the other elements then began. So essentially all the elements in the universe (Fig.15 shows their observed abundances), except Li, Be and B which are produced by spallation in cosmic rays [Aus81], were created in stars. To understand their creation, we therefore need reliable star models. That we are mainly concerned with heavy stars of mass 10–25 $M_\odot$ is seen from Fig. 16 which shows the production of nuclei per star weighted with the relative abundance of the different star masses.

Nuclei up to iron may be created by fusion during the phases of hydrostatic burning of heavy stars. Their relative abundances are then modulated [Wea80, Woo82] by explosive burning of these shells (see below) in supernova explosions. There are only two conceivable processes which may have led and lead to the formation of heavy elements (if we neglect the p-process which produces neutron-deficient nuclei to the left of the stability line, largely via (p,γ), (γ,n)

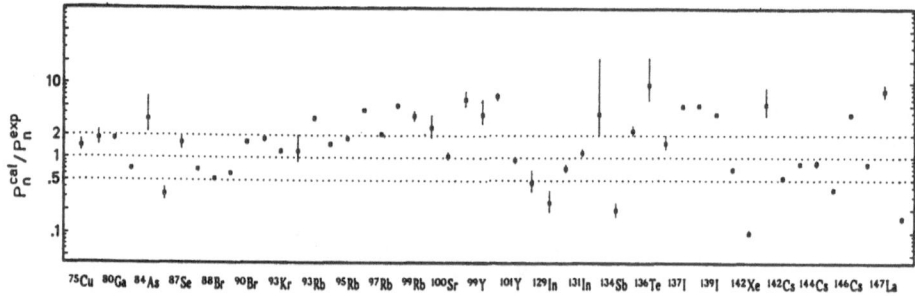

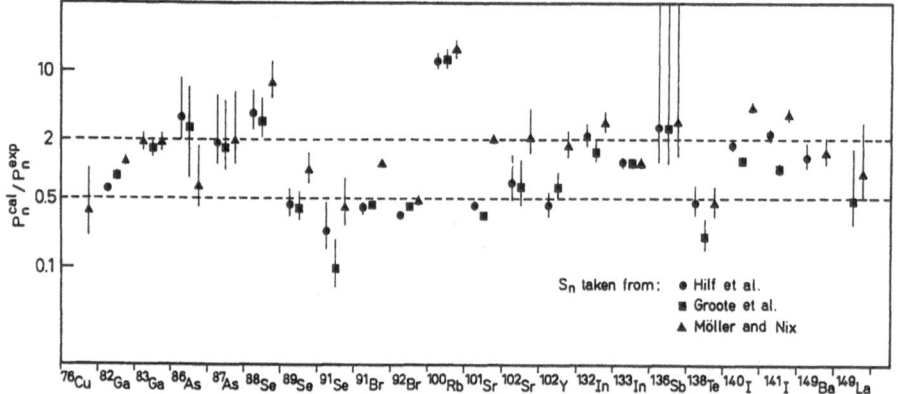

Fig. 14. Ratio of calculated [Hir91] and experimental [Lun89] P_n-values in the range of the fission products (for details see text)

and (γ,α) reactions – see [Bur57, Ray90]): slow and rapid neutron capture in the so-called s-(slow) and r-(rapid) processes (see Fig. 17).

The components due to these two processes are directly visible in the cosmic element distribution (Fig. 15). Slow neutron capture (in which the β decay half-lives are small in comparison with those for neutron capture, $(\tau_\beta \ll \tau_n)$) may explain the abundance maxima at nuclei with magic neutron numbers ($A \approx$ 90, 144, 208), in terms of the small neutron capture cross section for these nuclei. It is known that this process occurs, e.g. in the He burning shell of heavy stars, driven by the neutrons from the reaction $^{22}\text{Ne}(\alpha,n)$ (see Fig. 18), which produces neutron densities of order of magnitude 10^{10} cm^{-3}. The other component in the cosmic abundance distribution with maxima $10 - 15$ units of mass lower than the maxima of the s-component (e.g. at A $\approx$ 130, 195) can only be explained by a short-duration very high neutron concentration ($n_n > 10^{18}$ cm^{-3}) and a process in which ($\tau_\beta \gg \tau_n$)).

In what follows, we shall restrict ourselves to a discussion of the r-process (for the s-process, see e.g. [Käp89]). We shall see that precise knowledge (i.e. calculation) of the β decay of neutron-rich nuclei is crucial to an understanding of this process. Only once we understand the creation of the heavy elements,

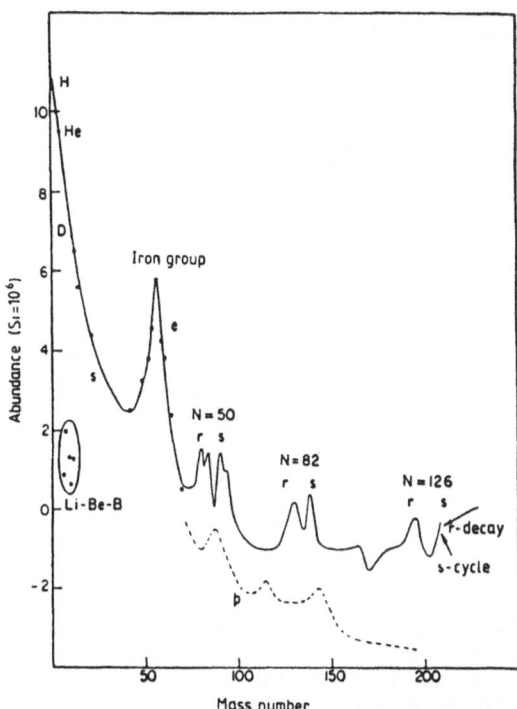

Fig. 15. Simplified diagram of the observed solar (cosmic) element abundances as a function of the mass number A (normalised to $Si = 10^6$). The distinct maxima near the magic neutron numbers $N = 50, 82, 126$ may be traced back to the r- and s- processes (see text)

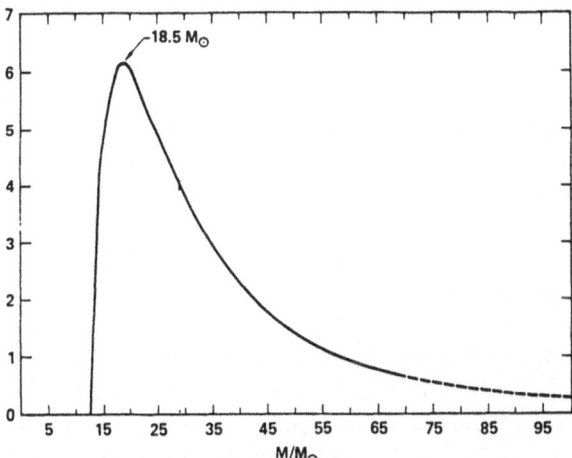

Fig. 16. Schematic diagram of the yield of nuclei with $Z > 2$ in supernova explosions, as a function of the star mass. The yields per star are weighted by the mass-dependent star creation rate (from [Mil79, Wea80]

including the so-called actinide chronometers Th, U, Pu, can we make cos-mochronological inferences about the age of our Milky Way.

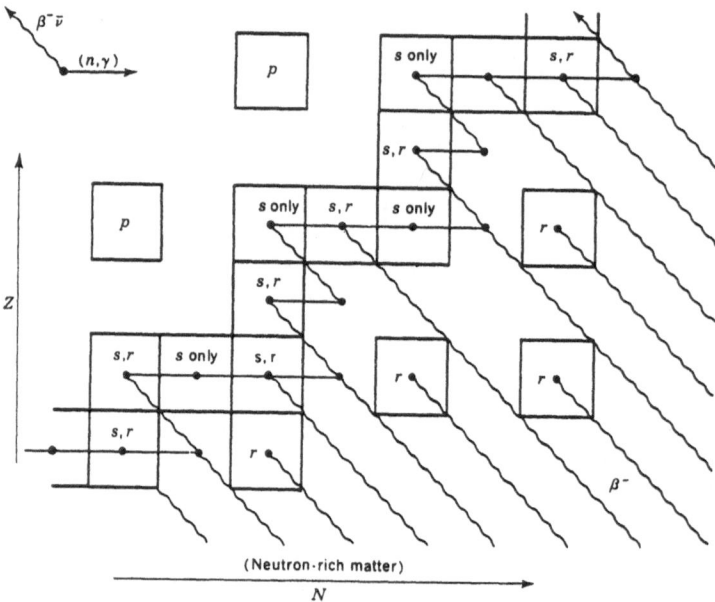

Fig. 17. Characterisation of a section of the nuclidic chart according to the generability of its nuclei in the r- and the s-process. The s-process path of (n,γ) reactions, followed by (relatively) rapid β decay ($\tau_\beta \ll \tau_n$), passes through nuclei denoted by "s '. Neutron-rich stable nuclei are generated by multiple β decay of the element distribution of the r-process path which is formed momentarily by multiple neutron capture ($\tau_\beta \gg \tau_n$) in explosive processes. Nuclei which are screened from the r-production process are denoted by "s only '. The rare proton-rich nuclei which are not reached by the r- and s-processes are denoted by "p" (after [Cla83])

3.1.1. The r-Process, Principle and Dependence on β-Decay:

What is known about the r-process? Its principle has been known for some 30 years (Fig. 19): the initial element distribution (in early models, mainly Fe; in more recent models, built up by hydrogen burning and s-process, see below) is exposed to a high neutron flux.

By rapid successive n capture and β decay, ($\tau_{n,\gamma}, \tau_{\gamma,n} \ll \tau_\beta < \tau_{r-process}$) the initial distribution is transformed into an r-element distribution, some 20-30 units to the right of the β stability line. The magic neutron numbers lead to the steps at $N = 82$ and 126, which act as bottlenecks due to the increased β half-lives closer to the β stability line, in which the element formation piles up. As the neutron flux decreases, this r-distribution of extremely n-rich nuclei decays by β decay back to the stability line. Projection of the abundances built up at these magic neutron numbers onto the β stability line gives the r-maxima to the left of the s-maxima in Fig. 15.

Both the element distribution on the r-path, and the resulting final distribution of stable elements are thus sensitive to the β decay properties of the neutron-rich nuclei involved in the process (about 6000 nuclei between the β stability line and the neutron drip line), see Fig. 19:

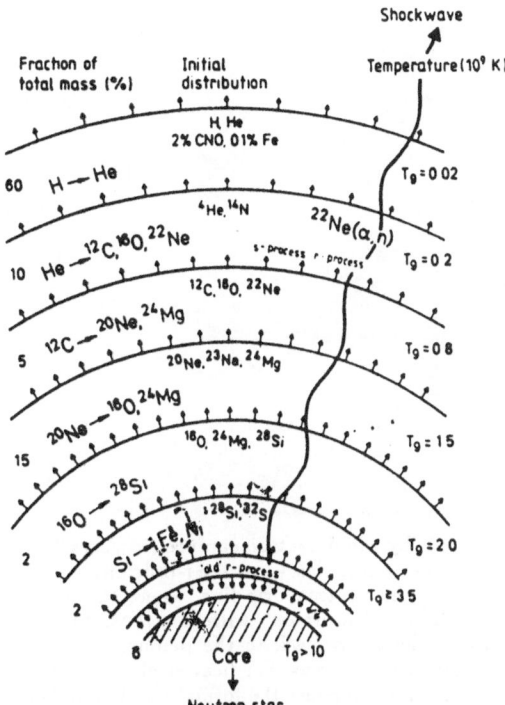

Fig. 18. Schematic diagram of the structure, composition and development of a heavy star (about 25 M$_\odot$). In the hydrostatic burning of the shells, elements of higher nuclear charge numbers, up to Fe, Ni, are produced from the initial distributions (whose main components are given). Graviational collapse of the core leads to the formation of a neutron star and ejection of $\approx$ 95% of the star mass (supernova explosion). The cast off outer shells are breached by a detonation shock front which induces explosive burning of the shells. A probable site for synthesis of heavy elements (r-process) is the explosive He burning

- The β half-lives in the r-path determine the rate of formation of the heavy elements, and the masses to which the r-path may extend. These β half-lives are particularly important if, as required, the r-process is treated dynamically, i.e. as an explosive process.
- The β half-lives between the r-path and the β stability line, and the rates of β-delayed neutron emission largely determine the distribution of stable nuclides on the β stability line after the β decay.
- The rates for β-delayed fission determine where the r-path breaks off at high mass numbers A, and also influence the production rates of heavy nuclei and, in particular, of the so-called cosmochronometers (see below) ^{235}U, ^{238}U, ^{244}Pu, ^{232}Th.

Most of the nuclei involved in the r-process (in Fig. 19 for instance all to the right of the foot of the "mountain chain" of stable elements) cannot, at present and for the foreseeable future, be produced in terrestrial laboratories – with the possible exception of some region of heavy nuclei which may become accessible via thermonuclear explosions. Therefore, in respect of beta decay properties, one has to rely on theoretical extrapolation, as described in the

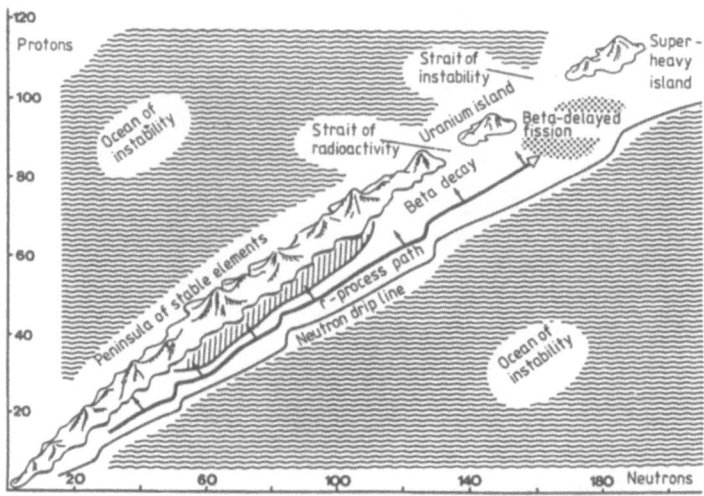

Fig. 19. Schematic diagram of element synthesis by rapid neutron capture (r-process). The nuclides created for a short time (0.4 s) in supernova explosions lie on the r-process path away from the peninsula of stable elements and near the neutron drip line, where the neutron separation energy is zero, i.e. where the nuclei are "saturated" with neutrons. After the neutron density decreases, the neutron-rich nuclei decay back to the peninsula of stable elements; the uranium island is also still attained in this way. For heavier elements, the r-process path is broken off by β-delayed fission, which impedes the formation of superheavy nuclei. The β decay properties of nuclei between the r-process path and the β stability curve (about 6000 nuclides, most of which cannot be studied in terrestrial laboratories) critically affect the resulting final distribution of stable nuclides (the shaded region of nuclides determines the residual heat formed from the β decay of fission products when a nuclear reactor is switched off; see [Hir91a, Kla88c])

above sections. Except for the sensitivity to β decay, which was emphasised much later [Kla76, Kla80, Kla83, Kla86a], these fundamentals of the r-process have been known for 30 years [Bur57]. The question however as to where in the universe conditions may be found under which such a process may actually occur, is not yet completely solved.

3.1.2 Site of the r-Process: It is immediately clear that the necessary neutron densities may only be attained in explosive processes, and that practically only supernova explosions should be given any consideration. Thus the question reduces to "where in the supernova".

The original idea (see [Bur57, Hoy60]) was that the r-process occurs in part of the core material, close to the forming neutron star, which is ejected together with the hydrostatic burning outer shells (see Fig. 18). Here the matter has undergone full photodisintegration from the shock wave, and is very neutron-rich due to electron capture. If this matter is ejected, then during expansion and cooling to $< 10^{10}$ K, n-rich nuclei of the "iron" group are initially synthesised in nuclear statistical equilibrium with free neutrons. After all reactions with charged particles are "frozen out", neutron capture and β decay continue and transfer iron group "seed nuclei" into the r-process nuclei.

Quantitative investigations of this so-called "classical" r-process scenario, by hydrodynamic explosion calculations using detailed supernova models, were carried out since 1976. However, these investigations led from the initial hopes to a damping of the expectations, and finally to the practical exclusion of this scenario (see [Hil82]). The main reason for this negative result was the initially disregarded existence of neutral weak currents [Fre74, Tub75]. These enable the coherent scattering of neutrinos formed in the neutron star by nuclei (mainly Fe), they thus increase the probability of neutrinos staying in the layer above the neutron star, and thus also increase the rate of the reaction $\nu + n \rightarrow e^- + p$. As a result, a much smaller neutron concentration (neutron/proton ratio) establishes in equilibrium than that required for an r-process under the conditions of the "classical" scenario. Other inconsistencies are described e.g. in [Hil78, Hil79, Kla81ab, Nor79]. In particular these include the fact that the mass cut which must be assumed in order to obtain solar r-abundances would lead to vast overproduction of r-elements.

Several alternatives to the "classical" r-process have been proposed and investigated, including, recently, a primordial r-process in an inhomogeneous big bang [App85, Kaj88]. Amongst these, the explosive helium burning [Hil77, Hil82, Kla81a,b, Kla82b, Kla83, Kla86a, Kla89b, Kla90b, Thi83, Tru78] seems currently to be the most consistent, despite some still disputed questions, such as the source of the neutrons in this process (see below). It is also the most quantitatively investigated r-process site as far as its dependence on β-decay properties and more general, nuclear physics parameters, is concerned. Its crucial dependence on the β-decay input data had been investigated first by [Kla81ab, Kla83, Thi83] who found that this scenario did not work when using the gross-theory half-lives, but yielded a good reproduction of the observed cosmic element distributions, when using the KMO microscopic half-lives. Since recent years have proved the KMO β-decay half-life predictions to be rather reliable (see the discussion in [Kla89a]), the essential conclusions we have drawn earlier [Kla81ab, Kla83, Kla86a, Thi83] on the astrophysical r-process remain valid (corresponding investigations using the second-generation half-life calculations have not been made up to now but are in progess).

Also the recent measurement of β half-lives of the (probable) r-process waiting point nuclei ^{80}Zn and ^{130}Cd do not lead to new conclusions, as claimed in [Mat90] since they just were a confirmation of the calculations within 10% (see [Kla89a, Kla90b]). They rather confirm earlier conclusions.

That the cosmic r abundances in their peaks are in good agreement with that expected from an (n,γ)-(γ,n) equilibrium (discussed recently by [Kra90]) had been shown already by [Kla81ab, Thi83]. Also the conclusion [Mat89] that the neutron density necessary to allow for this equilibrium is $n_n > 2 \cdot 10^{17}$ cm^{-3} was made already ten years ago in [Kla81ab].

All these points may justify to show here some of the results obtained for this site.

In explosive He-burning the neutrons come from the reaction ^{22}Ne(α,n) or less probably from ^{13}C(α,n) which yield in the He-burning shell (Fig. 18) during

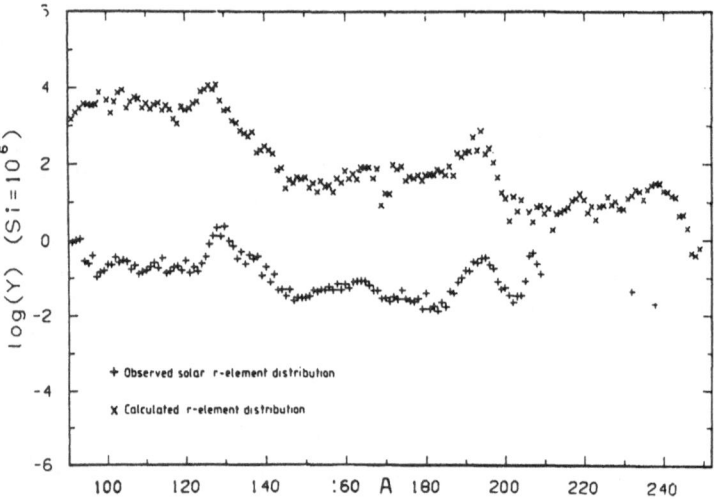

Fig. 20. The calculated distribution of stable elements after β decay of the temporary element distribution formed in the r-process, as a function of the mass number A (after [Thi83]). The diagram also shows the observed solar r-element distribution according to Cameron

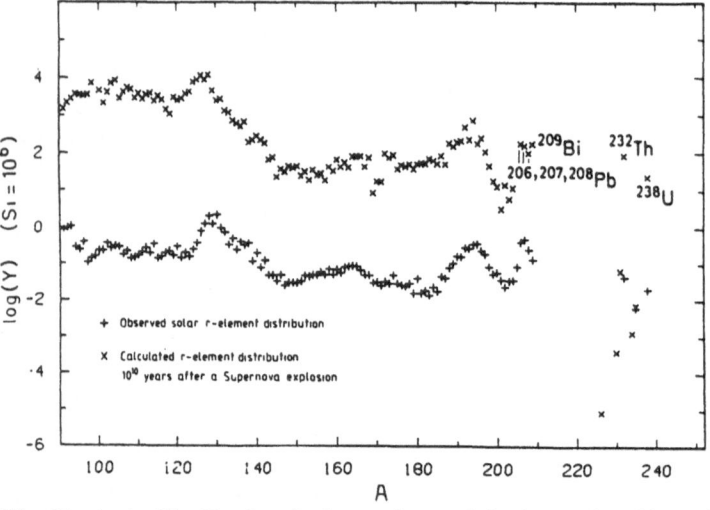

Fig. 21. As in Fig. 20, after further α decay of the heavy β stable nuclei over 10^{10} years (from [Thi83])

the short time ($\sim$0.5 sec) of the passage of the detonation shock front in the supernova explosion up to 10^{20} neutrons per cm^{-3} [Kla81a,b].

The initial element distribution in the He burning shell, which is exposed to the high neutron flux, may be assumed to be solar (i.e. as in stars of the population I), modified by the CNO- and s-processes. An r-process running in this scenario would represent more a conversion of an s-distribution into an r-distribution than a production of heavy elements starting from the Fe peak as in the r-process close to the neutron star described above.

The results of hydrodynamic explosion calculations for the element creation in explosive He burning are shown in Figs. 19-21 [Kla81ab, Kla83, Kla86a, Thi83]. The first figure shows schematically that the temporary element distribution formed in the r-path, around atomic number $Z \approx 90$ runs into the region of massive β-delayed fission which interrupts the r-path. This rules out the formation of superheavy elements ($Z > 102$) in the r-process, and thus probably also in nature. Fig. 20 shows the final distribution of β stable elements which arises after the re-decay of the distribution on the r-path by β decay, taking into account β-delayed neutron emission and fission. If we permit further decays by α emission and spontaneous fission for a period of time of order of magnitude of the age of the Galaxy, we obtain the distribution shown in Fig. 21: only the so-called cosmochronometers Th, U, Pu remain beyond Bi.

The calculation shows a remarkable agreement between the calculated cosmic r-element distribution and observation — however only for the microscopically calculated β decay data. The systematically longer β half-lives of the so-called gross theory [Tak73], which were previously used, would require a longer period of neutron irradiation, which is excluded in this scenario.

The calculated absolute abundances of r-elements after the explosion are, according to Fig. 21, a factor 10^3 larger than the observed solar r-abundances. This enrichment of r-material only occurs in a part of the He shell of the star (about 1/10). In relation to the total mass of the star of 25 M$\odot$, this corresponds to an r-material enrichment factor of 10–20 and is thus compatible with the enrichment factor for lighter elements derived by [Wea80, Woo82] from the explosive burning of the inner shells.

The consistency in the description of the creation of the heavy elements in the universe obtained using the microscopic β decay data goes still further. This r-process site should also help to explain observed isotope anomalies in meteorites, such as excess r-abundances in ^{107}Ag, ^{130}Te, Nd, ... The conjecture that these were introduced shortly before the condensation phase of the protosolar cloud by a nearby exploding supernova (which "triggered" the condensation of the solar system) has for a long time been faced by the problem that only the outermost layers of the supernova in the form of dust could penetrate into the protosolar cloud [Mar79]. Thus this explanation would require excess r- abundances in the outer shells of the supernova. Explosive He-burning would achieve precisely that.

Finally, we note that the above discussion of the site of the r-process is independent of unresolved theoretical problems of details associated with the generation of the supernova explosion by gravitational collapse. This is because our derivation involves only the fact of the explosion of the heavy star and the known portion of the energy released in the gravitational collapse which reappears as energy of the shock front. The hydrodynamic conditions, arising due to the shock front, in the He burning shell remote from the centre of the explosion may be described by assuming a point source — i.e. essentially independently of the details of the "ignition" of the explosion and of the specific star model.

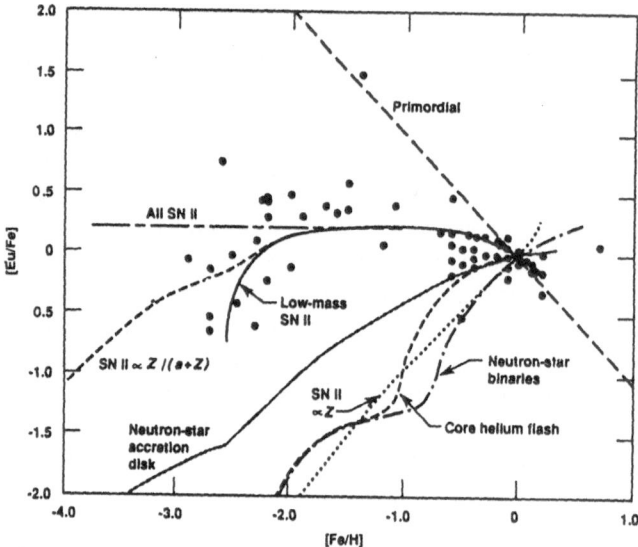

Fig. 22. Measured correlation of [Eu/Fe] with [Fe/H] (points) and correlations expected for various r-process sites (lines) according to [Mat90]

An interesting recent discussion of r-process sites was given by Mathews and Cowan [Mat90]. They showed (see Fig. 22) that correlations of heavy element abundances measured in recent years for the surfaces of metal-poor stars with respect to iron "seed" material rule out most of the r-process models or sites invented in the past (although the correlations might not yet be considered as extremely striking).

They rule out in particular a primordial r-process in an inhomogeneous big bang which may contribute only at a level of 10^{-3} of the solar abundance [Mat89]. They conclude that "the apparent decrease in [Eu/Fe] for [Fe/H] $<$ -2.5 suggests, that the r-process may be a primary process in low-mass type II supernovae or a secondary event with a neutron source that does not require pre-existing seed material". An example for the latter would be explosive He-burning (curve denoted SNII $\sim Z/(s+Z)$ in Fig. 22). Other investigations [Woo90] favour a primary process close the forming neutron star. They discuss jet-like ejections of r-nuclides at the poles of the rotating supernova core. However, the efficiency of such a mechanism remains to be shown. Also it has been argued recently [Thi90, Arn90], that in order to satisfy various observational constraints on the composition of the supernova SN 1987A, and in particular on its amount of ejected ^{56}N, no material enriched with r-nuclides can be ejected from the layers just outside the mass cut.

4. The r-Process and the Age of the Universe

The production ratios of the actinide cosmochronometers (^{232}Th, ^{235,238}U, ^{244}Pu) in the r-process allow one by comparison to ratios observed at the time of condensation of the solar system (in meteorites) to get information on the age of the galaxy. These production ratios are, in addition to half-lives and P_n values, particularly sensitive to β-delayed fission (see Figs. 11–13 and [Kla89c, Sta91]). Earlier work [Fow78, Fow85] on these chronometers ignored completely the effect of β-delayed fission and led to ages of the universe systematically too short, compared to independent other sources. Inclusion of β-delayed fission rates calculated from the KMO β-strength functions allowed one for the first time to reach consistency of actinide chronometer ages with ages from the most important other independent sources, the globular clusters and the Hubble constant [Kla83, Kla86a, Thi83]. The conclusions from the cosmochronometers are independent of the final site of the r-process. The arguments for this were already given by [Thi83]. For a detailed discussion we refer to [Gro90, Kla88b, Kla89a, Kla90a].

5. Age of the Universe, Cosmology and Neutrino Mass

Particular interest in improving the precision of the cosmochronometers comes from cosmology. The age of the universe is one of the boundary conditions for cosmological models and thus of importance to determine the value of the cosmological constant, which is of extreme importance for the structure of grand unified theories (see, e.g., [Abb88]), and the amount of non-baryonic dark matter in the universe. This has been pointed out by [Blo84] and by [Kla86b, Gro90] and will be outlined in the following.

The expansion of a homogeneous and isotropic universe with a Robertson-Walker metric can be described by the radius of curvature $R(t)$ of the three-dimensional space manifold at time t after the big bang:

$$\left(\frac{\dot{R}(t)}{R(t)}\right)^2 = \frac{8\pi G}{3}\rho(t) - \frac{kc^2}{R(t)^2} + \frac{1}{3}\Lambda c^2$$

$$\left(\frac{\ddot{R}(t)}{R(t)}\right)^2 = -\frac{4\pi G}{3}\left[\rho(t) + \frac{3p(t)^2}{c^2}\right] + \frac{1}{3}\Lambda c^2$$

Here $\rho(t)$ is the mean density of matter, $p(t)$ the radiation pressure in the universe; k is the metric parameter describing spherical, Euclidean and hyperbolic spaces for $k = +1, 0, -1$, respectively, and Λ is the cosmological constant which cannot be fixed by general relativity. In most cosmological models Λ was assumed to be zero (mainly from "esthetical" reasons). However, in the context of quantum field theory Λ can be interpreted [Zhe68] as energy density of the vacuum ϵ_V,

$$\Lambda = \frac{8\pi G}{c^4}\epsilon_V$$

and there is no reason for the latter to be zero. Very high vacuum energy densities (or equivalent vacuum matter densities $\rho_V = \epsilon_V/c^2$) are required in order for inflation [Gut81, Bra86] to develop at some very early time $t(t \sim 10^{-35}$ sec) after the big bang (one must have $\rho_V \geq \rho(t)$ at this time). Such large vacuum energy densities may result from the Higgs fields introduced in gauge theories. On the other hand contributions to the vacuum energy in our present world must be many orders of magnitude (more than 50 or even 100) smaller not to completely dominate the universe. From observation the value at present is [Abb88] $\Lambda \leq 3 \cdot 10^{-56}$ cm^{-2}.

A precise determination of the age of the universe helps to further fix the present value of Λ. To obtain a unique solution for R(t) of the above equation three boundary conditions are required (if $\rho(t)$ is assumed to be pressure-free matter density and p(t) is neglected (good approximation for $t > 10^6$ years)). One of them can be the age of the universe, the two others the Hubble constant $H_0 = \dot{R}/R$ today and the matter density ρ_0 of the universe today. Although the latter two quantities are fixed by observation still only with relatively wide limits ($H_0 = 40$–100 km/Mpc sec with the most probable value around $H_0 = 50$ km/Mpc sec [Tam86]), $\rho_0 \geq \rho_B = 0.5(+0.7, -0.3) \cdot 10^{-30}$ g cm^{-3} [Blo84]) we shall see that the age of the universe can play a critical role for the question whether the present value of Λ is zero or not. ρ_B here denotes the baryonic matter density in the universe (shining or not, i.e., including baryonic dark matter) which can be determined from primordial nucleosynthesis (see, e.g., [Blo84, Kol90]). Figs. 23, 24 show the relation between age of the universe and present matter density ρ_0 for various cosmological model types with Λ and k as parameters (H_0 is fixed in these figures to an average value of 75 km/Mpc sec, the case of $H_0 = 50$ km/Mpc sec is shown in [Kla86b]); q_0 in Fig. 24 denotes the deceleration parameter, $q_0 = -[\ddot{R}(t) \cdot R(t)/\dot{R}(t)^2]_{t=t_0}$, which is fixed by observation to $-1.27 < q_0 < 2$.

With Euclidean metric (favoured by inflation according to many authors, but questioned as necessary consequence of inflation by [Mad88]), and $\Lambda = 0$, the age would be only $t_0 = 8.7$ Gyr (for $H_0 = 75$ km/Mpc sec). The simultaneous requirements $\Lambda = 0$ and Euclidean metric require further a large amount of non-baryonic dark matter ρ_D in the universe: $\rho_0 = \rho_B + \rho_D > \rho_B$. If one allows, however, for $\Lambda > 0$ ($\rho_V \neq 0$) (Friedmann-Lemaitre models), the Euclidean models can give t_0 in the right range according to globular clusters and cosmochronometers and at the same time, the non-baryonic dark matter ρ_D required would be largely reduced. A positive vacuum energy density to a large amount "replaces" the dark matter needed to obtain Euclidean metric and $\Lambda = 0$.

We find [Kla86b] that for ages $t_0 > 13 \cdot 10^9$ years in the case of Euclidean metric solutions with $\Lambda = 0$ would be excluded for $H_0 = 50$ km/Mpc sec. For a Hubble constant $H_0 \geq 43$ km/Mpc sec the value of Λ would become positive for $t_0 > 15.2$ Gyr.

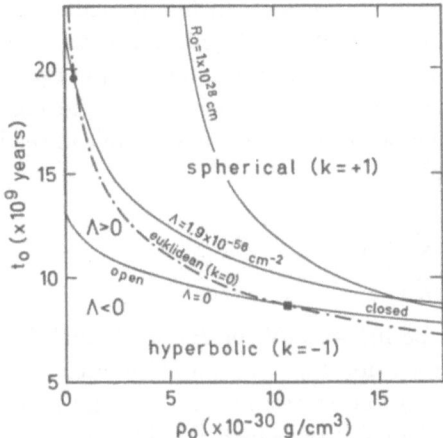

Fig. 23. Relation between the age of the universe t_0 and present cosmic matter density ρ_0 for various cosmological model types, with cosmological constant Λ and metric parameter k as parameters (H_0 is fixed to 75 km/Mpc sec). In Friedmann-Einstein models ($\Lambda = 0$) t_0 cannot exceed $13 \cdot 10^9$ years. Moreover this limiting value would only be obtained in strongly hyperbolic models without dark matter. Euclidean metric would restrict t_0 further to 8.7 Gyr. Larger values for t_0 can be obtained in Euclidean models if $\Lambda > 0$ (from [Kla86b])

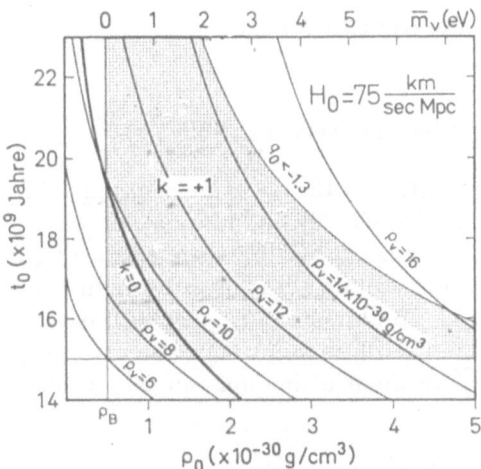

Fig. 24. The relation between t_0 and ρ_0 as in Fig. 23, however, the parameter range of interest shown in more detail. The hatched area corresponds to boundary conditions $\rho_0 \geq \rho_B$ $q_0 < -1.3$ and $t_0 \geq 15$ Gyr. In this area only models with $\rho_V > 0$ are allowed. For Euclidean metric ($k = 0$) this holds also for $H_0 = 50$ km/Mpc sec down to ages $t_0 > 13$ Gyr. An upper limit of the neutrino mass can be read from the upper scale, corresponding to attributing the non-baryonic dark matter $\rho_D = \rho_0 - \rho_B$ to neutrinos (from [Kla86b])

For a Euclidean universe with $t_0 \geq 17$ Gyr we obtain, assuming $H_0 = (50 - 100)$ km/Mpc sec, a value of $\Lambda = (0.47 - 1.9) \cdot 10^{-56}$ cm^{-2} or, correspondingly, $\rho_D = (2.5 - 10) \cdot 10^{-30}$ g cm^{-3}

From the amount of non-baryonic dark matter we can finally deduce an upper limit for the neutrino mass [Kla86b]. Adopting a Euclidean metric and an age of 15 Gyr, attributing of the non-baryonic dark matter completely to neutrinos (of course, there are many other candidates for dark matter, see, e.g., [Sch86, Smi90]) whose cosmic density is known to be $\sim$ 330 cm^{-3}, we can read an average neutrino mass (three flavours)

$$\overline{m_\nu} = \frac{1}{3} \cdot \sum_{i=1}^{3} m_\nu^i$$

from the upper scale given in Fig. 24 to be $\overline{m_\nu} \approx 2$ eV for $H_0 = 75$ km/Mpc sec. Correspondingly, we obtain $\overline{m_\nu} = 4$ eV for $H_0 = 50$ km/Mpc sec. If no assumption on the metric is made and only the restriction $H_0 \geq 50$ km/Mpc sec is used, the less stringent limit $\overline{m_\nu} = 18$ eV is obtained. Would the masses of ν_e, ν_μ, ν_τ be related as those of the e, μ and τ leptons (see, e.g., [Moh88]) a value of $m_{\nu_e} \leq 10^{-3}$ eV would be obtained. These numbers are consistent with our knowledge on the neutrino mass from other sources (double beta decay (see next section), tritium decay [Kün88], supernova SN1987A [Kol87]).

6. Double Beta Decay and Neutrino Mass

6.1. Neutrino Mass and Particle Theories (GUTs, SUSYs)

The neutrino mass is one of the key quantities for the structure of grand unified theories (GUTs, SUSYs, SUGRAs) [Kla88a, Gro90], and also, in this way, for cosmology. In GUTs, beyond minimal SU(5), e.g. SO(10), there exists a right-handed neutrino (the SU(5) singlet in Fig. 25), its Dirac mass being in lowest order of the order of the u quark mass. To obtain consistency with observation, the way of the see-saw mechanism [Gel79] is to introduce an even larger right-handed Majorana mass term m_R^M in the neutrino mass matrix which is, simplified to the one flavour case,

$$\left(\overline{\nu_L}\overline{(\nu_R)^C}\right) \begin{pmatrix} 0 & m_D \\ m_D & m_R^M \end{pmatrix} \begin{pmatrix} (\nu_L^C) \\ \nu_R \end{pmatrix} \qquad \begin{matrix} m_R^M \gg m_D \\ m_D \simeq \text{MeV} \end{matrix}$$

This leads to the phenomenologically required light (Majorana) neutrino

$$\nu_1 = \nu_L - \frac{m_D}{m_R^M}\nu_R \qquad \text{with} \qquad m_1 \approx \frac{\left(m^D\right)^2}{m_R^M} < (\text{eV})$$

and to a heavy (Majorana) neutrino with $m_?^M \simeq m_R^M$.

Introduction of m_R^M is, on the other hand, equivalent to (B-L) nonconservation – and equivalent to the occurrence of neutrinoless double beta decay.

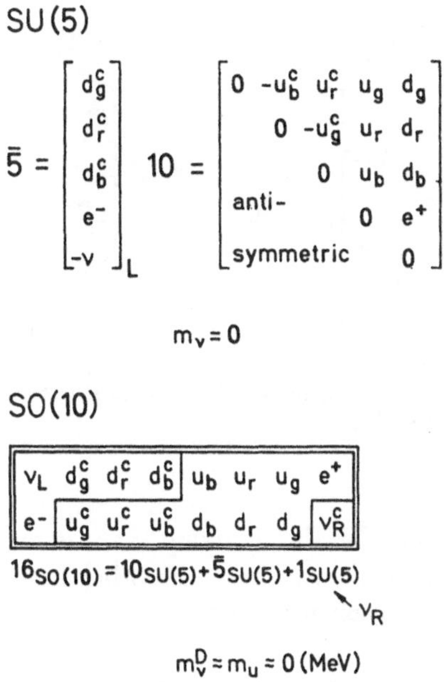

SU(5)

$$\bar{5} = \begin{bmatrix} d_g^c \\ d_r^c \\ d_b^c \\ e^- \\ -\nu \end{bmatrix}_L \qquad 10 = \begin{bmatrix} 0 & -u_b^c & u_r^c & u_g & d_g \\ & 0 & -u_g^c & u_r & d_r \\ & & 0 & u_b & d_b \\ \text{anti-} & & & 0 & e^+ \\ \text{symmetric} & & & & 0 \end{bmatrix}$$

$$m_\nu = 0$$

SO(10)

$$\begin{array}{|cccccccc|} \hline \nu_L & d_g^c & d_r^c & d_b^c & u_b & u_r & u_g & e^+ \\ e^- & u_g^c & u_r^c & u_b^c & d_b & d_r & d_g & \nu_R^c \\ \hline \end{array}$$

$16_{SO(10)} = 10_{SU(5)} + \bar{5}_{SU(5)} + 1_{SU(5)}$

$\searrow \nu_R$

$$m_\nu^D \simeq m_u \simeq 0 \,(MeV)$$

Fig. 25. Multiplets of the fermions of the first generation in the SU(5) and SO(10) model. The right-handed neutrino (left-handed antineutrino), not existing in SU(5), is a natural consequence in SO(10)

Since m_R^M is produced in GUTs when breaking left-right symmetry, its value and consequently the neutrino mass can range over a wide scale, depending on the energy scale at which left-right symmetry is broken [Lan88, Doi85]. This might happen in the range of the two examples

$$SO(10) \xrightarrow{M > M^x} SU(5) \xrightarrow{M^x} SU(3)_c \otimes SU(2)_L \otimes U(1) \qquad (a)$$

or

$$SO(10) \rightarrow SU(4)_{EC} \otimes SU(2)_L \otimes SU(2)_R \qquad (b)$$

extension of strong interaction   right-handed bosons $W_R^\pm$
(extended color) with leptons
as 4. color charge

between $\simeq 10^{15}$ GeV (a) or $\simeq 100$ GeV (Pati-Salam model, (b)). This is the energy range probed by $0\nu\beta\beta$ decay, which is reflected directly by the neutrino mass (see Tab. 1, from [Lan88]).

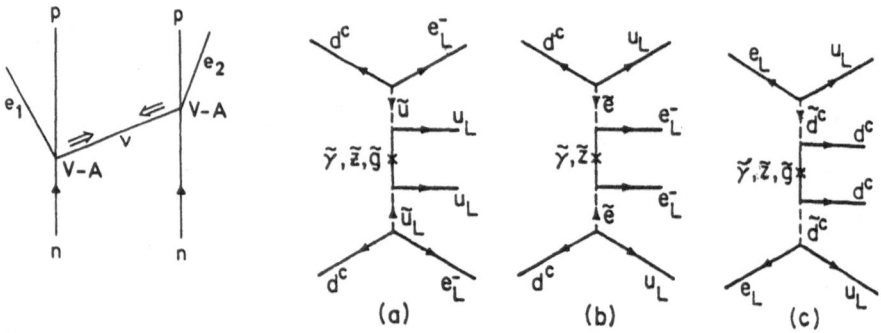

Fig. 26. Dominating transition modes for $0\nu\beta\beta$ decay in GUT models (left) and in supersymmetric extension of the standard model. In the latter the neutrino exchange of the GUT models is replaced by gaugino (gluino, photino, zino) exchange

Consequently the predictive power of GUTs for the neutrino mass is extremely weak: $m_\nu \simeq 10^{-11} - \simeq 1$ eV, (see Table 1), which on the other hand reflects the importance of experimental information for selecting the "right" GUT model.

In SUSY models such as a supersymmetric extension of the standard model neutrino exchange - dominating $0\nu\beta\beta$-decay in GUT models - would give only a small contribution to $0\nu\beta\beta$ decay, and the latter would be dominantly induced by gaugino (gluino, photino and zino) exchange [Moh86] (see Fig. 26).

Thus, also in the case that the neutrino mass would be so small that it is not in the range which can be probed by $\beta\beta$ decay, such experiments would remain in any case a sensitive probe of physics beyond the standard model.

Table 1. Models of neutrino mass, along with their most natural scales for the light neutrino masses.

Model	m_{ν_e}	$\langle m_{\nu_e} \rangle$	m_{ν_μ}	m_{ν_μ}
Dirac	1 - 10 MeV	0	100 MeV - 1 GeV	1 - 100 GeV
pure Majorana (Higgs triplet)	arbitrary	m_{ν_e}	arbitrary	arbitrary
GUT seesaw ($M \sim 10^{14}$ GeV	10^{-11} eV	m_{ν_e}	10^{-6} eV	10^{-3} eV
intermediate seesaw ($M \sim 10^9$ GeV)	10^{-7} eV	m_{ν_e}	10^{-2} eV	10 eV
$SU_{2L} \otimes SU_{2R} \otimes U_1$ seesaw ($M \sim 1$ TeV)	10^{-1} eV	m_{ν_e}	10 keV	1 MeV
light seesaw ($M \ll 1$ GeV)	1-10 MeV	$\ll m_{\nu_e}$	—	—
charged Higgs	< 1 eV	$\ll m_{\nu_e}$	—	—

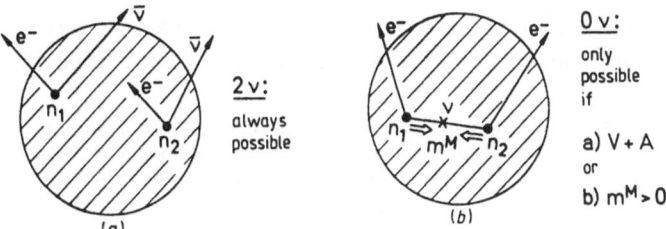

Fig. 27. 2ν and $0\nu\beta\beta$ decay, schematic

6.2. Double Beta Decay and Neutrino Mass

6.2.1. The Required Matrix Elements: There are two main decay modes in $\beta\beta$-decay, two-neutrino (2ν) decay

$$_Z^A X \rightarrow _{Z+2}^A X + 2e^- + 2\bar{\nu}$$

and neutrinoless (0ν) decay

$$_Z^A X \rightarrow _{Z+2}^A X + 2e^-$$

The first process is a usual effect of second order in the classical weak interaction, the second, however, requires as a prerequisite a non-vanishing Majorana neutrino mass or a right-handed weak interaction (Fig. 27).

In the framework of GUTs, or more general gauge theories, both of these conditions cannot be seen independently, but in any case a non-vanishing mass is required.

The half life for neutrinoless $\beta\beta$-decay $T_{1/2}$ is given by [Doi85, Mut88a]

$$\left[T_{1/2}^{0\nu}(0_i^+ \rightarrow 0_f^+)\right]^{-1} = C_{mm}\frac{\langle m_\nu\rangle^2}{m_e^2} + C_{\eta\eta}\langle\eta\rangle^2 + C_{\lambda\lambda}\langle\lambda\rangle^2 + C_{m\eta}\langle\eta\rangle\frac{\langle m_\nu\rangle}{m_e}$$

$$+ C_{m\lambda}\langle\lambda\rangle\frac{\langle m_\nu\rangle}{m_e} + C_{\eta\lambda}\langle\eta\rangle\langle\lambda\rangle$$

or, when neclecting the effect of right-handed weak currents, by

$$\left[T_{1/2}^{0\nu}(0_i^+ \rightarrow 0_f^+)\right]^{-1} = C_{mm}\frac{\langle m_\nu\rangle^2}{m_e} = \left(M_{GT}^{0\nu} - M_F^{0\nu}\right)^2 G_1\frac{\langle m_\nu\rangle^2}{m_e}$$

where G_1 denotes the phase space integral.

Thus a reliable neutrino mass can be extracted from a measured rate only if the nuclear matrix element $M^{0\nu} = (M_{GT}^{0\nu} - M_F^{0\nu})$ can be reliably calculated.

In the calculation of $2\nu\beta\beta$ decay one finds a strong sensitivity of the calculated matrix elements on details of the wave functions of initial and final states – and a cancellation reminiscent of a broken symmetry requiring $M^{2\nu} = 0$. In 0ν decay this sensitivity fortunately is not present, the essential reason being the excitation of states of various multipolarities in the intermediate nucleus in this case. These problems, which are decisive for $0\nu\beta\beta$ decay to be or not to be a tool for probing the neutrino mass have been understood only very recently. For details we refer to [Gro85, Gro86a, Kla84b, Kla90ad, Mut89, Sta90c,d].

Table 2. Present status of (effective) neutrino masses deduced from double beta decay experiments, for the nuclear matrix element of various nuclear structure calculations. Contributions of right-handed currents are neglected. Experimental $0^+ \rightarrow 0^+ 0\nu\beta\beta$ decay half-life limits are given in the last line.

Refs.	^{48}Ca	^{76}Ge	^{82}Se	^{128}Te	^{130}Te	^{100}Mo	^{136}Xe	^{150}Nd
Hax84	< 40	< 1.8	< 7.3	< 0.90	< 10	—	—	—
Gro86a*		< 0.7	< 2.9	< 0.43	< 5.4	< 8.7	< 13.3	< 3.2
Tom87		< 2.0	< 7.4	< 1.4	< 19	—	—	—
Mut89		< 2.1	< 7.4	< 1.2	< 18	< 54	< 36	< 3.8
Eng88**		< 7-17	< 32-64	< 2.2-3	< 29-39	< 66	< 61-80	—
$T^{0\nu}_{1/2}$	$> 2\,10^{21}$	$> 5\,10^{23}$	$> 1.1\,10^{22}$	$> 5\,10^{24}$	$> 1.5\,10^{21}$	$> 4.4\,10^{20}$	$> 1.7\,10^{21}$	$> 2.3\,10^{21}$

* no pp force **Not realistic (zero range) interaction

We have calculated for 2ν and 0ν decay with a realistic interaction (G matrix from Paris potential) the nuclear matrix elements for all potential $\beta\beta$-emitters with $A > 70$ [Mut89a,b, Sta90c] and particularly for the seven nuclei of main experimental interest, and extracted limits for a (Majorana) neutrino mass from experimental half-life limits. The results are shown in Table 2 (for references concerning the experimental limits see [Mut88a]). In the few cases, where we can compare to the Tübingen calculation [Tom87], also using a realistic interaction (G matrix from Bonn potential) we find good agreement (see Table 2). The 0ν results remain also stable against different choices of the renormalized effective interactions, e.g. Bonn-A and Reid potentials instead of the Paris potential [Sta90d]. A discrepancy occurs only when a not realistic interaction is used (as in [Eng88]).

Concluding, $0\nu\beta\beta$ decay remains the most powerful tool for probing the mass of the electron neutrino.

6.2.2. Future Experimental Possibilities. The Heidelberg-Moscow Experiment using enriched ^{76}Ge:

Double beta decay yields at present the sharpest limit on the Majorana mass of the neutrino (see Table 2). Among the direct (non-geochemical) experiments the most stringent limit is given by $0\nu\beta\beta$ decay of ^{76}Ge: $m_\nu < 1.4$ eV, corresponding to the claimed present half-life limit of $T^{0\nu}_{1/2} > 1.2 \cdot 10^{24}$ years [Cal90, Kir90].

In next generation $\beta\beta$ decay experiments ^{76}Ge will play also the most important role. Use of enriched ^{76}Ge (86%) [Kla90d, Hei89] instead of the presently used detectors from natural Ge (containing 7.8% of ^{76}Ge) could explore the half-life of $0\nu\beta\beta$ decay up to $\sim 10^{25}$ years and correspondingly the neutrino mass down to $\sim 10^{-1}$ eV, probing a class of left-right symmetric GUT models, with a right-handed Majorana mass term of about 1 TeV, based on $SU(2)_L \otimes SU(2)_R \otimes U(1)$ (see Table 1).

Fig. 28. The $\beta\beta$-laboratory of the HEIDELBERG-MOSCOW experiment in the GRAN SASSO near Rome

Fig. 29. The worldwide first enriched HP-^{76}Ge detector (with Si cup) in its extreme low-level shielding in the GRAN SASSO Laboratory

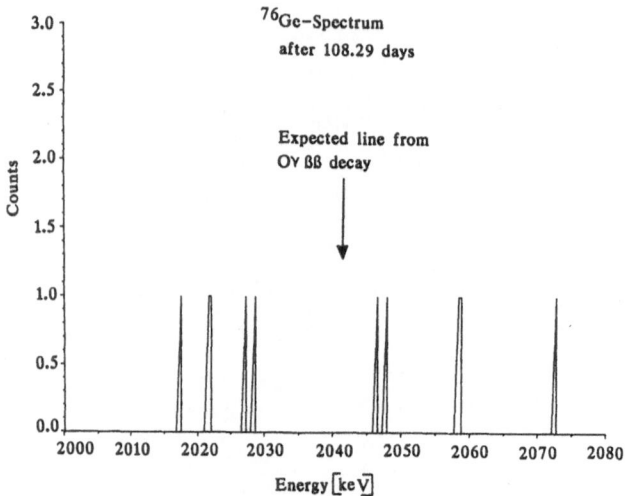

Fig. 30. The measured spectrum with the first HP-^{76}Ge detector after 108 days of measurement in the range of the hypothetical $0\nu\beta\beta$-line at 2041 keV

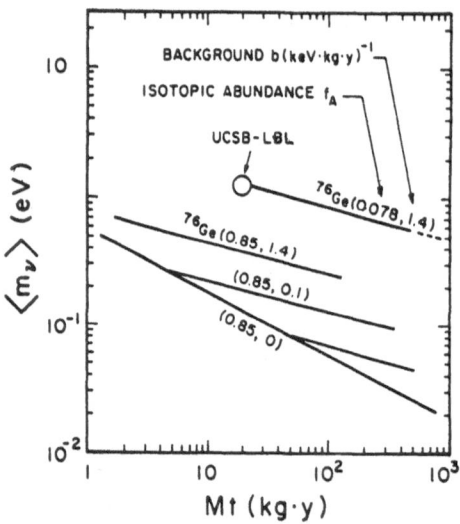

Fig. 31. The ranges of neutrino mass which can be probed with Ge-detectors of different enrichment in ^{76}Ge, as function of the product detector mass times measuring time (kg y) (from [Moe90])

The Heidelberg-Moscow $\beta\beta$ experiment [Hei89, Kla90d] makes use of 16.9 kg of ^{76}Ge metal enriched to 86%, corresponding to 14.5 kg of the isotope ^{76}Ge. The full amount of ^{76}Ge has been transferred from Moscow to Heidelberg. Up to now one enriched detector of ~ 1 kg and 2 crystals of 3.4 kg each (the largest ever produced p-type Ge crystals) have been produced. The detector is running in the GRAN SASSO Underground Laboratory in Italy since end of July 1990. Figs. 28, 29 show the Heidelberg-Moscow $\beta\beta$-Laboratory built generously by the INFN in the GRAN SASSO, and the first detector in its extreme low-level shielding from electrolytic copper and special lead. The results of the first 108 days of measuring are $T_{1/2} > 5.66 \cdot 10^{23}$ (90% confidence limit) or $> 1.19 \cdot 10^{24}$ years (68% confidence limit). The corresponding limits for the neutrino mass are $m_\nu < 2.0$ and < 1.4 eV. The background level which characterizes the quality of the setup is 0.45 ± 0.14 events/kg year keV (Fig. 30) in an 80 keV interval around the hypothetical 2041 keV $0\nu\beta\beta$-line (10 events after 108 days). The prospects of the experiment, compared to standard experiments using natural Ge are illustrated in Fig. 31.

The new experiment indeed will represent a major step forward to higher sensitivity and lower neutrino mass. The time scale for the full experiment is approximately 5 years.

It may be of interest to note that the HEIDELBERG-MOSCOW $\beta\beta$-experiment showed for the first time that the technology of production of enriched HP Ge-detectors can be handled. This opens – by use of ^{73}Ge and ^{70}Ge detectors – exciting perspectives for dark matter search and galactic γ-spectroscopy, among others of nucleosynthesis products, by satellite missions such as NAE (Nuclear Astrophysics Explorer [Nuc89]) and INTEGRAL (International Gamma-Ray Astrophysics Laboratory [Int90]). We refer for these questions to [Geh90, Kla90ce].

7. Conclusion

The distribution of nuclear beta strength, governing beta decay far from stability as well as double beta decay of the neutron-richest "stable" nuclei, is of decisive importance for exciting problems of astrophysics and cosmology. Both are beautiful examples of the close interconnections between nuclear physics and the latter disciplines, and also particle physics, and of the growing together of micro- and macrophysics observed in this century.

References

Abb88   L. Abbott: Sci. American **5**, 82 (1988)
App85   J.H. Applegate, C.J. Hogan: Phys. Rev. **D31**, 3037 (1985)
Arn90   M. Arnould, M. Rayet: Ann.Phys.Fr. **15**, 183 (1990)
Aus81   S.M. Austin: Progr.Part.Nucl.Phys. **7**, 1 (1981)
Ben88   E. Bender, K. Muto, H.V. Klapdor: Phys. Lett. **208B**, 53 (1988)
Ber69   E.Ye. Berlovich, Yu.N. Novikov: Phys.Lett. **B29**, 155 (1969)
Blo84   H.J. Blome, W. Priester: Naturwissenschaften **71**, 456, 515, 528 (1984)
Bra86   R.H. Brandenberger: In *Proc.Int.Symp.on Weak and Electromagnetic Interactions in Nuclei*, ed. by H.V. Klapdor (Springer, Heidelberg 1986) p. 991
Bur57   E.M. Burbidge, G.R. Burbidge, W.A. Fowler, F. Hoyle: Rev. Mod. Phys. **29**, 547 (1957)

Cal90   D.O. Caldwell et al.: In *Proc.14th Europhys.Conf.on Nucl.Phys.*, (Bratislava, 1990)

Cla83   D.D. Clayton: *Principles of Stellar Evolution and Nucleosynthesis*, (Univ. of Chicago Press, 1983)

Doi85   M. Doi, T. Kotani, E. Takasugi: Progr.Theor.Phys.Suppl.83, 1 (1985)

Eng88   J. Engel, P. Vogel, M.R. Zirnbauer: Phys. Rev. C37, 731 (1988)

Fow78   W.A. Fowler: In *Proc. of the Welch Foundation Conferences on Chemical Research*, ed. by W.D. Milligan, Cosmochemistry Vol. 21 (1978) p. 61

Fow85   W.A. Fowler and C.C. Meisl: In *Cosmogonical Processes*, ed. by W.D. Arnett, C.J. Hansen, J.W. Truran, S. Tsuruta (Singapore, VNU Press), 1985 p. 83

Fre74   D.Z. Freedman: Phys. Rev. D9, 1389 (1974)

Geh90   N. Gehrels: Nucl. Instr. Meth. A292, 505 (1990)

Gel79   M. Gell-Mann, P. Ramond, S. Slansky: In *Supergravity*, ed. by P. van Nieuwenhuizen, D.Z. Freedman (North-Holland, Amsterdam 1979)

Gil86   P.L. Gill, R.F. Casten, D.D. Warner, A. Piotrowski: Phys. Rev. Lett. 56, 1874 (1986)

Gre88   R.C. Greenwood et al.: In *Proc. Int. Conf. on Nucl. Data for Science and Technology*, (Mito, Japan,1988)

Gra90   M. Graefenstedt et. al.: Z.Phys. A 336, 247 (1990)

Gro76   H. von Groote, E.R. Hilf, K. Takahashi: At. Data Nucl. Data Tables 17, 418 (1976)

Gro85   K. Grotz, H.V. Klapdor: Phys. Lett. 153B , 1 (1985);
         K. Grotz, H.V. Klapdor: Phys. Lett. 157B, 242 (1985)

Gro86a   K. Grotz, H.V. Klapdor: Nucl. Phys. A460, 395 (1986)

Gro86b   K. Grotz, H.V. Klapdor, J. Metzinger: Phys. Rev. C33, 1263 (1986)

Gro90   K. Grotz, H.V. Klapdor: *The Weak Interaction in Nuclear, Particle and Astrophysics*, Adam Hilger (Bristol, New York 1990), and
         Die schwache Wechselwirkung in Kern-, Teilchen- und Astrophysik (Teubner, Stuttgart 1989)

Gut81   A.H. Guth: Phys. Rev. D23, 347 (1981) ;
         K. Sato: Phys. Lett. 99B, 66 (1981) ;
         A.D. Linde: Phys. Lett. 108B, 389 (1982)

Hal67   J.A. Halbleib, R.A. Sorensen: Nucl. Phys. A98, 542 (1967)

Hax84   W.C. Haxton, G.J. Stephenson: Progr. Theo. Nucl. Phys. 12, 409 (1984)

Hei89   Heidelberg-Moscow-Cooperation: H.V. Klapdor-Kleingrothaus, G. Heusser, A.Piepke, A. Buchner, A. Müller, U. Schmidt-Rohr, H. Strecker (MPI); S.T. Belyaev, A. Balish, A. Gurov, A. Demehin, I. Kondratenko, V.I. Lebedev (Kurchatov Institute): In *Proc. Internat. Sympos. on Weak and Electromagn. Interactions in Nuclei* (Montreal, 1989)

Hil76   E.R. Hilf, H.V. Groote, K. Takahashi: CERN-Report 76-13, 142 (1976)

Hil77   W. Hillebrandt, F.K. Thielemann: Mitt. Astron. Ges 43, 243 (1977)

Hil78   W. Hillebrandt: Space'Sci. Rev. 21, 639 (1978)

Hil79   W. Hillebrandt: In *Proc. Fourth EPS General Conf.* (1979) p. 255

Hil82   W. Hillebrandt: Phys. Bl. 38, 189 (1982)

Hir91a   M. Hirsch, A. Staudt, H.V. Klapdor-Kleingrothaus: to be published

Hir91b   M. Hirsch, A. Staudt, K. Muto, H.V. Klapdor-Kleingrothaus: Nucl. Data Tables, to be published

Hof86   R. Hoff: In *Proc. Int. Symp. on Weak and Electromagnetic Interactions in Nuclei*, ed. by H.V. Klapdor (Springer, Heidelberg 1986) p. 207

How80   W.M. Howard, P. Möller: At. Data Nucl. Data Tables 25, 219 (1980)

How87   W.M. Howard, P. Möller, G. Mathews, B. Meyer: In *Symp. on The Origin and Distribution of the Elements at the 194th American Chemical Society National Meeting*, (New Orleans, USA 1987)

Hoy60   F. Hoyle, W.A. Fowler: Astrophys. J. 132, 565 (1960)

Int90   INTEGRAL Proposal, 1990

Kaj88   T. Kajino, G.J. Mathews, G.M. Fuller: In *Proc. Int. Symp. Heavy Ion Physics and Nuclear Astrophysical Problems, Tokyo 1988* (World Scientific, Singapore 1989) p. 51

Käp89   F. Käppeler, H.Beer, K. Wisshak: Rep. Progr. Phys. 52, 945 (1989)

Kir90	I.V. Kirpichnikov: In *Proc. 14th Europhys. Conf. on Nucl. Phys.*, (Bratislava, 1990)
Kla76	H.V. Klapdor: Phys. Lett **65B**, 35 (1976) ; H.V. Klapdor: CERN Report **76-13**, 311
Kla80	H.V. Klapdor, C.O. Wene: J. Phys. **G6**, 1061 (1980)
Kla81a	H.V. Klapdor, T. Oda, J. Metzinger, W. Hillebrandt, F.K. Thielemann: Z. Phys. **A299**, 213 (1981)
Kla81b	H.V. Klapdor, T. Oda, J. Metzinger, W. Hillebrandt, F.K. Thielemann: CERN Report **81-09**, 341 (1981)
Kla82a	H.V. Klapdor, J. Metzinger, T. Oda: Z. Phys. **A309**, 91 (1982)
Kla82b	H.V. Klapdor: Phys. Bl. **38**, 182 (1982)
Kla83	H.V. Klapdor: Progr. Part. Nucl. Phys. **10**, 131 (1983)
Kla84a	H.V. Klapdor, J. Metzinger, T. Oda: At. Data Nucl. Data Tables **31**, 81 (1984)
Kla84b	H.V. Klapdor, K. Grotz: Phys. Lett. **142B**, 32 (1984)
Kla85	H.V. Klapdor: Fortschr. d. Physik **33**, 1 (1985)
Kla86a	H.V. Klapdor: Progr. Part. Nucl. Phys. **17**, 419 (1986)
Kla86b	H.V. Klapdor, K. Grotz: Astrophys. J. **301**, L39 (1986)
Kla88a	H.V. Klapdor (ed.): *Neutrinos*, (Springer, Heidelberg 1988)
Kla88b	H.V. Klapdor: Invit. Talk at *Internat. Symp. Heavy Ion Physics and Nuclear Astrophysical Problems, Tokyo, 1988*, Proc. (World Scientific, Singapore 1989) p. 127
Kla88c	H.V. Klapdor, J. Metzinger: In *Proc. Int. Conf. Nuclear Data for Science and Technology Mito, Japan, 1988*, ed. by S. Igarasi (Japanese Atomic Energy Research Institute) p. 827
Kla89a	H.V. Klapdor: Invited talk presented at *Internat. Conf. on Selected Topics in Nuclear Structure*, (Dubna, USSR 1989)
Kla89b	H.V. Klapdor: *Der Betazerfall der Atomkerne und das Alter des Universums*, Rheinisch-Westfäl. Akad. d. Wiss. Vorträge **365** (Opladen Westdeutscher 1989) p.73
Kla89c	H.V. Klapdor-Kleingrothaus: Invit. Talk at *Intern. Conf. "Fiftieth Anniversary of Nuclear Fission"* , (Leningrad, 1989)
Kla90a	H.V. Klapdor-Kleingrothaus: Invit. Talk at *Intern. School-Seminar on Heavy Ion Physics*, (Dubna, 1989) Proc. **D7-90-142** 440-461 (1990)
Kla90b	H.V. Klapdor-Kleingrothaus: Invit. Talk at *7th Int. Symp. Capt. γ-Ray Spectroscopy*, (Asilomar, California, 1990)
Kla90c	H.V. Klapdor-Kleingrothaus: Summary Talk given at *14th Europhysics Conf. on Nucl. Phys.*, (Bratislava, 1990)
Kla90d	H.V. Klapdor-Kleingrothaus: Invit. Talk at *14th Europhys. Conf. on Nucl. Phys.*, (Bratislava, 1990)
Kla90e	H.V. Klapdor-Kleingrothaus: Invit. Talk at *Int. Symp. on Gamma-Ray Line Astrophysics*, (Paris-Saclay, 1990)
Kol87	E.W. Kolb et al.: Phys. Rev. **D35**, 3598 (1987)
Kol90	E.W. Kolb, M.S. Turner: *The Early Universe*, (Addison-Wesley, 1990)
Kra86	K.L. Kratz, H. Gabelmann, W. Hillebrandt, B. Pfeiffer, H.L. Raven, F.K. Thielemann: Z. Phys. **A325**, 489 (1986)
Kra90	K.L. Kratz et al.: Z. Phys. **A336**, 357 (1990)
Kün88	W. Kündig et al.: In *Neutrino Physics*, ed. by H.V. Klapdor, B. Povh, (Springer, Heidelberg 1988) p.126; M. Fritschi et al. Phys. Lett. **173B**, 485 (1986)
Lan88	P. Langacker: In *Neutrinos*, ed.by H.V. Klapdor (Springer, Heidelberg 1988), p. 71.
Lun86	E. Lund, K. Aleklett, B. Fogelberg, A. Sangariyavanish: Phys. Scripta **34**, 614 (1986)
Lun89	E. Lund, G. Rudstam: University of Upsala, Research Report **NFL-60** (1989)
Mad88	M.S. Madsen, G.F.R. Ellis: University of Cape Town, Preprint **88/3**, (1988)
Mar79	S.H. Margolis: Astrophys. J. **231**, 236 (1979)
Mat89	G.J. Mathews, J.J. Cowan: In *Proc. Int. Symp. on Heavy Ion Physics and Nuclear Astrophysical Problems*, (Tokyo, World Scientific, 1989), p. 143
Mat90	G.J. Mathews, J.J. Cowan: Nature **345**, 491 (1990)
Mil79	G.E. Miller, J.M. Scalo: Astrophys. J. Suppl. **41**, 513 (1979)

Moe81   P. Moeller, J.R. Nix: At. Data Nucl. Data Tables **26**, 165 (1981)
Moe90   M. Moe: In *Proc. 14th Intern. Conf. Neutrino Physics and Astrophysics (Neutrino '90)*, ed.by K. Winter (CERN, Geneva, 1990)
Moh86   R.N. Mohapatra: Phys.Rev. **D34**, 3457 (1986)
Moh88   R.N. Mohapatra: In *Neutrino Physics*, ed.by H.V. Klapdor and B. Povh (Springer, Heidelberg 1988), and
In *Neutrinos*, ed.by H.V. Klapdor (Springer, Heidelberg 1988)
Mut88a   K. Muto, H.V. Klapdor: In *Neutrinos*, ed.by H.V. Klapdor (Springer, Heidelberg 1988), p. 183
Mut88b   K. Muto, H.V. Klapdor: Phys. Lett. **201B**, 420 (1988)
Mut89a   K. Muto, E. Bender, H.V. Klapdor: Z. Phys. **A333**, 125 (1989)
Mut89b   K. Muto, E. Bender, H.V. Klapdor: Z. Phys. **A334**, 177 and 187 (1989)
Nor79   E.B. Norman, D.N. Schramm: Astrophys. J. **228**, 881 (1979)
Nuc89   Nuclear Astrophysics Explorer Final Concept Study, Phase A Study Team, July 1989, Contract NAS5-30338
Ray90   M. Rayet, N. Prantzos, M. Arnould: Astron. Astrophys. **277**, 271 (1990)
Sch86   D.N. Schramm: In *Proc. Int. Symp. on Weak and Electromagnetic Interactions in Nuclei*, ed. by H.V. Klapdor (Springer, Heidelberg 1986) p. 1032
Smi90   P.F. Smith, J.D. Lewin: Phys. Rep. **187**, 203 (1990)
Sta89   A. Staudt, E. Bender, K. Muto, H.V. Klapdor: Z. Phys. **A334** 47 (1989);
Sta90a   A. Staudt, E. Bender, K. Muto, H.V. Klapdor: At. Data Nucl. Data Tables **44**, 79 (1990)
Sta90b   A. Staudt, M. Hirsch, K. Muto, H.V. Klapdor-Kleingrothaus: Phys. Rev. Lett. **65**, 1543 (1990)
Sta90c   A. Staudt, K. Muto, H.V. Klapdor-Kleingrothaus: Europhys. Lett. **13**, 31 (1990)
Sta90d   A. Staudt, T.T.S. Kuo, H.V. Klapdor-Kleingrothaus: Phys. Lett. **242B** 17 (1990)
Sta91   A. Staudt, H.V. Klapdor-Kleingrothaus: submitted for publication
Tac88   T. Tachibana, S. Ohsugi, M. Yamada: In *AIP Conf. Proc. 164 "Nuclei far from Stability"* (1988) p. 614;
M. Yamada: In *Proc. Int. Conf. on Nuclear Data for Science and Technology*, (Mito, Japan,1988) p. 841
Tak73   K. Takahashi, M. Yamada, Z. Kondoh: At. Data Nucl. Data Tables **12**, 101 (1973)
Tam86   G.A. Tammann: In *Proc. Int. Symp. on Weak and Electromagnetic Interaction in Nuclei*, ed.by H.V. Klapdor (Springer, Heidelberg 1986) p. 1016
Thi83   F.K. Thielemann, J. Metzinger, H.V. Klapdor: Z. Phys. **A309**, 301 (1983);
F.K. Thielemann, J. Metzinger, H.V. Klapdor: Astron. Astrophys. **123**, 162 (1983)
Thi90   F.K. Thielemann, M. Hashimoto, K. Nomoto: Astrophys. J. **349**, 222 (1990)
Tom87   T. Tomoda, A. Faessler: Phys. Lett. **199B**, 475 (1987)
Tru78   J.W. Truran, J.J. Cowan, A.G. Cameron: Astrophys. J. **222**, L63 (1978)
Tub75   D.L. Tubbs, D.N. Schramm: Astrophys. J. **201**, 467 (1975)
Wap88   A.H. Wapstra, G. Audi, R. Hoekstra: At. Data Nucl. Data Tables **39**, 281 (1988)
Wea80   T.A. Weaver, S.E. Woosley: Ann. N.Y. Acad. Sci **336**, 335 (1980)
Wen75   C.O. Wene: Astron. and Astrophys. **44**, 233 (1975)
Wen76   C.O. Wene, S.A.E. Johansson: CERN-Report **76-13**, 584 (1976)
Wil83   B.H. Wildenthal, M.S. Curtin, B.A. Brown: Phys. Rev. **C28**, 1343 (1983)
Woo82   S.E. Woosley, T.A. Weaver: In *Supernovae: A Survey of Current Research*, ed.by M.J. Rees, R.J. Stoneham (Reidel, Dordrecht 1982)
Woo90   S.E. Woosley: 1990, priv. comm.
Zhe68   Ya.B. Zheldovich: Usp. Fiz. Nauk **95**, 209 (1968)

Subject Index